LIGHT-WEIGHT MATERIALS FOR TRANSPORTATION
AND BATTERIES AND FUEL CELLS
FOR ELECTRIC VEHICLES

EUROPEAN MATERIALS RESEARCH SOCIETY SYMPOSIA PROCEEDINGS

Volume 45: Molecular Electronics (eds. Ch. Ziegler, W. Göpel and G. Zerbi)
Volume 46: Amorphous Insulating Thin Films II (eds. R.A.B. Devine, W.L. Warren, J. Kanicki and M. Matsumara)
Volume 47: Photon-Assisted Processing of Surfaces and Thin Films (eds. U.K.P. Biermann, J. Dieleman and P. Hess)
Volume 48: Photorefractive Materials (eds. G. Roosen, F. Agulló-López and O. Schirmer)
Volume 49: Organic Materials for Electronics (eds. J.L. Brédas, W.R. Salaneck and G. Wegner)
Volume 50: High Temperature Electronics (eds. K. Fricke, V. Krozer, J.P. Collinge and J.L. Robert)
Volume 51: Porous Silicon and Related Materials (eds. R.Hérino and W. Lang)
Volume 52: Small Scale Structures (eds. N.F. de Rooij, J.-M. Moret, H. Schmidt, W. Göpel, A.L. Greer, K. Samwer and C.V. Thompson)
Volume 53: Ion Beam Processing of Materials and Deposition Processes of Protective Coatings (eds. P.L.F. Hemment, J. Gyulai, R.B. Simonton, I. Yamada, J.-P. Thomas, P. Thévenard, W.L. Brown, P.B. Barna, Y. Paulau and G. Wahl)
Volume 54: Selected Topics in Group IV and II-VI Semiconductors (eds. E. Kasper, E.H.C. Parker, R. Triboulet, P. Rudolph and G. Müller-Vogt)
Volume 55: Laser Ablation (eds. E. Fogarassy, D. Geohegan and M. Stuke)
Volume 56: C, H, N and O in Si and Characterization and Simulation of Materials and Processes (eds. A. Borghesi, U.M. Gösele, J. Vanhellemont, A.M. Gué and M. Djafari-Rouhani)
Volume 57: Porous Silicon Material, Technology and Devices (eds. R. Hérino, W. Lang and H. Münder)
Volume 58: Organic Materials and Fullerenes (eds. D. Bloor, G. Leising, G. Zerbi, C. Taliani, P. Bernier and R. Taylor)
Volume 59: Frontiers in Electronics: High Temperature and Large Area Applications (eds. J. Camassel, K. Fricke, V. Krozer, J.L Robert, B. Drévillon, B. Equer, I. French and T. Kallfass)
Volume 60: Semiconductors and Organic Materials for Optoelectronic Applications (eds. R.L. Aulombard, B. C. Cavenett, B. Gil, R. Triboulet, G. Leising and F. Stelzer)
Volume 61: Group IV Heterostructures, Physics and Devices (Si, Ge, C, -Sn) (eds. J.-M. Lourtioz, G. Abstreiter and B. Meyerson)
Volume 62: Magnetic Ultra Thin Films, Multilayers and Surfaces (eds. M.A.M. Gijs, F. Petroff)
Volume 63: High Temperature Superconductor Thin Films: Growth Mechanisms-Interfaces-Multilayers (eds. H.-U. Habermeier and M.L. Hitchman)
Volume 64: Laser Processing of Surfaces and Thin Films (eds. C.N. Afonso, E. Matthias and T. Szörényi)
Volume 65: New Trends in Ion Beam Processing of Materials and Beam Induced Nanometric Phenomena (eds. F. Priolo, J.K.N. Lindner, A. Nylandsted Larsen, J.M. Poate, E.E.B. Campbell, R. Kelly, G. Marletta, and M. Toulemonde)
Volume 66: Advanced Materials for Interconnections (eds. Th. Gessner, J. Torres and G. Crean)
Volume 67: New Developments in Porous Silicon. Relation with other Nanostructured Porous Materials (eds. L.T. Canham and D. Bellet)
Volume 68: Fullerenes and Carbon Based Materials (eds. P. Delhaes, and H. Kuzmany)
Volume 69: Recent Developments in Thin Film Research: Epitaxial Growth and Nanostructures, Electron Microscopy and X-Ray Diffraction (eds. G. Ritter, C. Matthai, O. Takai, A.M. Rocher and A.G. Cullis)
Volume 70: Computational Modeling of Issues in Materials Science (eds.H. Dreyssé, Y. Kawazoe, L.T. Wille and C. Demangeat)
Volume 71: Light-Weight Materials for Transportation and Batteries and Fuel Cells for Electric Vehicles (eds. T. Hartkopf A. Moretti, J.G. Wurm, M. Wakihara, R. Ciach and H. Wallentowitz)
Volume 72: Biomaterials and Biodegradable Polymers and Macromolecules: Current Research and Industrial Applications (eds. F. Burny, D. Muster and A. Steinbüchel)
Volume 73: Coatings and Surface Modifications for Surface Protection and Tribological Applications (eds. J.P. Rivière, L. Pranevicius J.-M. Martinez-Duart and A. Grill)
Volume 74: III-V Nitrides Semiconductors and Ceramics: from Material Growth to Device Applications (ed. B.K. Meyer)
Volume 75: Materials, Physics and Devices for Molecular Electronics and Photonics (eds. J. Zyss and F. Garnier)

LIGHT-WEIGHT MATERIALS FOR TRANSPORTATION AND BATTERIES AND FUEL CELLS FOR ELECTRIC VEHICLES

PROCEEDINGS OF SYMPOSIUM J
ON
LIGHT-WEIGHT MATERIALS FOR TRANSPORTATION

AND

PROCEEDINGS OF SYMPOSIUM E
ON
MATERIAL ASPECTS FOR ELECTRIC VEHICLES
INCLUDING BATTERIES AND FUEL CELLS

OF THE 1997 ICAM/E-MRS SPRING CONFERENCE
STRASBOURG, FRANCE, JUNE 16-20, 1997

ELSEVIER
AMSTERDAM–LAUSANNE–NEW YORK–OXFORD–SHANNON–TOKYO

Published by:
Elsevier
Elsevier Science SA
Avenue de la Gare 50
1003 Lausanne
Switzerland

ISBN: 0-444-20515-2

Reprinted from

MATERIALS AND DESIGN Vol. 18 (4-6)
and
JOURNAL OF POWER SOURCES Vol. 72 (1)

The manuscripts for the Proceedings were received by the Publisher between

24 September and 10 October 1997 (Symposium J)
and
7 November 1997 and 16 January 1998 (Symposium E)

∞ The paper used in this publication meets the requirements of ANSI/NISO Z39.48-1992 (Permanence of Paper).

Transferred to digital printing 2006
Printed and bound by CPI Antony Rowe, Eastbourne

PART I

Symposium J
on
Light-Weight Materials
for Transportation

LIGHT-WEIGHT MATERIALS FOR TRANSPORTATION

PROCEEDINGS OF THE 1997 ICAM/E-MRS SPRING MEETING
SYMPOSIUM J ON LIGHT-WEIGHT MATERIALS FOR TRANSPORTATION
STRASBOURG, FRANCE, JUNE 16-20, 1997

R. CIACH
Institute of Metallurgy and Materials Science, Krakow, Poland

H. WALLENTOWITZ
Institut für Kraftfahrwesen, RWTH, Aachen, Germany

Sponsors

This conference was held under the auspices of:

The Council of Europe
The Commission of European Communities

It is our pleasure to acknowledge with gratitude the financial assistance provided by:

Banque Populaire	(France)
Elsevier Science	(The Netherlands)
Office du Tourisme, Strasbourg	(France)

Materials & Design
Vol. 18, Nos. 4/6, 1997

Elsevier Science Ltd

Contents

Materials & Design, Vol. 18, Nos. 4/6, pp. ix, 1997

0261-3069/98 $19.00 + 0.00

PII: S0261-3069(97)00100-3

Editorial and conference report

This special issue is based on a Symposium on 'Light-Weight Materials for Transportation' held in Strasbourg from 16 to 20 June 1996, as part of the International Conference on Advanced Materials ICAM '97 and the European Materials Research Society Spring Meeting E-MRS '97. The venue for the conference was the Congress Centre — Palais de la Musique et des Congrès. Forty oral contributions and some posters were presented during the Symposium by the authors from Austria, Brasil, China, France, Germany, Great Britain, Israel, Japan, Yugoslavia, Korea, Latvia, Poland and USA. Among the Symposium participants there were scientists from academic institutions and industrial organisations involved in the research and production, so it was a very good forum for the meeting and discussion between scientists and the industry representatives, both groups interested in light materials development and implementation.

These modern materials are the basis for further progress in the industry and in our life. Among them the light advanced materials with desired ratios of weight/properties and cost/properties are of special value for transportation for almost all applications.

Progress in this area depends on co-operation and development of metallurgy, casting and solidification techniques (like uni- and directional solidification or die-casting), plastic and superplastic deformation, heat and surface treatment. Dealing with common alloys we have well defined materials with a wide data base at our disposal, whereas designing materials based on composites still requires a thorough research in order to establish data bases and that results not only in high costs, but also gives inefficient designs and less-than-optimal structures. But, however difficult and problematic the composites are they bear the inherent potential of new materials. That is why the problem should be thought over from the beginning in terms of function and design which should include concepts for successful manufacturing.

On the other hand, materials science in the field of light materials is now transforming from an empirical approach to a more quantitative scientific stage. The revolution in materials has begun with the emergence of supercomputer simulation and computer-enhanced quantitative microscopic image analysis.

'Light-Weight Materials for Transportation' of 1997 was held when the advanced materials science and industry are undergoing a thorough transformation but at the same time the commercialisation of scientific and technological achievements is expected to go to a large extent. The advanced materials applied previously in the defence and aerospace area should expand over the commercial market including the air transportation and civil engineering. The new generation of modern cars and trains as well as aircraft (Boeing 777) can be good examples for the application of new materials.

I hope that the challenges met by the Organising Committee will be taken up, for the benefit of the participants and the development of light materials science, technology and applications, as a result of the articles presented during the Symposium as well as the fruitful discussion stimulated by them.

Guest Editor:
Professor Ryszard Ciach,
Vice President,
European Materials Research Society,
c/o Institute of Metallurgy and Materials Science,
Polish Academy of Sciences,
25 Reymonta St,
30-059 Kraków,
Poland

ELSEVIER

Materials & Design, Vol. 18, Nos. 4/6, pp. 203–209, 1997
© 1998 Published by Elsevier Science Ltd
Printed in Great Britain. All rights reserved
0261-3069/98 $19.00 + 0.00

PII: S0261-3069(97)00049-6

Technical Report

New cars — new materials

Arno Jambor*, **Matthias Beyer**

Daimler-Benz AG, Adv Development Cars EP/VF;HPC:G250,D 71059 Sindelfingen, Germany

Received 10 July 1997; accepted 30 July 1997

Due to more demanding requirements of car occupants in relation to comfort and safety enhancing measures, the weight of cars has been increasing, and as a result additional difficulties have been encountered in making lighter cars. In the development of every new car there is a search for new ways to combine the demands of the customers with reducing the weight of new cars. Further progress in optimizing steel body design can only take place gradually. Reinforced steel or tailored blanks are already in common use today. Even further reductions can be achieved by design in aluminium, magnesium or plastics. At Daimler–Benz, for example, the hard-top of the SL-sports-car is made of aluminium and the petrol tank partition panel of the SLK-roadster is made of die-cast magnesium. Lightweight design and, consequently, fuel saving will only be successfully realized, if proper materials are selected for appropriate parts. © 1998 Published by Elsevier Science Ltd. All rights reserved.

Keywords: cars; materials; design; parts

Introduction

The entire automobile industry is under considerable pressure to reduce the fuel consumption and therefore the emissions of their products. Over the last few decades, traffic density has continued to grow, because both the vehicle population and the mileage driven per vehicle have risen considerably. At the same time, even more demanding requirements in relation to comfort and safety-enhancing measures have tended to increase the weight of vehicles.

Total energy consumption of a motor vehicle

The aim of all car makers is to reduce fuel consumption. It is known that the energy consumption over the full life cycle of a vehicle is essentially determined by the fuel consumed during active use (*Figure 1*). The chart shows the percentage of energy consumed during the life cycle of a car. This figure has to be reduced. These reductions can be achieved by various measures (*Figure 2*):

- Disciplined driving style
- Improvement of drive efficiency and tyre rolling resistance
- Improvement and optimisation of the ancillary components
- Reduction of drag
- Reduction of vehicle weight

Within this presentation, the remainder of this paper will concentrate briefly on the effect of vehicle weight on fuel consumption followed by examples, which show how we can reduce the vehicle weight.

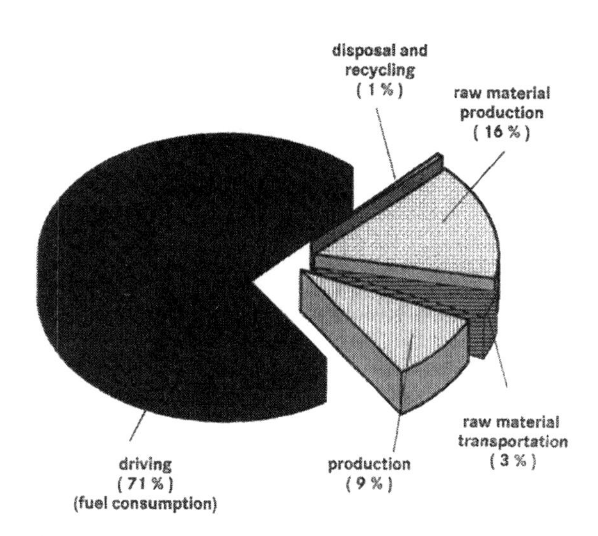

Figure 1 Energy consumption during life cycle

Influence of vehicle weight on fuel consumption

Road traffic generally moves at permanently changing speeds. The influence of mass acceleration can be clearly seen from the example of the fuel consumption figures calculated according to the ELA (European Legislative Average). Tyre rolling resistance also depends on mass (*Figure 3*).

The influence of vehicle mass on fuel consumption depends more on the kind of the engine than on the

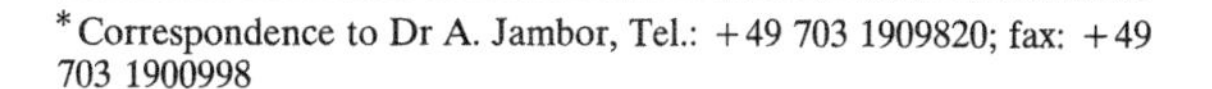

*Correspondence to Dr A. Jambor, Tel.: +49 703 1909820; fax: +49 703 1900998

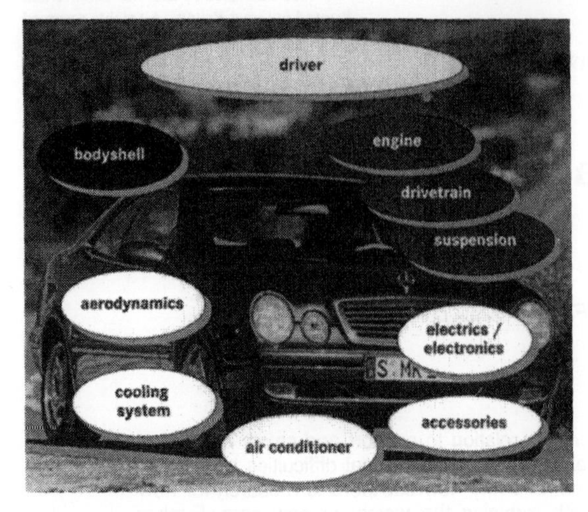

Figure 2 Impact on fuel consumption

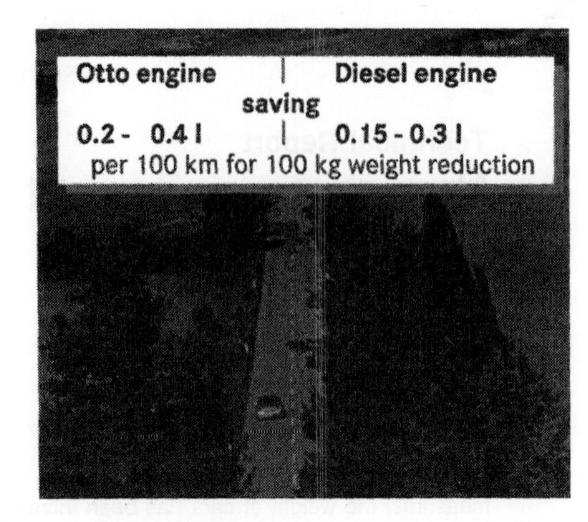

Figure 4 Impact of weight reduction

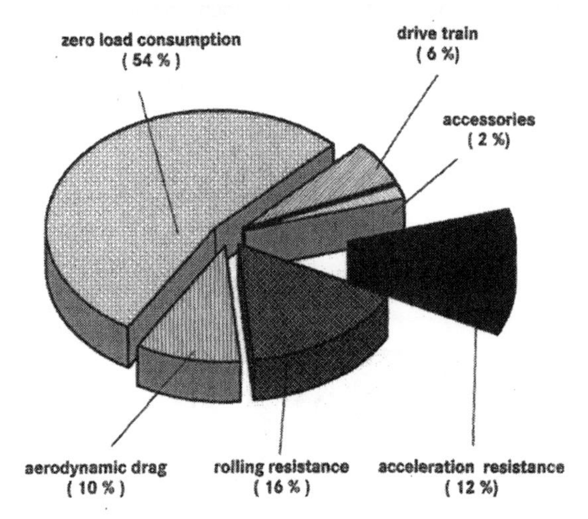

Figure 3 Impact of vehicle weight on NEDC (with C 180)

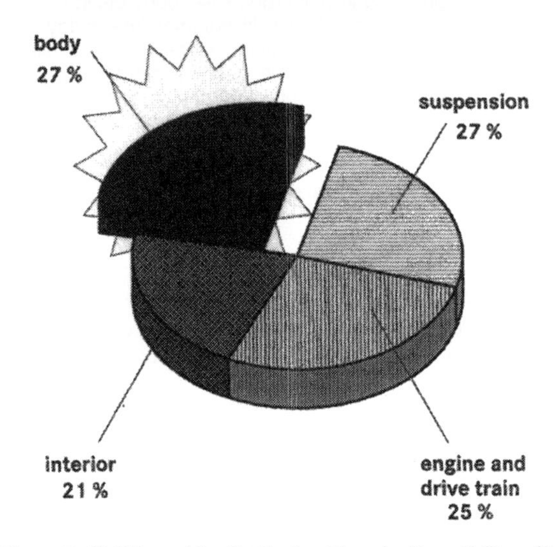

Figure 5 Vehicle weight distribution Mercedes-Benz C-Class (C 180)

category of the car. If we assume, that the axle-drive ratio is adapted for same flexibility (60–120 km h^{-1}), the Daimler–Benz product range has the following values according to the ELA (*Figure 4*)

Lightweight vehicle construction

Vehicle weight factors

The overall weight of our cars is distributed about 50% to the engine drive train and running gear and 50% to the body. The chart shows the weight distribution of a Mercedes-Benz C-class (*Figure 5*).

The increased use of lightweight materials in the engine and drive train brings a potential weight saving of 1–2% in relation to the gross vehicle weight. For the running gear, the potential weight saving from the use of these materials is about 6%. Greater potential weight savings can be obtained in the vehicle structure, with the body in white providing the largest contribution through the use of new technologies or new materials. Some examples will be considered in detail in this paper.

Lightweight design concepts in the body

The possibilities for reducing the weight of the vehicle body start with an optimised all-steel body in white—with a potential weight saving of about 7% (in relation to the body-in-white)—and span all the way to the all-aluminium car. Here, weight reductions of 30–50% are possible. More extreme lightweight designs can only be obtained by using fibrous composite materials (*Figure 6*).

Between the extremes of all-steel and all-aluminium, there are solutions that combine steel with lightweight materials. It should be noted that costs do not increase in relation to the use of lightweight materials. Reducing these costs is a primary aim.

Lightweight design requirements

A lightweight design must also meet the following

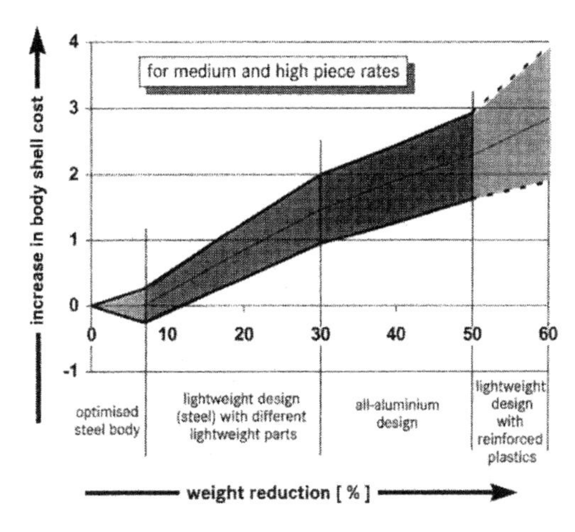

Figure 6 Cost curve of various lightweight design concepts

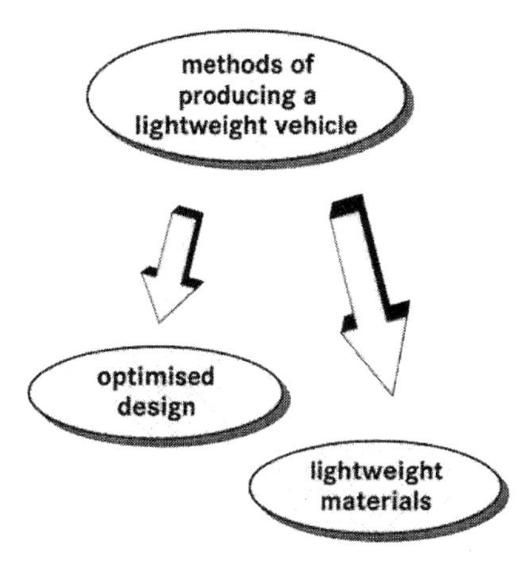

Figure 8 Methods of producing a lightweight vehicle

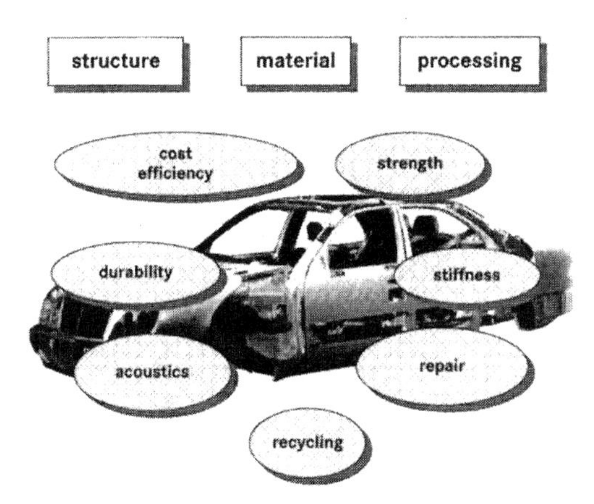

Figure 7 Body shell requirements

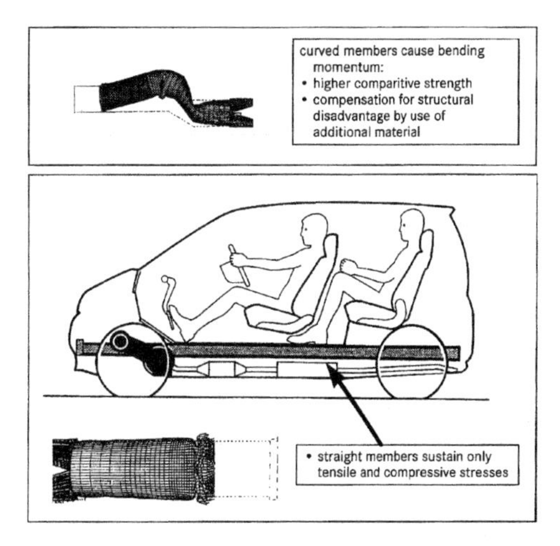

Figure 9 Straight members in new A-class

criteria (*Figure 7*):

- Low-cost production in high quantities
- Requirements of strength, stiffness and crash resistance
- Durability
- Recycling
- Repair concept
- Acoustic properties

The requirements regarding acoustic properties, stiffness and crash safety are now sufficiently satisfied, as has been proved by well-known models of aluminium construction. These questions will therefore not be considered further in this paper.

Lightweight design methods

Vehicle weight can be reduced by an optimised design and by lightweight materials (*Figure 8*).

Lightweight design shapes

Lightweight design shape means a design that is optimised for the expected load and for the material. This can be illustrated by two examples of the recently presented A-class (*Figure 9*).

The concept of this car gave us the possibility to design the front side member straight and without offsets, so that in the event of a frontal crash it is optimally axial-loaded. By contrast, bent members are always subject to bending forces, and the generated strains have to be compensated by extra material (*Figure 10*).

A second example is provided by flat panels with cambered impressing. This type of panel was used for the floorpan of the A-class. These cambered impressions increase the impedance of the vehicle floor in the low frequency range. In consequence, the thickness of the fused sheets otherwise required can be reduced. The result is a reduction of 50% in weight.

Lightweight design materials

Steel materials. Steel, as the traditional automobile material, has long since proved its worth. New high-strength sheet metal is now more and more being used for parts exposed to high stresses.

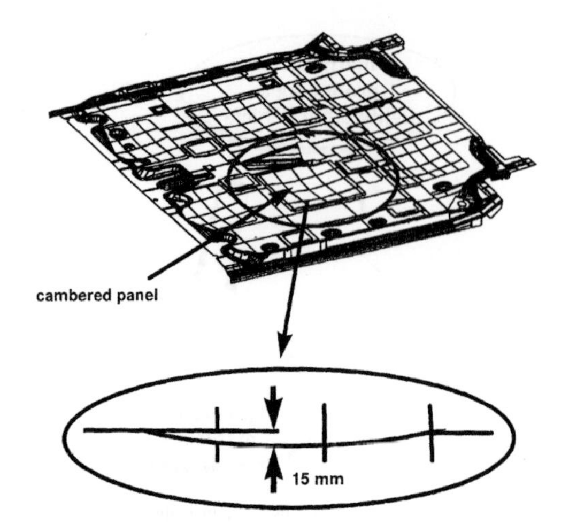

Figure 10 Cambered panels

High-strength panels. These micro-alloyed panels enable us to reduce the panel thickness and achieve reductions in body weight and/or improve the vehicle properties (strength) at reasonable costs. Modern strength ratings of up to 540 MPa compared to 180 MPa for conventional body panels permit thinner panels in strength-relevant and crash-relevant areas (*Figure 11*).

At present, the proportion of micro-alloyed high-strength steel panels in the E-class is 20%. The increased use of these panels is at present restricted by their limited formability. However, new steel materials combining the qualities of higher strength and improved formability are currently being tested. These new materials are essentially dual and triple-phase steels. In addition to their improved formability, increased strength is a prime aim. These steels will increase the proportion of high-strength steels in future models, and therefore the proportion of lightweight design. One example is a tunnel panel that could not previously be pressed in high-strength steel (*Figure 12*).

Continuing research into the above materials is highly promising. It shows, that the thickness of the panels can be reduced by 0.1–0.2 mm with the result of an additional weight reduction of 5–7 kg at the E-class.

Aluminium as lightweight construction material. The lightweight construction potential of aluminium is generally well-known, and so there is no need to explain it in detail here. The disadvantages in terms of processing, recycling and costs are also outside the scope of this paper. After a brief history, only the current state of development at Daimler–Benz will be described here.

The use of aluminium at Daimler–Benz. Our sports cars were already making extensive use of aluminium (and magnesium) before the war and in the post-war years. One representative example is the 300 SLR, a racing car dating back to January 1956 (*Figure 13*).

Another example from the past, this time a standard production model, is the 230 SL dating from 1963. This vehicle had an inner door panel of die-cast aluminium. The weight of the part was 6.6 kg, representing a weight saving of about 5 kg compared to a steel construction (*Figure 14*).

In today's Daimler–Benz model range, the hard-top of the SL sports car is made of aluminium. The weight of the roof structure has been reduced by 52%. The customer has the advantage of a light and therefore easily removable hard top. This all-aluminium design is cost-intensive. The costs are about 30 DM per kg of weight reduction compared to steel (*Figure 15*).

Interior parts are made of aluminium, too. For example, the structure of the new A-class seats is assembled of aluminium parts: the seat back is a tubular space frame, the seat rails are made of extrusion parts and the seat cushion pan consists of a panel.

Convertible in aluminium design (study). In order to research lightweight aluminium design at reasonable costs using the latest technical knowledge, a lightweight vehicle design study was carried out by Daimler–Benz on the basis of the current Roadster SL. The development goal was to prove that all specific requirements for the body in white—as already mentioned—could

NEW CARS

NEW MATERIALS

material	characteristics	scope	yield point
St14	low strength high elongation at break high formability	large series, outside panels	140 - 210 MPa
IF-steel (interstitial free) IF high strength Z St E 260 Z St E 300 Z St E 340 Z St E 380 Z St E 420	low strength high elongation high strength steel elongation ca. 20%	best for deep-drawing body structure	120 - 160 MPa 240 MPa 260 - 340 Mpa 300 - 380 Mpa 340 - 440 Mpa 380 - 500 Mpa 420 - 540 MPa
phosphate alloy steel Z St E 220 P Z St E 260 P Z St E 300 P	high strength steel elongation 28%	body structure	 220 - 280 Mpa 260 - 320 Mpa 300 - 360 MPa
BH-steel (bake-hardening) Z St E 180 BH Z St E 220 BH Z St E 260 BH Z St E 300 BH	hardening while stove-enamelling, low transformation force before hardening, high strength after hardening, high elongation	body structure	 180 - 240 MPa 220 - 280 Mpa 260 - 320 Mpa 300 - 360 MPa
dual-phase steel DP 500 DP 600			 230 - 290 Mpa 290 - 350 MPa
trip-steel transformation induced plasticity	optimal relation between strength and elongation		230 - 460 MPa

Figure 11 Material properties of various steels

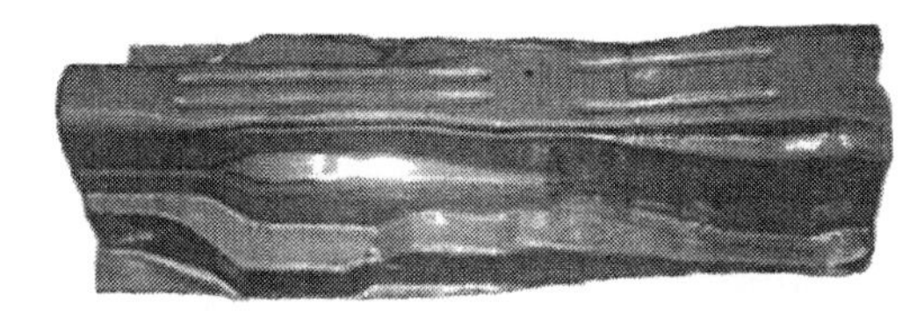

high strength steel (DP-steel)

- weight reduction of up to 1 kg
- reduction sheet thickness up to 0.2 mm

first try-out of deep-drawing with dual-phase steel
optimisation of geometry with deep-draw simulation

Figure 12 Tunnel E-class

Aluminium body in white with tubular space frame

Figure 13 Mercedes-Benz 300 SLR 1956

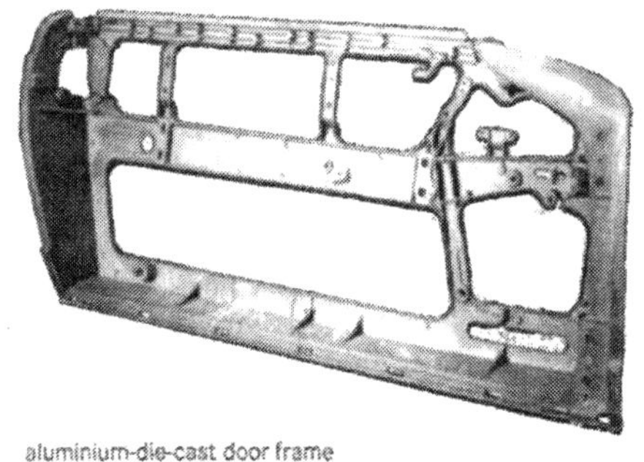

aluminium-die-cast door frame

Figure 14 Mercedes-Benz 230 SL 1963

be met by an all-aluminium concept also for a convertible (*Figure 16*).

The body in white consisted of panels, extrusion parts and die-cast components. Die-cast parts have the advantage of high integration: there are fewer individual parts and therefore less costly joining work and an ideal design e.g. with ribbing in the stiffness-relevant

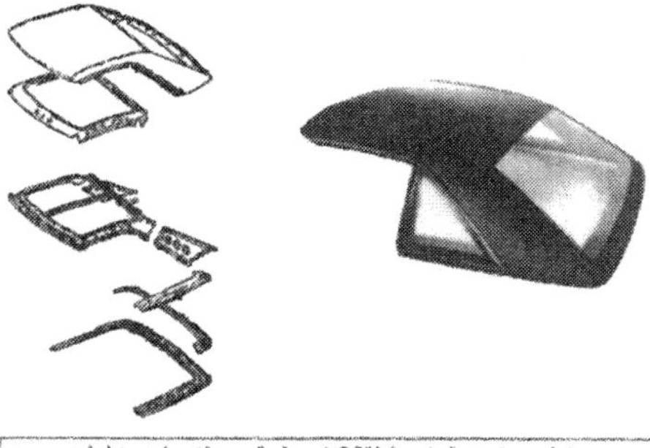

- weight reduction of about 50% (metal parts only)
- cost increase of 30 DM/kg weight reduction

Figure 15 SL-hardtop in aluminium

weight (complete BIW):	
steel:	415 kg
aluminium:	250 kg
weight reduction:	40 %

material:	
sheet inside:	Al Mg 5
sheet outside:	Al Mg 0.4 Si 1.2
die-casting:	Al Si 10
extrusion part:	Al Mg Si 0.7

Figure 16 Aluminium body in white SL (study)

zones of the body. The following tests were performed on the cars:

- Static and dynamic stiffness measurements
- Frontal crash
- Frontal offset crash at 65 km h^{-1} with 40% overlap against a deformable barrier
- Rear impact crash
- Comprehensive road testing

The result of this study is that the high demands of Daimler–Benz regarding passive safety, stiffness, strength and vibration comfort can be fully satisfied in all functional criteria—even in convertibles–by an aluminium body designed to meet the requirements of the material. The overall weight reduction compared to the steel body in white of the SL including doors, bonnet and boot lid, is 40%. Despite this remarkable weight reduction, the costs prevented the industrialization yet (*Figure 17*).

Magnesium. Magnesium is an even lighter construction material than aluminium. The use of magnesium in sheeted panels is not feasible at present. The sheet has to be deformed at high temperatures (over 300°C).

Figure 17 Aluminium body in white SL (study); crash test

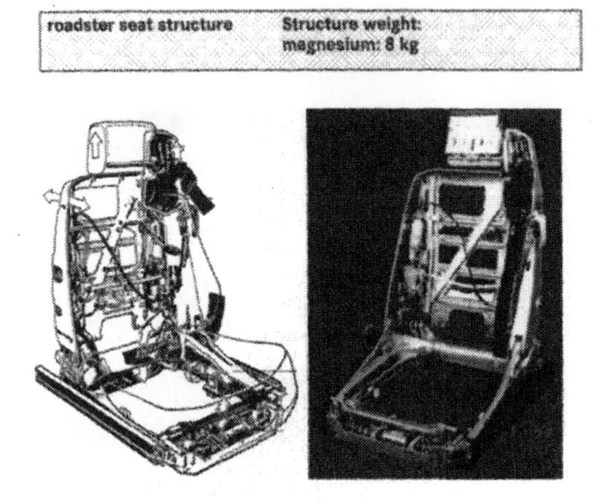

Figure 18 Magnesium die-casting for structure parts

Figure 19 Magnesium die-casting

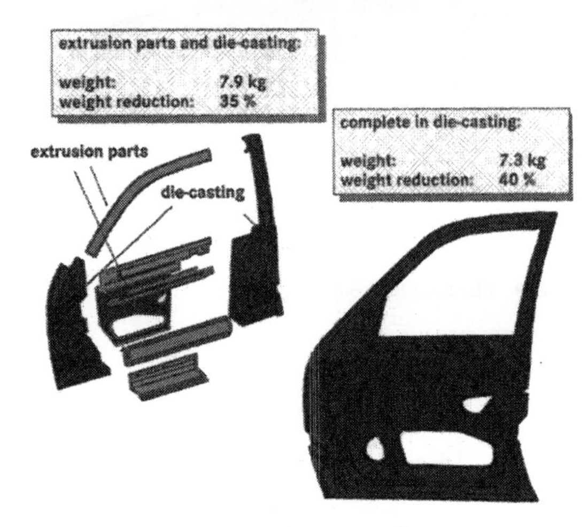

Figure 20 Door in magnesium (study C-class)

Therefore, heated tools are necessary. This makes the process expensive. Consequently, magnesium can only reasonably be used in die-cast production.

A well-known example of the use of magnesium for structural components is the seat frame of the present Roadster (start of production 1989): this seat frame is entirely made of die-cast magnesium, and the complete seat structure weighs only about 8 kg, even though the attachment of the restraint systems to the backrest imposes very strict strength requirements (*Figure 18*).

A second standard production part made of magnesium is the petrol tank partition panel in our new SLK sports car. The original steel panel weighed 6.7 kg and was replaced, for weight reasons, by an aluminium panel with the weight of 4.0 kg. Since a magnesium die-cast version allowed a further weight reduction to 3.2 kg, the decision was made, despite the higher costs, to introduce this lightest version in series production (*Figure 19*).

Other possible applications for magnesium are in the doors. Due to the possibility of casting thin-walled parts (1–1.5 mm wall thickness), the door inner panel could be built of die-cast parts combined with extrusion parts or die-cast as a single piece. A weight reduction of 40% is imaginable (*Figure 20*).

One problem with the use of magnesium is the corrosion that occurs in contact with steel or other materials. In this case, the materials must be kept separate, e.g. by plastic intermediate layers or special coatings.

Plastics. The outstanding advantages of plastics are their low specific weight. Known or possible uses include both the outer body skin and the load-bearing structure, mainly with the use of fiber-reinforced plastics. An example of an outer skin part at Daimler–Benz is the plastic wing of the A-class. The material is an unreinforced high-quality thermoplastic polyamide blend (PPO/PA) (*Figure 21*).

The weight reduction compared to a steel wing is 45%. Another essential advantage is the increased customer benefit due to the reduced risk of minor damage. The costs of the plastic wing are on a similar level to the steel variant. Other applications for plastic outer skin components include doors, hatches and lids. Necessary extra strength can be achieved by a two-shell construction.

The hatch of the A-class was built using this concept. The outer shell (unreinforced thermoplastic) is divided into a lower covering section and a rear roof spoiler. In

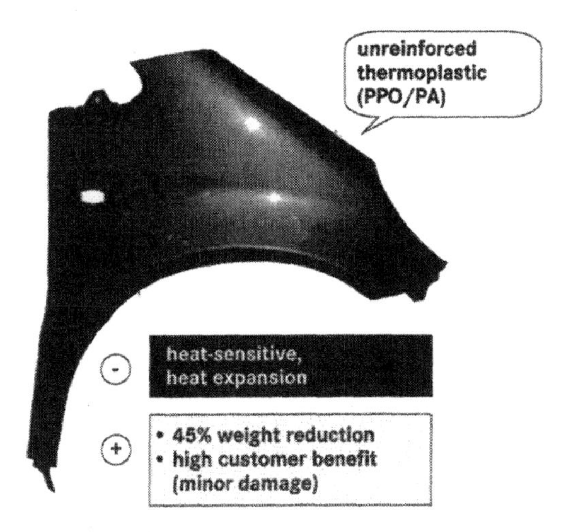

Figure 21 Wing A-class

Figure 23 LK-GTR in carbon fibre

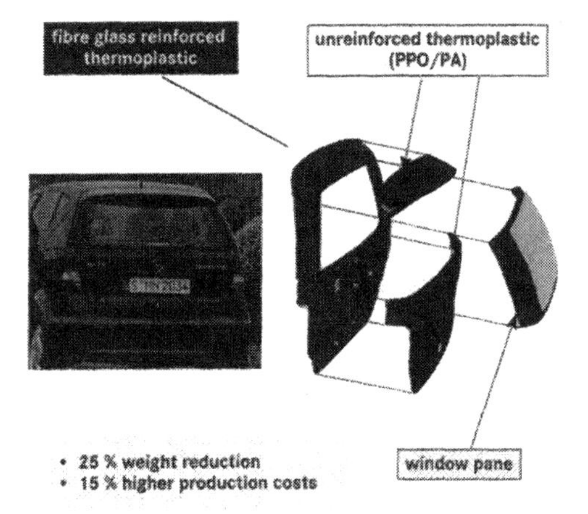

Figure 22 Hatch A-class

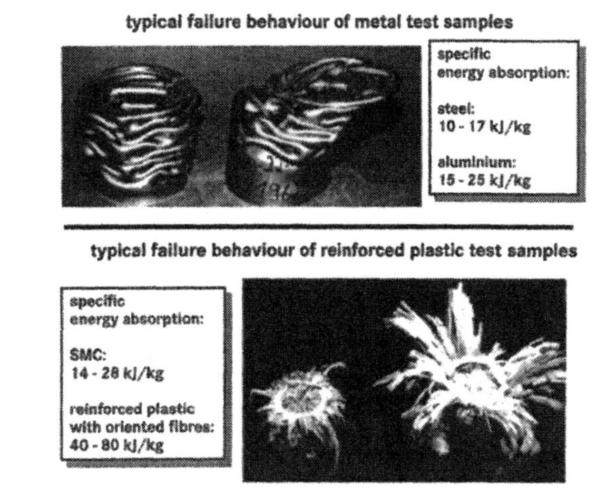

Figure 24 Crash behaviour reinforced plastic

the inner part made of GMT, the formability of the material is fully exploited by the integration of the lock fixing, hinges, rear wash/wipe fixing and number plate lamp housing (*Figure 22*).

The door is fully assembled as a module by the supplier and is painted off-line. This concept offsets a part of the higher material costs compared to steel. However, the manufacturing costs are about 15% higher than for the steel variant. The weight reduction of the plastic door is 3 kg (about 25%).

Structural components. For load-bearing body in white structures, only fibre-reinforced composite plastics with appropriately oriented reinforcing fibres offer suitable lightweight construction potential, even compared to aluminium. The technical feasibility and advantages of carbon fibre reinforced plastics are already well-known from applications in aviation and space travel as well as in motor-racing.

As long as high stiffness remains a major concern, carbon fibres should be preferred, whereas strength requirements can also be very well met by glass or—this is perhaps new—natural fibres. With this technology we can achieve a weight reduction of 50%. Fibre reinforced plastics, depending on the structure and orientation of the fibres, not only have high stiffness and strength but also a much higher energy absorption potential than metals, so that in principle they can even be used as lightweight materials in crash-relevant structural areas (*Figures 23* and *24*).

Conclusion

In the course of this paper, I have shown with the aid of a few examples the possibilities that we now have for using materials other than steel in order to reduce vehicle weight. These methods must be improved for future vehicles in order to meet our high demands and specifications, whether self-imposed or laid down by law. We must combine our efforts to attain these goals to preserve the car itself in its full fascination.

Materials & Design, Vol. 18, Nos. 4/6, pp. 211–215, 1997

0261-3069/98 $19.00 + 0.00

PII: S0261-3069(97)00084-8

Microstructure and properties of a new super-high-strength Al–Zn–Mg–Cu alloy C912

Y. L. Wu[a,*], F. H. Froes[a], A. Alvarez[a], C. G. Li[b], J. Liu[c]

[a]*Institute for Materials and Advanced Processes (IMAP), University of Idaho, Moscow, ID 83844-3026, USA*
[b]*Chinese Materials Research Society, BIAM, P.O. Box 81, Beijing 100095, China*
[c]*Alcoa Technical Center (ATC), Pennsylvania, PA 15069-0001, USA*

Received 22 July 1997; accepted 30 July 1997

Using scanning electron microscopy (SEM) and transmission electron microscopy (TEM), the microstructure of a new super-high-strength Al–Zn–Mg–Cu alloy (C912) has been investigated. Compared with some other high-strength aluminum alloys, the C912 alloy exhibits higher strength and good stress-corrosion resistance and its specific strength is even higher than some Al–Li alloys. Its potential for use in the Chinese AE100 airplane is discussed.

Keywords: Al–Zn–Mg–Cu alloy; high-strength aluminum alloys; stress-corrosion resistance; heat treatment; airframe

Introduction

To meet the increasing demands of the aerospace industry, much more advanced structural materials are needed. Aluminum alloys, as the main structural materials for airplanes, are developing in low-density, heat-resistant and high-strength arenas. Because decreasing density and improving strength can all reduce the structural weight (*Figure 1*), many new low-density and high-strength powder metallurgy aluminum alloys are being developed, such as Al–Li alloys, RS/PM 7090, 7091 and 7093 alloys. However, their high cost limits their application in aerospace. In the past 10 years, some new ingot metallurgy aluminum alloys with higher strength and lower cost and some new heat-treatment tempers have been developed by Alcoa, such as the 7150, 7055 alloys and the T77 temper. As the alloys 7150-T77 and 7055-T77 all have higher strength than 7075-T6 and improved resistance to SCC which is similar to the 7150-T76, they have been used for the upper wing structure of the C-17 military transport aircraft and B-777 commercial airplane, respectively[2]. This article investigates the microstructure and properties of a new super-high-strength Al–Zn–Mg–Cu alloy (C912). By comparison with some other super-high-strength aluminum alloys, its application prospect in airplanes is given.

* Correspondence to Y. L. Wu

Experimental procedure

The C912 alloy belongs to 7xxx series aluminum alloys and has the following nominal composition (in wt.%): 8.7 Zn, 2.6 Mg, 2.5 Cu, < 0.05 Fe, < 0.05 Si. The semi-continuously DC casting ingots and the extrusions were produced at BIAM (Beijing Institute of Aeronautical Materials, Beijing, China).

Solution heat treatment was carried out at 455°C for all specimens. There were three aging regimes: (a) CS1, 121°C/24 h, near the peak-aged condition; (b) CS2, 121°C/24 h + 240°C/48 s + 121°C/24 h, a retrogression re-aging process; and (c) CS3, 99°C/8 h + 163°C/26 h, an over-aged condition. Round tensile bars and compress column were tested to obtain the mechanical properties of the C912 alloy. According to HB5254-83 specification (equivalent to ASTM G49-76), tensile specimens were used to investigate the susceptibility of the alloy in different temper conditions to stress-corrosion cracking (SCC). The specimens were immersed in 3% NaCl + 0.5% H_2O_2 solution at 35°C for testing. Tensile tests of replicate specimens exposed with no applied stress, in conjunction with stressed specimens (exposed for the same times), were used to provide the $(1-\alpha)$ parameter for stress-corrosion susceptibility evaluating:

$$(1-\alpha) = (\sigma_{b1} - xx\%\sigma_{0.2})/(\sigma_b - xx\%\sigma_{0.2})$$

where σ_{b1} is the ultimate tensile strength of the exposed specimens with no applied stress, $xx\%\sigma_{0.2}$ is the

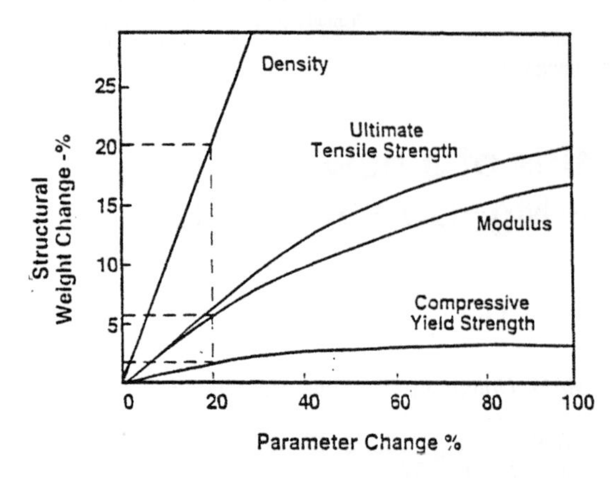

Figure 1 Effect of property improvement on weight saving in airplanes[1]

applied stress on the specimens, σ_b is the ultimate tensile strength of the unexposed specimens and $\sigma_{0.2}$ is the yield tensile strength of the unexposed specimens.

If the σ_{b1} is greater than the applied stress (xx% $\sigma_{0.2}$), the sample has SCC tendency. The $(1-\alpha)$ means the percentage of SCC fracture in the whole fracture:

$1-\alpha=$ 0 means that the specimen is not susceptible to SCC,

$1-\alpha=$ 1 means that the specimen is fully susceptible to SCC.

C-ring specimens were used to evaluate the stress corrosion resistance of the alloy. The specimens were prepared from the ∅ 25-mm extruded bars, loaded in the transverse direction and exposed in 3.5% NaCl solution under alternate immersion condition in accordance with the HB5259-83 specification (equivalent to ASTM G45-75 and G38-73).

Microstructural examination were performed with JSM-35 scanning electron microscopy (SEM) and H-800 transmission electron microscopy (TEM).

Results and discussion

Mechanical properties

Some mechanical property values of the C912 alloy in different temper conditions are listed in *Table 1*. It shows that the C912 alloy has very high strength and good ductility in the CS1 and CS2 temper conditions and there is a large strength difference (approx. 24%) between the CS2 and CS3 temper conditions for the alloy. This determines that the use of the C912-CS3 alloy is limited.

The differences of strength among these tempers can be explained by the microstructure variation. The C912-CS1 alloy has small, closely-spaced grain-boundary precipitates (η, $MgZn_2$) and a matrix consisting primary of very fine GP zones and some η'($MgZn_2$) precipitates (*Figure 2 (a), (b)*). The C912-CS2 alloy involves larger grain-boundary precipitates spaced further apart and GP zones in the matrix due to the retrogression treatment (*Figure 2 (c), (d)*). Compared with the C912-CS2 alloy, the C912-CS3 alloy has no GP zones and larger η' and η particles and interparticle spacing in the grain (*Figure 2 (e), (f)*).

Table 1 Typical mechanical properties of the C912 alloy on room temperature[a]

Temper	Extrusion size (mm)	UTS (MPa)	YS (MPa)	El (%)	E (GPa)	CYS (MPa)	Ec (GPa)
CS1	∅ 25	706	679	12	71	703	74
CS2	∅ 25	674	662	12	—	—	—
CS3	∅ 25	501	453	12	—	—	—
CS1	11 × 55	677	634	12	—	—	—
CS2	11 × 55	649	610	9	—	—	—
CS3	11 × 55	499	441	11	—	—	—

[a]Tested in the longitudinal direction.

As fully coherent with the matrix, GP zones of small size in the CS1 temper can be sheared by dislocations easily, which results in a shear fracture of the alloy (*Figure 3(a)*). Its fracture surface consists of many shear regions interspersed with areas containing dimples and secondary cracks (*Figure 3 (b)*). In this condition, the highest strength comes from the very high volume fraction of GP zones. In CS2 temper, the alloy is slightly over-aged, the GP zones become smaller and coarse and the volume fraction of semi-coherent η' phase increases which retards the planar slip — looping and bypassing occur. Then the cup and cone fracture features can be observed (*Figure 3(c)*). Its central fracture surface is made up of lots dimples and some shear regions (*Figure 3(d)*). The lower volume fraction of coherent particle (GP zones) and higher volume fraction of semi-coherent phase make the strength decrease. In CS3 temper, the alloy is over-aged deeply, there are no coherent GP zones in the matrix and the volume fraction of incoherent η phase increases. Therefore the alloy in CS3 temper exhibits a typical cup and cone fracture (*Figure 3(e)*) and there are no shear regions in its central fracture surface (*Figure 3(f)*). The larger interparticle spacing and the coarse size of the precipitates (η and η') in the grain result in a great decrease in strength.

For comparison, mechanical properties of some aluminum alloys are listed in *Table 2*. The C912 alloy exhibits similar properties to the Alcoa new aluminum alloy 7055 (and 7150) and the Russian alloy B96C. They all have very high strengths and good ductilities. Compared with some Al–Li alloys (2090 Al–Cu–Li, 8090 Al–Li–Cu–Mg) and PM alloys, the C912 alloy not only has higher strength, but also possesses higher specific strength. This implies that the C912 alloy has great potential for weight saving.

Stress-corrosion cracking

Table 3 lists the susceptibility of the C912 alloy to the SCC. In the CS1 temper, the C912 alloy has serious SCC tendency. After RRA heat treated (CS2), the C912 alloy gains favorable improvement in SCC resistance and in the CS3 temper, the alloy has no SCC tendency.

There are several SCC mechanisms that may account

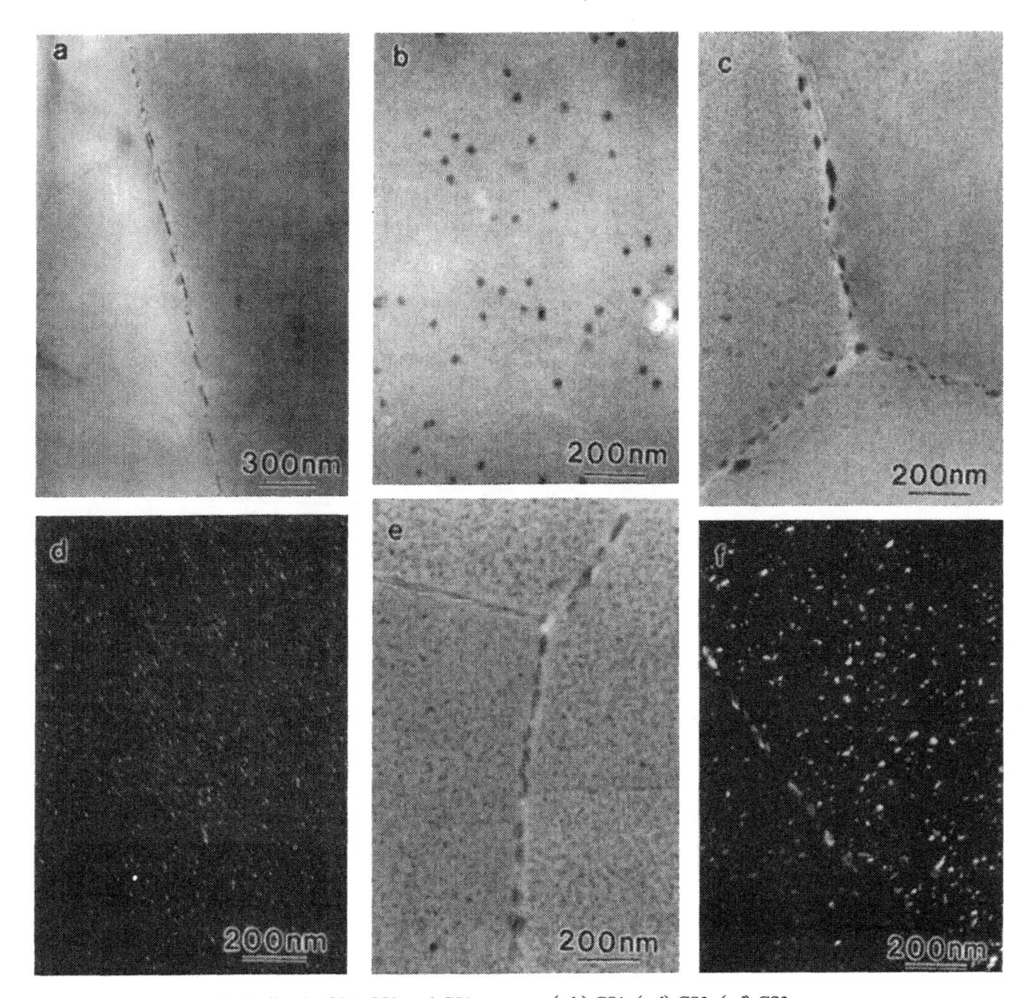

Figure 2 Microstructures of the C912 alloy in CS1, CS2 and CS3 tempers: (a,b) CS1; (c,d) CS2; (e,f) CS3

for this improvement, such as active path dissolution, hydrogen-assisted cracking and dislocation-assisted cracking[3–5].

In the active path dissolution mechanism, the crack velocity depends mainly on the anodic dissolution rate, i.e. the dissolution rate of grain-boundary precipitates. A larger size and spacing of grain-boundary particles will decrease the dissolution rate[3,6]. Comparing the C912-CS1 with the C912-CS2 and CS3 in *Figure 2*, the grain-boundary precipitates in the C912-CS2 and CS3 alloys really have larger size and more smooth shape. The spacing of the grain-boundary precipitate also increases. Therefore the lower dissolution rate of particles and the slow propagating rate of cracks could be expected.

When hydrogen atoms segregate to the region in front of the crack tip, they reduce the bonding force of metal and decrease the fracture stress, result in the crack developing easier. To understand the hydrogen-assisted cracking, it is helpful to utilize the concept of reversible and irreversible traps. The reversible trap is one at which hydrogen has a short residence time, while for an irreversible one, the release of the hydrogen atom is highly improbable once it is captured[4]. Many studies have shown that incoherent precipitates larger than a critical size (~ 20 nm) can act as hydrogen atom irreversible traps[4,7] and it has been observed that these traps serve as nucleation sites for molecular hydrogen[7,8]. In the C912-CS1, CS2 and CS3 alloys, the sizes of η phases in the grain boundary are all larger than 20 nm, meaning that in the SCC process, these particles all can act as nucleation sites for molecular hydrogen. Considering the η particle size and shape, the presence of hydrogen in the grain boundary of the C912-CS1 alloy could be small and sharp, but larger and more smooth in the C912-CS2 and CS3 alloys. The later cases would decrease the stress concentration at the crack tip and then decrease the propagation rate of the crack, causing crack blunting.

When studying the SCC improvement of the C912 alloy, it is also very important to consider the contribution of matrix precipitates. As mentioned earlier, in the CS2 and CS3 tempers, the matrix precipitate size is greater than that in the CS1 temper, which makes the planar slip more difficult. This results in the decrease of stress concentration at the grain boundary. In addition, following the reversible and irreversible trap theory, the η' and η particles in the matrix can act as reversible traps of hydrogen atoms, due to their sizes being less than 10 nm. Therefore it could be expected

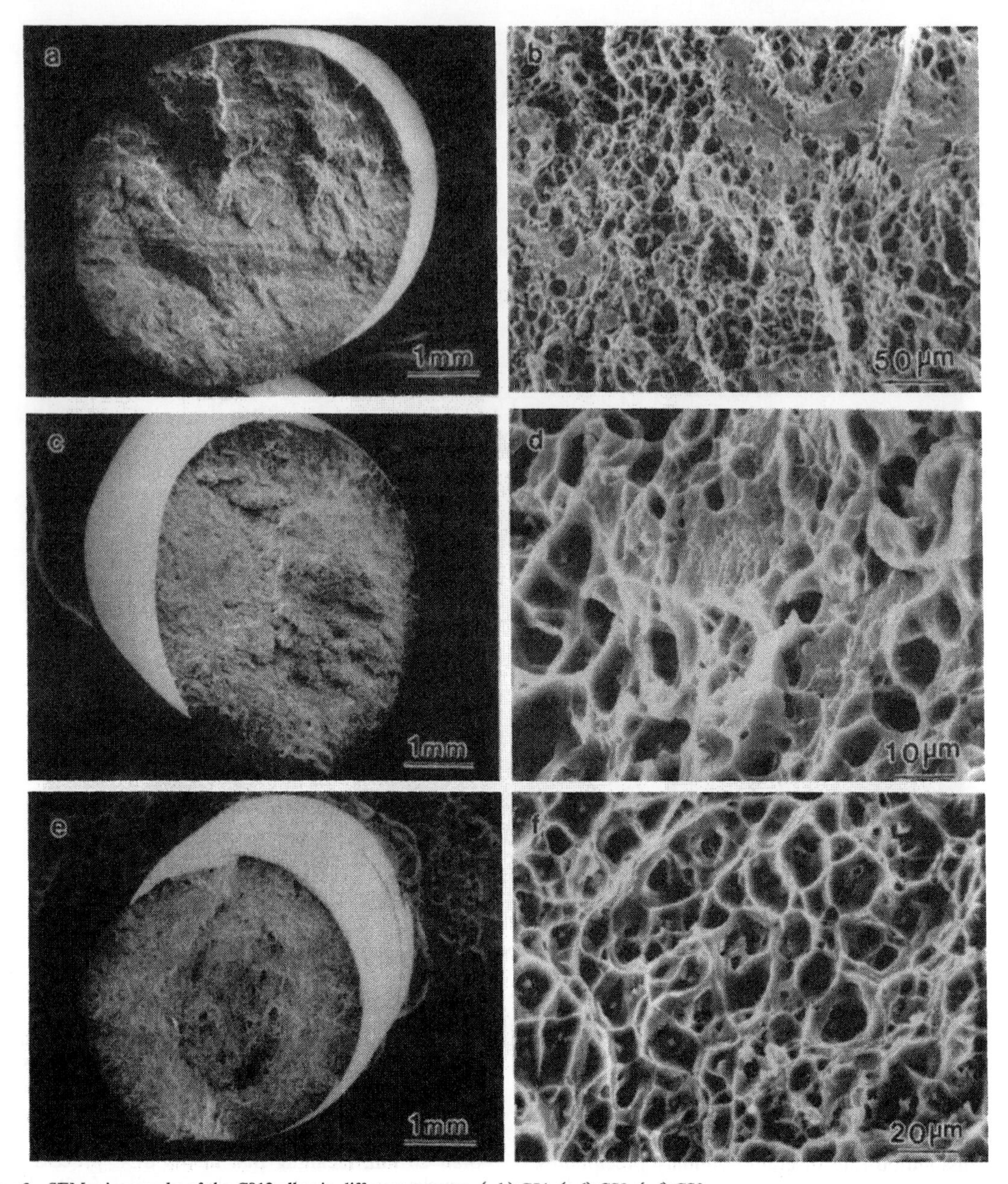

Figure 3 SEM micrographs of the C912 alloy in different tempers: (a,b) CS1; (c,d) CS2; (e,f) CS3

that the hydrogen atoms will move easily from the matrix to the grain boundary through the dislocations. As the larger η' and η particles retard the movement of dislocation, the concentration of hydrogen at the grain boundary decreases. The decreasing concentration of stress (dislocation) and hydrogen at the grain boundary improves the SCC resistance of the materials[9,10].

For comparison, C-ring SCC test data of the C912 alloy and some other 7xxx aluminum alloys are listed in *Table 4*. In the peak-aged condition (T6 and CS1), the C912 alloy has higher SCC threshold than that of the 7075, 7178 and 7079 alloys and in the over-aged condition, the C912 alloy shows a similar SCC resistance to the 7075-T76, -T73 alloys.

Future use in airplanes

The Aviation Industries of China (AVIC) is moving aggressively into the commercial airplane arena. A 100-seater commercial airplane (the AE100) with up to 2700 units to be built is in the planning. An artist's rendition of the airplane and some suggested alloys for its optional performance and long structure-life are shown in *Figure 4*.

In the design of the upper and lower skin-stringer, compressive yield strength and modulus of elasticity in compression and the tensile strength, tensile yield strength and tensile modulus are the static material properties, respectively. The corrosion resistance, fatigue resistance are also very important[2]. Considering

Table 2 Mechanical properties and specific strength of some aluminium alloys extrusion

Alloy/temper	UTS (MPa)	El (%)	ρ (g/cm^3)	UTS/ρ
IM/X2094-T6	720	3.4	2.7	266.7
IM/C912-CS1	706/677	12	2.86	246.8/236.7
IM/7055-T77	662	10	2.85	232.3
IM/7150-T77	648	12	2.82	229.8
IM/C912-CS2	674/649	12/9	2.86	235.7/226.9
IM/B96c-1-T2	650	8	2.89	224.9
IM/2090-T8	565	7	2.59	218.1
IM/B96c-3-T2	610	10	2.87	212.5
IM/8090-T8	515	4	2.54	202.8
PM/7090-T7	627	10	2.85	220.0
PM/7091-T7	593	12	2.82	210.3

Table 3 Tensile stress corrosion test values for the C912 alloy

Temper	Applied stress (MPa)	Failure time[a] (h)	Blank sample's UTS (MPa)	$1-\alpha$
CS1	482	136	652	0.84
CS2	457	412	572	0.60
CS3	331	73	331	0.00

[a]Mean values of five samples.

Table 4 C-ring SCC results of some Al–Zn–Mg–Cu alloys[11]

Alloy/temper	Direction of applied stress	SCC threshold (MPa)
7075-T6	LT	220
7075-T76	LT	340
7075-T73	LT	330
7178-T6	LT	170
7178-T76	LT	360
7079-T6	LT	240
C912-CS1	LT	280
C912-CS2	LT	330

the very high tensile strength, compressive yield strength and good corrosion resistance, the plate and forging of the C912 alloy will find their uses mainly in the upper (and lower) wing structure of the AE100 airplane and the extrusion of the C912 alloy will be used in the fuselage structure, such as stringers and seat tracks. Using its higher strength, higher weight saving will be gained.

Conclusion

(1) The C912 alloy is a new developed super-high-strength IM/Al–Zn–Mg–Cu alloy. In the CS1 temper (near peak-aged), its ultimate tensile strength and compressive yield strength at room temperature are all higher than 700 MPa.

Figure 4 Appearance of AE100 plane and some suggested alloys

(2) CS2 temper (retrogression and re-aging) is suitable for the C912 alloy to get the combination of high strength and good corrosion resistance. RRA processed the C912 alloy have a precipitate structure similar to that for a CS1 temper, but the grain boundary possessed a larger and more widely separated coarse phase, which is characteristic of a CS3 temper. The combination of these microstructural features are thought to be responsible for the favorable combination of properties.

(3) The improvement in SCC resistance correlates very well with the grain-boundary precipitate size and the volume fraction of matrix precipitates. The stress-corrosion characteristics in the C912 alloy mainly come from (a) the hydrogen-assisted cracking mechanism, (b) the lower stress-concentration at the grain boundaries and (c) the slower anodic dissolution rate.

(4) With the combination of very high strength, good corrosion resistance and high specific strength (even higher than some Al–Li alloys), the C912 alloy can be widely used in the commercial airplanes (such as the AE100) for weight saving.

References

1 Bobeck, G. E. and Froes, F. H., *Light Metal Age*, 1990, **6**, 5
2 Starke, E. A. Jr. and Staley, J. T., Application of Modern Aluminum Alloys to Aircraft. ATC report, Alcoa, 1995, p. 3
3 Poulose, P. K. et al., *Metallurgical Transactions*, 1974, **5A**, 1393
4 Pressouyre, G. M. and Bernstein, I. M., *Metallurgical Transactions*, 1978, **9A**, 1571
5 Talianker, M. and Cina, B., *Metallurgical Transactions A*, 1989, **20A (10)**, 2087
6 Kent, K. G., *Journal of the Institute of Metals*, 1969, **93**, 127
7 Scammans, G. M. et al., *Corrosion Science*, 1976, **16**, 443
8 Christodoulou, L. and Flower, H. M., *Acta Metallurgica*, 1980, **29**, 481
9 Hisamichi Kimura et al., *Materials Transactions*, JIM, 1995, **36 (8)**, 1004
10 Jacobs, A. J., *Metal Progress*, 1966, **5**, 80
11 Speidel, M. O., *Metallurgical Transactions A*, 1975, **6A**, 631

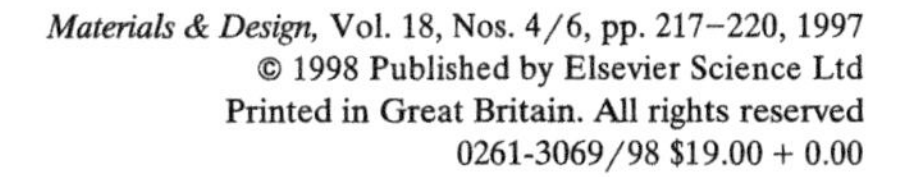
Materials & Design, Vol. 18, Nos. 4/6, pp. 217–220, 1997
© 1998 Published by Elsevier Science Ltd
Printed in Great Britain. All rights reserved
0261-3069/98 $19.00 + 0.00

PII: S0261-3069(97)00050-2

Aluminium foams for transport industry

J. Baumeister*, **J. Banhart, M. Weber**

Fraunhofer-Institute for Applied Materials Research, PowderTechnology, Lesumer Heerstraße 36, 28717 Bremen, Germany

Received 16 June 1997; accepted 18 July 1997

Foamed materials are widespread in transportation industry applications. While polymeric foams have been applied for many years foamed metals are now beginning to move into the focus of interest. A powder metallurgical method which allows the production of aluminium foams with porosity levels up to 90% is described. The foams typically have closed pores and densities ranging from 0.4 to 1 g cm^{-3}, so that this foamed metals float on water. The unique mechanical properties of metal foams are described. The density dependence of metal foam properties is shown with the Young's modulus, flexural strength and compression strength as examples. A non-linear dependency of these properties on the density is found and discussed. The discussion then focuses on the energy absorption properties of aluminium foams and tools to select appropriate foams for a given energy absorption task. © 1998 Published by Elsevier Science Ltd. All rights reserved.

Keywords: foamed materials; aluminum; transportation industry

Introduction

As a rule, the growing demands for active and passive safety of vehicles, particularly in the automotive industry, lead to an increase of vehicle weight. But this is contrary to further demands, e.g. for a possibly lower fuel consumption. For this reason, materials of low specific weight and high energy absorbing capacity are of special interest. Foamed organic materials have low specific weight, but the energy amounts convertible to deformation energy are also relatively low, as polymer foams have small strength only. Applying aluminium foams, energy absorbing devices with a corresponding higher energy level can be realised.

Manufacturing of aluminium foams

Using a powder metallurgical process, metallic foams can be produced in an elegant way[1–4]. For this, customary powders of aluminium or aluminium alloys are mixed by applying conventional techniques, for example in a tumbler mixer, with low quantities of an also powdered foaming agent. Thus, a homogeneous distribution of the gas releasing substance in the powder mixture is realised. Afterwards, this powder mixture is compacted to a semi-finished product of nearly vanishing porosity. Depending on the intended application, different compaction techniques are used. As a rule, direct powder extrusion is recommended, whereas uniaxial hot pressing is often used for test series in laboratories. Other methods such as powder rolling or hot isostatic pressing have turned out to be feasible as well, but they are more complicated and therefore only used for special applications.

Provided that correct process parameters have been chosen, the result of the compression process is a foamable semi-finished product that expands during a final heat treatment at a temperature above the melting point of the corresponding alloy. This way the material develops its highly porous structure consisting of closed cells. This implies that within the foamable semi-finished product each particle of the foaming agent must be embedded in a gas-tight metallic matrix. Otherwise, the evolving gas could escape prior to the beginning of the expansion through existing interconnected pores and, thus, would no longer be effective in producing and developing pores. The result of a successful foaming process is seen in *Figure 1*. The expansion has been driven to its maximum thus yielding a multitude of irregularly shaped cells with thin walls.

The foamable semi-finished product can be worked into sheets, rods, profiles, etc. by applying conventional techniques such as rolling, forging or extrusion prior to the actual foaming. It is possible to manufacture relatively complex shaped parts by filling adequately shaped hollow moulds with the foamable material and then heating up both the mould and the foamable material to the required temperature.

Properties of aluminium foams

Foamed materials in general and aluminium foams in particular show a number of interesting features due to their porous structure and open a wide range of applications. Metal foams combine properties which arise from the metallic nature of the matrix with the be-

*Correspondence to J. Baumeister. Tel.: +49 042 16383181; fax: +49 042 16383190; e-mail: bt@ifam.fhg.de

Figure 1 Cellular structure of an aluminium foam (area of photograph 60 × 50mm)

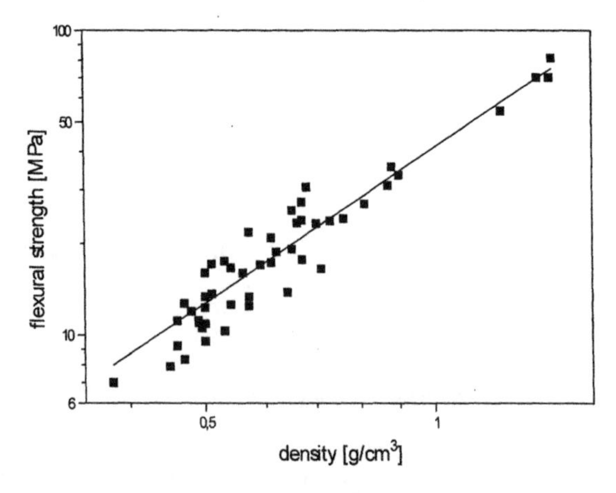

Figure 2 Flexural strength of AlMg3 foams

haviour due to their morphology. The following list includes the most interesting properties of aluminium foams which are of importance for users in transport applications. Also, some references are given:

- Nearly closed porosity
- Low specific weight[4]
- High energy absorption capacity during plastic deformation[5]
- High specific stiffness[11]
- Reduced thermal and electrical conductivity[6]
- Good mechanical[9] and acoustic damping[12]
- Not inflammable
- Recyclable
- Good machinability

An example of some aluminium foams properties is given in *Table 1* for two foams made of different alloys and having different densities.

An obvious fact is that most properties depend strongly on the density of the foam and the properties of its matrix material. Foams with a higher density, e.g. have higher compression strength and foams made of high strength aluminium have higher compression strength than those made of low strength aluminium.

Many properties obey a power law of the type:

$$A(\rho) = c \cdot \rho^n \tag{1}$$

Here A is the quantity of interest, c is a constant, ρ is the density and n is an exponent which is usually in the range between 1.5 and 2. Examples of such dependence for the compression strength, conductivity and Young's modulus have been measured by various authors[4,6–8,12]. *Figure 2* shows the flexural strength of AlMg3 foams as a function of density. In this case the exponent is about 1.7.

Energy absorption of aluminium foams

Up to now, polymer foams or honeycomb structures are used in energy absorbing structures. The possibility of controlling the stress-strain: behaviour by an appropriate selection of matrix material, cellular geometry and relative density makes foams an ideal material for such applications[10]. Decisive for the quality of packing protections or energy absorbers is the feature of

Table 1 Properties of aluminium foams (typical values, all measured at 20°C) (for comparison: conventional massive aluminium)

Alloy		Al99.5 foam	AlCu4 foam	Al99.5 massive
General data				
Foaming agent	—	TiH_2	TiH_2	—
Heat treatment of foam	—	None	Hardened	—
Density	$g\ cm^{-3}$	0.4	0.7	2.7
Mean pore diameter	mm	4	3	—
Mechanical properties				
Compression strength	MPa	3	21	—
Energy absorption at 30% strain	$MJ\ m^{-3}$	0.72	5.2	—
	$kJ\ kg^{-1}$	1.8	7.4	—
Young's modulus	GPa	2.4	7	67
Dynamical loss factor (1 kHz)	1	$25 \cdot 10^{-4}$	—	$< 5 \cdot 10^{-4}$
Electrical and thermal properties				
Electrical conductivity	$m\ (\Omega \cdot mm^{-2})$	2.1	3.5	34
Specific electrical resistivity	$\mu\Omega \cdot cm^{-1}$	48	29	2.9
Thermal conductivity	$W\ (m \cdot K)^{-1}$	12	—	235
Thermal expansion coefficient	$1/K$	$23 \cdot 10^{-6}$	$24 \cdot 10^{-6}$	$23.6 \cdot 10^{-6}$

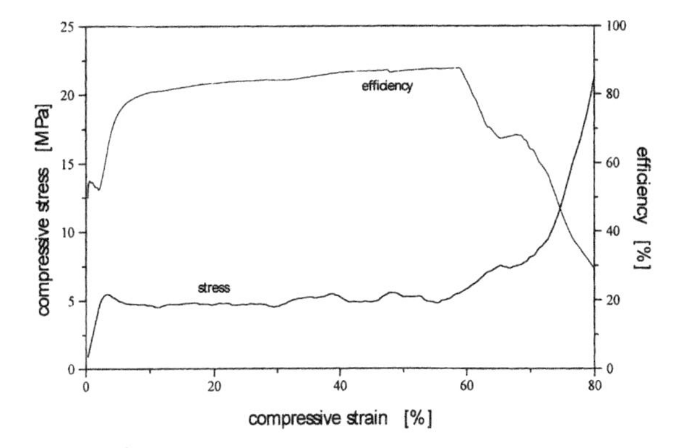

Figure 3 Compression stress and energy absorption efficiency of an AlSi12 foam ($\rho = 0.36$ g cm^{-3})

being able to absorb energy without the maximum stress or the highest occurring acceleration exceeding the upper limit at which damages or injuries occur. Compared to foamed organic materials, metallic foams are more advantageous if, due to a small available design space, a higher deformation stress with the same or uprated energy absorption is requested.

Figure 3 shows the deformation behaviour of foamed aluminium under compressive load. The energy per volume absorbed by the material corresponds directly with the area under the respective stress-strain curve. The foam shows a constant deformation stress and therefore can absorb much more deformation energy than a piece of massive aluminium when both are loaded up to a given limited stress level. The major part of the absorbed energy is irreversibly converted into plastic deformation energy which is a further advantage of foamed aluminium. At the same stress level the dense matrix material is deformed in the regime of reversible linear-elastic stresses and releases most of the stored energy after the load has been removed.

This elastic behaviour of a dense material is especially disturbing in applications where a controlled impact energy absorption without repercussion is requested as for example in automobile crush zones.

As aluminium foams can be produced in a relatively wide spectrum of densities and properties[4], the question arises which foam is the most suitable one for a given energy absorption task. For the selection of the appropriate energy absorbing materials, analytical methods can be applied which relate the demands and the foam parameters. The analytical techniques presented briefly in the following section describe the interrelationship between foam features and energy absorbing characteristics of aluminium foams taking data as a basis which has been determined in quasistatic compression tests of AlSi12-alloy foams. For a more detailed description of the typical testing parameters and the experimental set-up (see ref. 6).

Efficiency of energy absorption

The energy absorption efficiency compares the deformation energy absorbed by a real material or component with that of an 'ideal' energy absorber. An 'ideal' absorber shows a rectangular march of the load-compression curve, i.e. it reaches directly the maximum admissible strain and keeps it constant during the whole deformation process. The efficiency η is defined as ratio of the actually absorbed energy after a compression strain s and the energy absorption of the ideal absorber:

$$\eta = \frac{\int_0^s F(s')\mathrm{d}s'}{F_{\max}(s)s} \tag{2}$$

where $F_{\max}(s)$ is the highest force occurring up to the deformation s.

As all real materials show a varying stress under compression, the calculated efficiency also changes during the deformation process and therefore depends on the nature of the load-compression curve. Depending on density and alloy composition, the foamed aluminium reaches efficiency values of up to 90%, especially for the first 60% of deformation. The relative density, the cellular morphology, the foam homogeneity as well as density gradients influence considerably the length of the plateau during compression. In the area of densification, the efficiency decreases with increasing stress. It can be said that foams can be loaded optimally only until the end of the plateau area in the stress-strain curve. *Figure 3* shows the stress-strain curve of an AlSi12 foam with the numerically determined efficiency (right scale in the diagram).

The energy absorption efficiency is a valuable parameter in characterising a given test specimen with respect to its energy absorbing features and allows to draw conclusions from general foam properties to the deformation behaviour. However, for the selection of an appropriate material for a given energy absorption problem, the efficiency alone is not sufficient.

The energy absorption capacity

Especially concerning the construction of vehicles, the space and weight required for additional structural components are of high importance. The absorbed impact energy per initial volume of the energy absorber is therefore of special interest and is shown in Fig. 4 as a function of foam density. The three curves show the

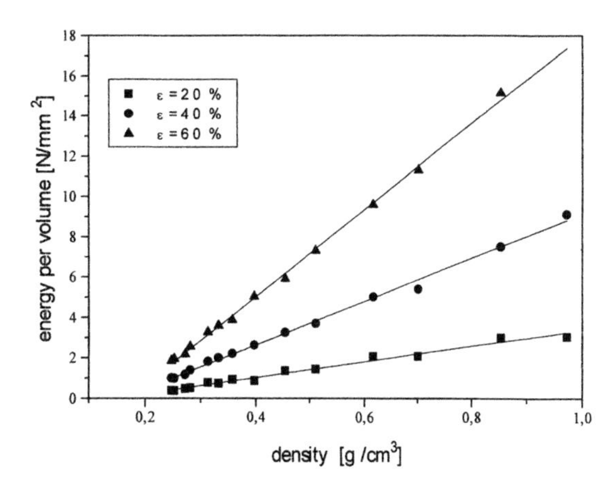

Figure 4 Energy per unit volume absorbed by various AlSi12 foams after compression strains ϵ of 20%, 40% and 60%

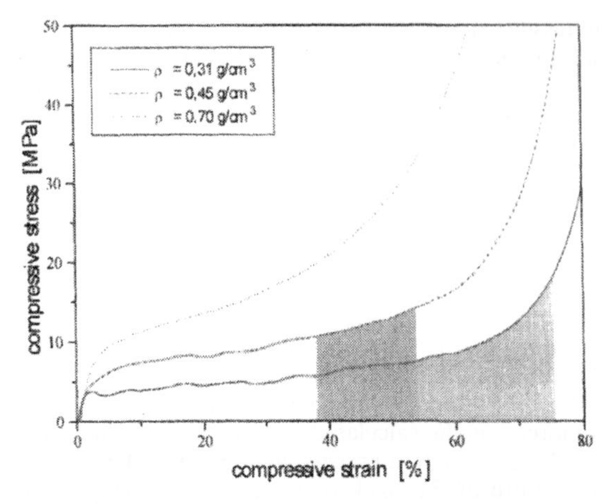

Figure 5 Compression behaviour of three AlSi12 foams of various densities. The various shaded areas correspond to the same absorbed amount of energy W^*

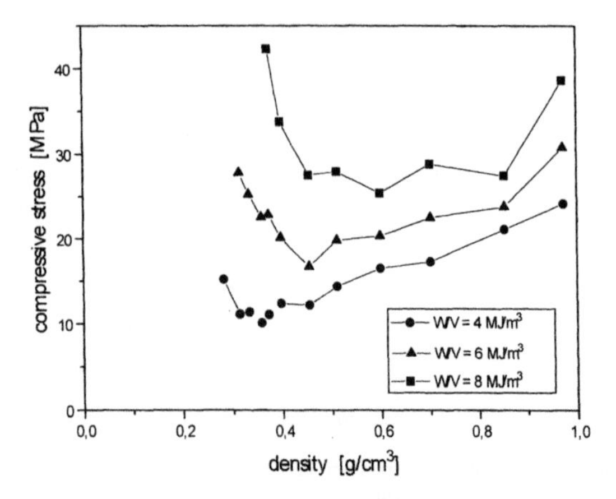

Figure 6 Maximum stress occurring when a given deformation energy is absorbed by foams of various densities

absorbed deformation energies after compression strains ϵ, of 20%, 40% and 60% and are calculated from the data of compression tests[6].

The energy absorption diagram

Maiti et al.[10] have developed the present methods and proposed energy absorption diagrams for the determination of optimised energy absorbers. Here it is presumed that a near-ideal foam absorbs a given energy at a minimum stress. *Figure 5*, showing the compression behaviour of three foams of various densities, explains these facts in detail.

The shaded areas correspond to an equal amount of energy W^* absorbed by the three foams. The right margin of each of the shaded areas marks the compression which is necessary to absorb this amount of energy. In the case of the lowest density, the stress-strain curve has already passed through the regime of constant stress before the energy W^* has been absorbed and therefore, the stress reaches high values. On the other hand, the foam of the highest density hardly shows a plateau area with constant stress at all and has a higher maximum stress, too. In contrast, for the given impact energy W^*, the foam of the medium density is loaded exactly up to the end of the plateau area. Therefore, it shows the lowest peak stress up to full energy absorption. In this way, for each given impact energy a foam of a specific density can be determined showing the lowest possible maximum stress during deformation.

Finally, *Figure 6* shows the maximum stress occurring when a given deformation energy is absorbed by foams of various densities. With decreasing energy absorption, the minimum of the curves in *Figure 6* is shifted to lower foam densities. Again it can be seen, how for various given energy densities and maximum permitted impact stress levels the appropriate foam can be selected.

Prospects

As described above, there are various techniques for the selection and evaluation of energy absorbers made of foamed aluminium. But it has to be emphasised that a simple foam structure does not necessarily present an optimum energy absorption element. By integrating such elements into the whole body structure, these elements could be tailor-made in so far that the deformation behaviour of the whole structure permits an efficient dissipation of energy. For example, it is possible to use integral foams or composite materials made of foamed aluminium and conventional materials to improve the very good energy absorption capacity of the aluminium foam in this way. This does not only apply to the frontal collision but also to the side impact protection. By selecting the alloy and possibly also the heat treatment condition, it is possible to influence besides the failure mode (brittle/ductile) also the stress level. For the development of highly efficient energy absorbers, aluminium foams represent an ideal technological basis.

References

1 Baumeister, J., German Patent DE, 1990, 4018360
2 Baumeister, J., US Patent, 1992, 5 151 246
3 Baumeister, J., European Patent EP 0460392A1, 1996
4 Banhart, J., Baumeister, J., Weber, M., In *Proceedings of the European Conferenc on Advanced Materials* (PM '95). Birmingham, 1995, p. 201
5 Baumeister, J., Banhart, J., Weber, M., In *Proceedings of the International Conference on Materials by Powder Technology*. Dresden, 1993, p. 501
6 Weber, M., Thesis, Technical University Clausthal, 1995
7 Thornton, P.H., Magee, C.L., *Metallurgical Transactions*, 1975, **6A**, 1253
8 Prakash, O., Sang, H., Embury, J.D., *Materials Science and Engineering*, 1995, **A199**, 195
9 Banhart, J., Baumeister, J., Weber, M., *Materials Science and Engineering*, 1996, **A205**, 221
10 Maiti, S.K., Gibson, L.J., Ashby, M.F., *Acta Metallurgica*, 1984, **32**, 1963
11 Baumeister, J., Banhart, J., Weber, M., Kunze, H.-D., *Powder Metallurg. International*, 1993, **25**, 182
12 Simancik, F., Degischer, H.P., Wörz, H., *Proceedings of 4th European Conference on Advanced Materials and Processes*. Padua, 1996, p. 191

Materials & Design, Vol. 18, Nos. 4/6, pp. 221–226, 1997
© 1998 Published by Elsevier Science Ltd
Printed in Great Britain. All rights reserved
0261-3069/98 $19.00 + 0.00

PII: S0261-3069(97)00054-X

Innovative light metals: metal matrix composites and foamed aluminium

H. P. Degischer[a,b]

[a]*Institute of Material Technology, Vienna University of Technology, A 1040 Vienna, Austria*
[b]*Austrian Research Centre Seibersdorf Ltd., Centre of Competence on Light Metals, A 5282 Ranshofen, Austria*

Received 8 August 1997; accepted 20 August 1997

Low weight is required especially for those means of transport, in which material properties have to be evaluated with respect to their specific mass. The possibility of increasing the specific properties of recyclable light metals are described: reinforcements by ceramic particulates, by continuous ceramic or carbon fibres, or by the reduction of weight by foaming the metal. Examples of castings, extrusions and forgings of particulate reinforced (< 30 vol.%) aluminium alloys are given and their advantages including stiffness and wear resistance are presented. The technique of selective reinforcements by co-extrusion of particulate reinforced alloys together with conventional alloys is described. High volume fractions (> 40 vol.%) of reinforcements can be produced by gas pressure infiltration of either particulate or fibre preforms. In the case of aluminium matrix, the specific strength can be increased by a factor of up to 15, and the specific stiffness by a factor of up to 7, whereas for carbon fibre reinforced magnesium the specific strength can be increased even more. The anisotropy of fibre reinforced metal matrix composites is discussed as well as the possibilities to use cross ply preforms. The technique of foaming aluminium alloys yields materials with a specific mass in the range of 0.3–1.0 g/cm^3. Such structures with essentially closed pores exhibit higher specific stiffness for beams and membranes than massive metal. The measurement and definition of stiffness and strength values appropriate for aluminium foams are presented by referring to compression tests. © 1998 Published by Elsevier Science Ltd. All rights reserved.

Keywords: metal matrix composites; foaming aluminum; strength values; stiffness

Introduction

Low weight is required especially of the means of transport and highly accelerated machinery components. Weight savings can be achieved, if the required service properties of the material related to its specific mass can be improved. Ashby[1] compares the different mechanical and physical properties of engineering materials by plotting them against density or by defining performance parameters, which are normalised with respect to mass density. As shown in *Figure 1*, there are two possibilities to increase, for example, the specific stiffness of light metals:

- Reinforcement by ceramics (including carbon fibres), which exhibit high stiffness and strength at low density, by producing metal matrix composites (MMC)[2].
- Weight reduction by foaming[3], which increases bending stiffness and resistance to buckling due to the increased thickness of a component yet weighing less: for instance foamed aluminium.

Metal matrix composites

There are two different classes of metal matrix composites[4,5]:

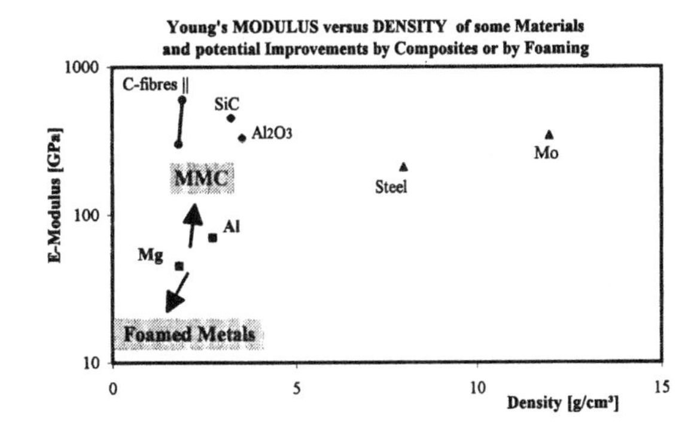

Figure 1 Tendencies to improve the stiffness vs. density ratio by metal matrix composites including foams

1. Discontinuously reinforced matrix, where either particulates, platelets (SiC, Al_2O_3, B_4C) or short fibres of high aspect ratio (usually Al_2O_3–SiO_2) are embedded within high strength alloys; conventional means of shaping: like casting, forging, extrusion, can be applied.
2. Continuous fibre reinforced metals, where either monofilaments of approximately 0.1 mm in diameter (e.g. SiC) or multifilament tows of at least some

hundreds of fibres of about 0.01 mm in diameter (e.g. SiC, Al_2O_3, carbon) are embedded within a rather ductile matrix; such components have to be manufactured by a net shape technique to preserve the continuous fibres.

There are two characteristic processing problems for such metal matrix composites:

1. The melt of light metals does not wet the reinforcing material. Therefore the two components have to be forced together (e.g. by pressure infiltration of molten metal into reinforcement preforms, or by powder metallurgical compaction in the solid state, or by stirring discontinuous reinforcement into the melt). The reactivity of the melt with certain atmospheres in the preform can stimulate pressureless infiltration due to the chemical activity[6].
2. Secondly, aluminium and magnesium are highly reactive elements, especially when liquid. As soon as bonding is achieved, care has to be taken to avoid chemical reaction between the matrix and the reinforcing phase, because the reaction products usually degrade the mechanical properties of the composite.

Typical micrographs of stirred-in particulates and of high volume fraction infiltrated particulates and continuous fibres are shown in *Figure 2*.

Particulate reinforced light metals (PRM)

There are four different production pathways[7] which determine the properties of PRM:

1. The powder metallurgical route to mix the ceramic powder with metal powder and to consolidate the mixture by extrusion, powder forging or hipping. Particulates smaller than 10 μm in diameter can be introduced up to 35 vol.% into the metal matrix. The reactivity of the components is less interfering because of solid state processing (e.g. SiC particulates embedded into Al wrought alloys).
2. The stir casting technique, where the ceramic powder is mixed into the molten matrix by stirring (Al: DURALCAN[8], Mg: MELRAM[9]). Due to the non-wettability only particulates larger than 10 μm can be introduced up to approximately 20 vol.%. The reinforcing systems have to be chosen carefully to avoid dissolution of the particulates in the melt (e.g. SiC particulates can be stirred into Al-cast alloys of high Si content, whereas Al_2O_3 particles can be introduced into Al wrought alloy melts).
3. Reinforcements by in situ reaction products can be achieved in certain phase systems similar to the carbide formation in tool steels (e.g. TiB_2 in Al–Cu-alloys[10]).
4. High volume fractions of particulate reinforcement can be achieved by infiltration of preforms either by pressure assisted casting techniques[11] or by chemically activated infiltration[6].

In contrast to dispersion strengthened metals, where the inclusions are smaller than 1 μm, the reinforcement of conventional light alloys by particulates can be explained by[4].

- An increase in stiffness due to the rule of mixture of the inverse Young's moduli of the components.
- An increase in yield strength of the matrix alloy due to the local strain hardening around the ceramic particles, which is produced during cooling from elevated temperatures due to the mismatch in thermal expansion of the components.
- A hardness increase due to the large enough, hard ceramic particles, which yield a considerable increase in wear resistance.

One of the major drawbacks of the reinforcement is the reduced ductility, especially of castings. The particulates tend to segregate along the grain boundaries, especially at slower cooling rates like sand casting,

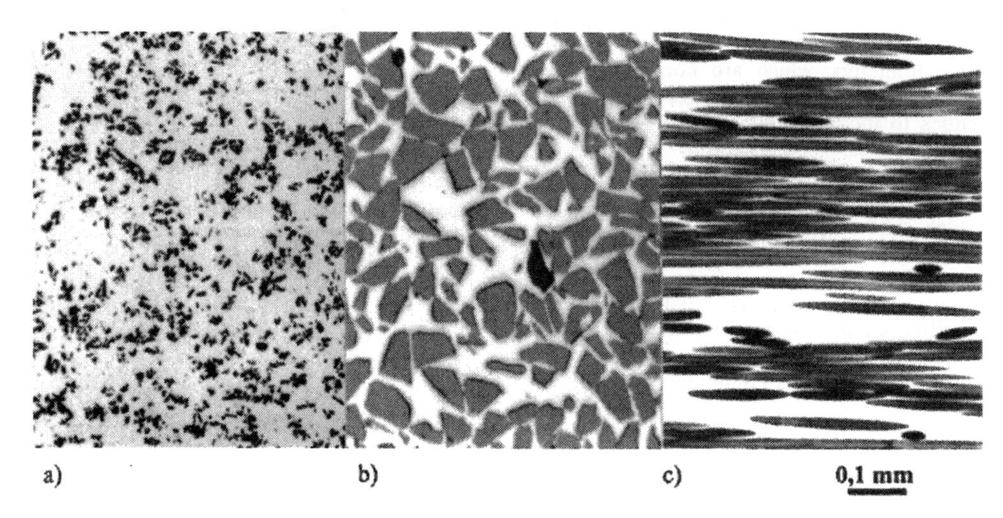

Figure 2 Optical micrographs at the same magnification of examples of different classes of metal matrix composites: (a) 10 vol.% of irregularly shaped alumina particulates embedded in an aluminium wrought alloy by stir casting; (b) 50 vol.% volume fraction SiC preform infiltrated by an Al-alloy; (c) 55 vol.% of unidirectional continuous alumina-fibres (Altex™) infiltrated by pure aluminium

Figure 3 Examples of particulate reinforced aluminium test components: (a) castings of brake and engine components, extruded cylinder liners, forged connecting rods; (b) extruded profiles, some of which are selectively reinforced by co-extrusion (PRM at crown of snow cat traction element, along inside of cylinder liner tube)

which causes embrittlement. On the other hand, the particulates conserve small grains of recrystallised hot extruded micro-structures even during prolonged heating. As mentioned, particulate reinforced aluminium and magnesium alloys can be worked (cast, extruded, forged, machined) like conventional alloys, but some additional precautions have to be applied. Polycrystalline diamond tools should be used for machining of these materials[12]. The potential applications are engine parts, brake components, transmission beams, stiffeners and sporting goods. Some examples of castings, forgings and extrusions are shown in *Figure 3*.

Higher volume fractions of particulates like SiC, as shown in *Figure 2b*, are embedded in aluminium by infiltrating preforms. The main purpose is to adjust the thermal expansion of conducting substrates to Si semiconductors of high power modules[13].

The reinforcement by short fibres is achieved by the shear lag of the load along the fibres. The strengthening by short alumina fibres becomes effective at elevated temperatures[4]. Selective reinforcement of components of combustion engines (like piston crowns[14] and cylinder liners) takes advantage of the increased strength and fatigue resistance at temperatures of approximately 200°C, of the increased stiffness and of the reduced thermal expansion due to the volume fraction of fibres without increasing the weight of the material.

Continuous fibre reinforced light metals (CFRM)

There are processing techniques[4,5], which start with coating the fibres with the matrix metal and consolidating packed bundles by diffusion bonding. Flat and tubular shapes can be produced by this method. More complicated, three-dimensional components of highest properties can be produced by gas pressure infiltration of wound or braided preforms[15]. Densely packed unidirectional ceramic fibres' (10–15 μm in diameter) preforms achieve 50–60 vol.%, whereas the thinner carbon fibres (approx. 7 μm in diameter) can be wound to 60–70 vol.%. The metal matrix provides the load transfer between the fibres and should be ductile to prevent local stress concentrations, which cause premature fracture. The advantages of metal matrices with respect

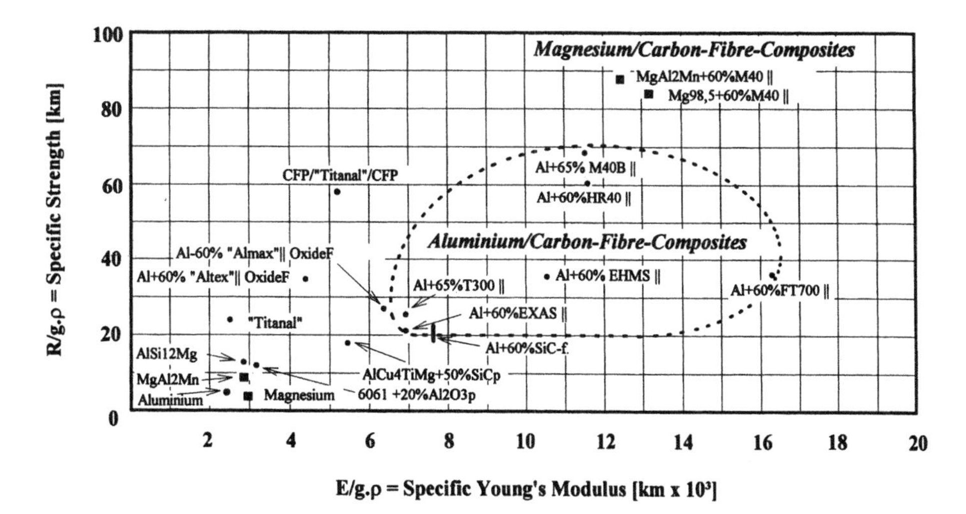

Figure 4 Comparison of specific properties of aluminium and magnesium matrix composites[11] indicating the increase of stiffness and strength with respect to the matrix (longitudinal CFRM properties)

Figure 5 Examples of continuous fibre reinforced Al test samples and a ceramic fibre bobin

to polymer matrix are the increased transverse strength and stiffness and the higher thermal stability.

The reactivity of the metals with ceramic and carbon fibres has to be taken into account for the adequate processing parameters:

- The contact time between liquid aluminium or Mg–Al-alloy melts and SiC-fibres or carbon fibres has to be minimized to avoid interfacial carbide formation, which would degrade the strength of the composite considerably.
- The contact time between liquid magnesium and all ceramic fibres has to be minimized, because Mg attacks SiC as well as alumina. On the other hand, Mg does not react with C-fibres, which causes debonding during cooling, therefore additions of alloying elements like Al are necessary to achieve sufficient transverse properties.

Although continuous fibre reinforced aluminium and especially magnesium achieve the highest specific strength and stiffness as shown in *Figure 4*, the high cost of the fibres and of the processing hinders their technical application. There are some chances to use these composites in small volumes for selected areas of components, where high loads are exerted, e.g. as inserts in castings. Some test samples are shown in *Figure 5*. Research on the processing of shaped parts of aluminium and magnesium matrix composites has resulted in some low weight components for aerospace and sports, where high strength and high stiffness, increased wear, temperature and fatigue resistance are required[4].

Figure 6 Examples of shapes produced by foaming powder metallurgically processed Al-precursor material, some of which are cut to reveal the pore structure

Figure 7 Examples of pore structures of different foamed Al-alloys (revealed by machining)

Foamed aluminium

A remarkable reduction of the weight of metals—down to one-tenth of their mass density—can be achieved by foaming[3,16]. Beams and plates made of foamed metals reach higher specific stiffness values than those made of massive metal due to their favourable thickness to weight ratio. The foaming of powder metallurgically produced precursor material within moulds allows to produce three-dimensional parts of complex shapes with a more or less tight surface skin[17,18], some of

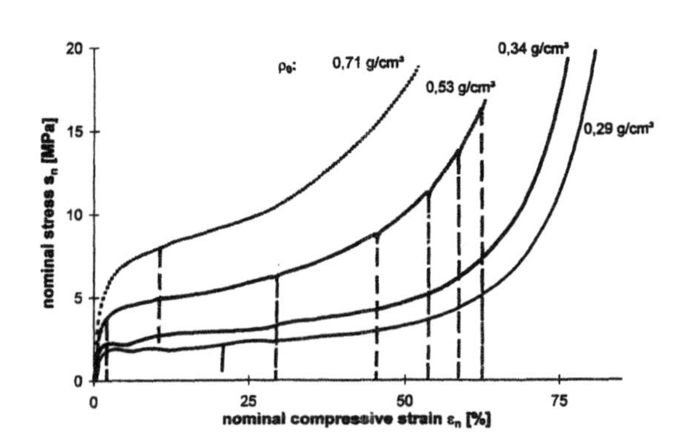

Figure 8 Compression curves of pure Al foams of indicated apparent mass densities (tested without surface skin) with some unloading–loading loops (dashed lines) for stiffness measurements

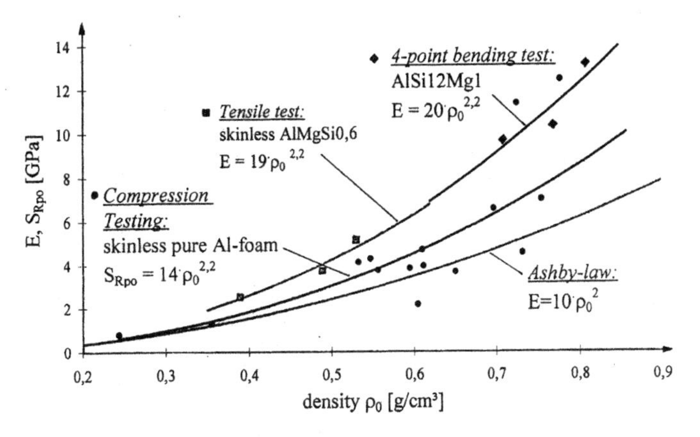

Figure 9 Stiffness of foamed Al determined by unloading–loading hysteresis at yield stress during the indicated tests. The fitted power law lines are compared with the Ashby formula[3]

which are shown in *Figure 6*. The size of the pores varies between fractions of 1 mm up to several millimetres as shown in *Figure 7*. The pores are closed cells during the foaming procedure, but some of them may crack during the cooling process. Extremely light aluminium-based material with densities between 0.3 up to approximately 1 g cm^3 can be produced of cast and wrought alloys[19]. The mechanical properties of foams have to be defined specifically taking into account the appropriate measuring procedures[20]. The compression curves in *Figure 8* are superimposed for different apparent foam densities showing the plateau stress increasing with density. The diagram in *Figure 9* shows the dependence of the stiffness of the porous structure determined by load hysteresis (as indicated by the dashed lines in *Figure 8*) on the average foam density, which obeys a power law of $n_E = 2.2$. The plateau strength extrapolated to zero deformation during compression as a function of apparent density (*Figure 10*) can be described as well by a power law with $n_R = 1.5–1.7$. If age hardening alloys are used, T5 treatments increase the strength of the foams as shown in *Figure 10* for a low alloyed AlMgSi-type.

Foamed metals may introduce innovations for vehicles, mechanical engineering, design, architecture and applied arts taking advantage of new property combinations: low mass density + stiffness + crash energy absorption + thermal stability + non inflammability + electromagnetic shielding + acoustic protection + decorative surface + corrosion resistance. Due to the early state of processing technology for foamed aluminium parts the development of small series productions of high value goods is envisaged first.

Conclusion

Both types of heterogeneous materials, MMC and metal foams, cannot be treated as semi-products, but have to be manufactured according to the special purpose of the component. Furthermore, due to their high cost they will be used selectively in combination with conventional materials and the benefits of weight gaining and higher performance have to be considered for the whole system, which has to be designed appropriately to be able to exploit the advantages of these advanced materials.

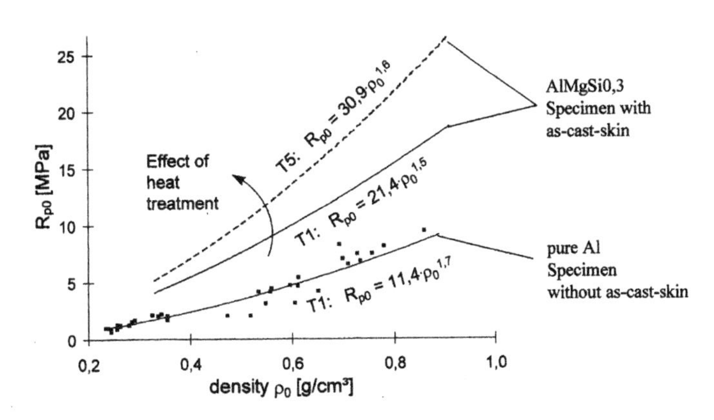

Figure 10 Plateau stress determined from compression tests of different Al-matrix alloy conditions extrapolated to the stress axis for different apparent densities

References

1 Ashby, M. F., *Materials Selection in Mechanical Design*. Pergamon Press, Oxford, 1992
2 Ashby, M. F., *Acta Metallurgical*, 1993, **41**, 1313–1335
3 Ashby, M. F., *Metallurgical Transactions*, 1983, **14A**, 1755–1769
4 Clyne, T. W. and Withers, P. J., *Introduction to Metal Matrix Comp*. Cambridge University Press, Cambridge, 1993
5 Kainer, K. U., *Metallurgical Verbundwerkstoffe*, DGM, Oberursel, 1994
6 Schiroky, G. H., Miller, D. V., Aghajananian, M. K. and Fareed, A. S., *Key Engineering Materials* 1997, **127 / 131**, 141–152
7 Masounave, J. and Hamel, F. G., *Fabr. of Part. Reinforcement Metal Comp*. ASM, Metals Park, 1990
8 DURALCAN-Composites, *Particulate Reinforced Aluminium Alloys*. San Diego, 1992
9 Magnesium Elektron Leaflet, *MELRAM*. Manchester, 1995
10 London and Scandinavian, Information sheet, London, 1996.
11 Degischer, H.P. in Ref. [5], pp. 129–168
12 Degischer, H. P. and Feuchtenschlager, F., *VDI-Bericht*, 1996, **1276**, 449–461
13 Electrovac, ELCOM-Product Information, Klosterneuburg, Austria, 1996
14 Henning, W. and Neite, G., in Ref. [5], pp. 169–191
15 Degischer, H. P., Schulz, P. and Lacom, W., *Key Engineering Materials*, 1997, **127 / 131**, 99–110.
16 Banhart, J., *Metallschäume*, Metall Innovation Technologie, Bremen, 1997.
17 Banhart, J., Baumeister, J. and Weber, M., *Aluminium*, 1994, **70**, 209–212
18 Degischer, H. P. and Simancik, F., In *Proceedings of Environmental Aspects in Materials Science*, ed. H. Warlimont. DGM-Oberursel, 1994, pp. 137–140
19 *Alulight*, Mepura-Prospekt, Ranshofen, 1996
20 Gradinger, R., Simancik, F., Degischer, H. P., in Int. Conf. WT–M & MT–FM · QM, eds. Felber. Vienna Univ. Techn., 1992, **2**, 701–712

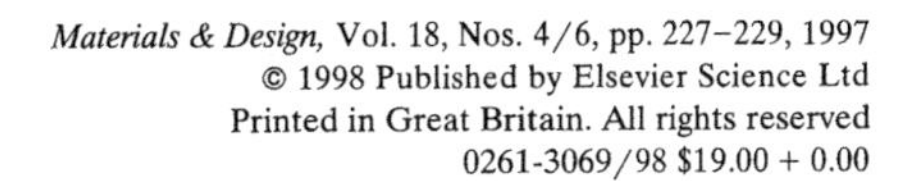

Materials & Design, Vol. 18, Nos. 4/6, pp. 227–229, 1997
© 1998 Published by Elsevier Science Ltd
Printed in Great Britain. All rights reserved
0261-3069/98 $19.00 + 0.00

PII: S0261-3069(97)00055-1

Grain growth and grain orientation distribution of Al–Mg alloys

Y. Umakoshi*, K. Matsumoto, T. Shibayanagi, H. Morisada

Department of Materials Science and Engineering, Faculty of Engineering, Osaka University, 2-1, Yamada-oka, Suita, Osaka 565-0871, Japan

Received 14 August 1997; accepted 20 August 1997

Grain growth behaviour of Al–Mg alloys containing 0.3, 2.7 and 5 mass% Mg was investigated focusing on the spatial distribution of grain orientation and grain boundary character. In Al–0.3 mass% Mg alloy the cube texture developed at the first stage and then the texture declined accompanied with abnormal grain growth of non-cube grains at the second stage. The development of cube grains was suppressed by an increase of solute Mg atoms. The texture change depended strongly on spatial distribution of grain boundary character and cube clusters. © 1998 Published by Elsevier Science Ltd. All rights reserved.

Keywords: Al-Mg alloys; grain orientation; grain boundary

Introduction

Aluminium alloys are used in car bodies and parts to save energy and minimize environmental pollution because of the light-weight and superior recyclability. The poor press-formability is one of the major problems to be solved before industrial application. The main cause of this poor formability results from an inadequate crystallographic orientation distribution in the recrystallized grains. The texture is controlled during annealing after cold rolling by several important factors such as distribution of grain orientations and grain boundary character, depending on initial orientation of grains before cold rolling, primary and subsequent grain growth process. A process to obtain the good balance of recrystallized texture is required to improve the press formability. The texture is known to be very sensitive to initial distribution of grain size, grain orientation, grain boundary character (GBC)[1–7].

In this paper, we describe several controlling factors for recrystallization texture in Al–Mg alloys focusing on the effect of geometrical distribution of initial grain orientation and GBC, and the growth behaviour of cube-oriented clusters.

Experimental procedure

Two Al–Mg alloy ingots containing 0.3 and 2.7 mass% Mg were made by melting pure Al (99.99%) and Mg (99.98%) and then homogenized at 773 K for 86.4 ks. The ingots were hot-rolled at 773 K to 72% of their original thickness. After annealing at 623 K for 370 s, they were cold-rolled to a final thickness of 0.9 mm with 90% reduction in thickness. Subsequent annealing for 3.6 ks was carried out at 558 and 573 K for the primary recrystallization of the Al–0.3 mass% Mg and Al–2.7 mass% Mg alloy, respectively. The further annealing was performed at 773 K for an adequate time from 2×10^2 to 2.6×10^4 s to examine the grain growth process.

Al–5 mass% Mg alloy was also prepared to investigate the effect of cube-clusters on the grain growth process with the same initial grain size. After cold-rolling, the Al–5 mass% Mg alloy sheets were recrystallized at 773 K for 20 s and at 673 K for 1.8 ks to receive two kinds of samples with different initial grain orientation and distribution, but with the same average grain size of about 70 μm. The further annealing was carried out at 773 K for 1×10^3 to 1×10^5 s to examine the effect of distribution of cube-clusters on the grain growth behaviour.

Microstructures were examined using an optical microscope and a scanning electron microscope (SEM), JSM-840A. The macro texture was measured in the normal direction (ND) plane using X-ray diffraction and SEM-Electron Channelling Pattern (ECP) techniques. Microtextures and/or the spatial distribution of the crystallographic orientation in grain growth were measured on the rolling direction (RD) plane and the ND plane using SEM-ECP technique.

Grain boundary structure was characterized by Σ values determined from the ECPs of two adjacent grains. Brandon's condition was used as the criterion of the exact coincidence orientation relationship[8].

Results and discussion

Primary recrystallization texture was mainly composed of cube {001} ⟨100⟩, rotated (R)-cube {001} ⟨310⟩ and other non-cube components containing a high volume

*Correspondence to Prof Y. Umakoshi. Tel.: +81 6 8797494; fax: +81 6 8797495; e-mail: umakoshi@mat.eng.osaka-u.ac.jp

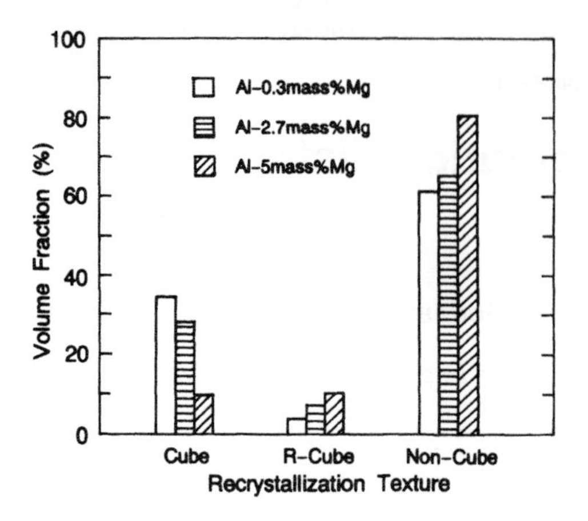

Figure 1 Volume fraction of cube, R-cube and non-cube components in primary recrystallized grains of Al–Mg alloys

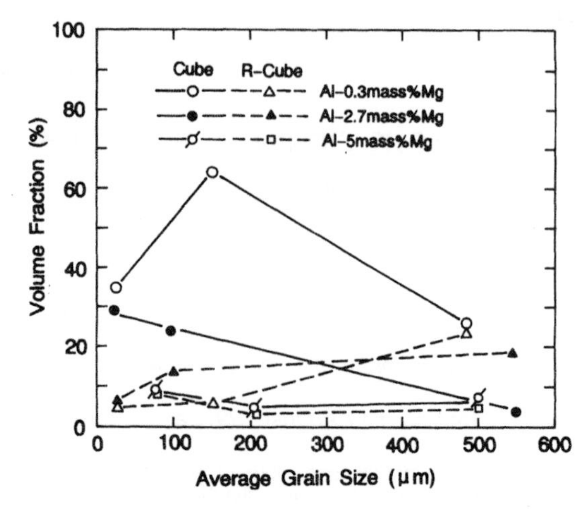

Figure 2 Variation of volume fraction of cube and R-cube components in A–Mg alloys with the average grain size

fraction of grains with {123} ⟨634⟩ orientation. *Figure 1* shows the volume fraction of each texture component observed in the primary recrystallized specimens measured on the ND plane by the SEM-ECP technique. In Al–0.3 mass% Mg alloy, the primary recrystallized texture mainly consisted of a cube component, while the rather high volume fraction of grains with the R-cube component was observed in Al–2.7 mass% Mg and A–5 mass% Mg alloys. The volume fraction of cube component decreased and the R-cube component increased with increasing Mg concentration.

The change of texture was examined using X-ray diffraction method and SEM-ECP technique for each grain and there was no significant difference in grain orientations obtained by both methods. *Figure 2* shows variation in volume fraction of grains with cube and R-cube orientations during grain growth for the three alloys. The volume fraction of grains with cube component in Al–0.3 mass% Mg alloy remarkably increased at an early stage of annealing and then turned into a slight decrease as grain growth proceeded, while a monotonic decrease in the volume fraction was observed in Al–2.7 mass% Mg and Al–5 mass% Mg alloys. In contrast, the volume fraction of R-cube component monotonically increased as the grain growth proceeded. The frequency of cube and R-cube components in three alloys showed similar tendency to the change in volume fraction of their components.

Grain orientation distribution and grain growth process are very sensitive to GBC and GBC distribution (GBCD) during the grain growth. *Figure 3* shows the change in frequency of $\Sigma 1$, coincidence boundaries with $\Sigma 3$ to $\Sigma 51$ and random boundaries in Al–0.3 mass% Mg and Al–2.7 mass% Mg alloys. The $\Sigma 1$ boundary increased in the frequency at the initial stage in Al–0.3 mass% Mg alloy, followed by a decrease at the second stage, while random boundaries showed the reverse tendency. No significant change was observed for coincidence boundaries with $\Sigma 3$ to $\Sigma 51$. In Al–2.7 mass% Mg alloy, the frequency of $\Sigma 1$ boundary maintained a low level during grain growth and there was no significant change in the frequency of coincidence and random boundaries.

The GBCD in Al–0.3 mass% Mg and Al–2.7 mass% Mg alloys can be evaluated by the classification of grain

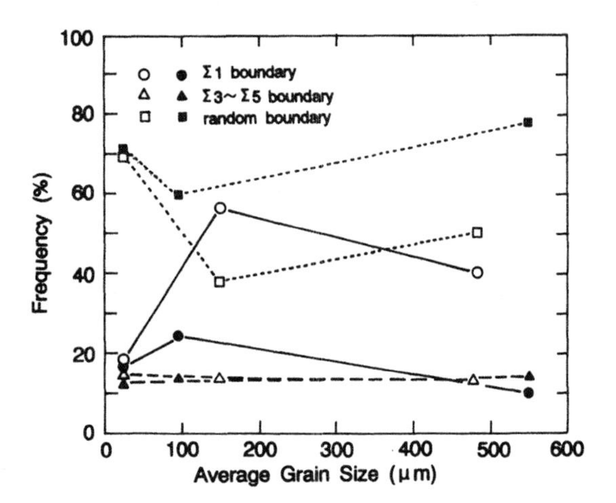

Figure 3 Change in frequency of grain boundary character in Al–Mg alloys as a function of the average grain size. Open and closed symbols represent data of Al–0.3 mass% Mg and Al–2.7 mass% Mg alloys, respectively

boundaries forming between adjacent cube/cube (C/C), cube/non-cube (C/N) and non-cube/non-cube (N/N) grain groups. After primary recrystallization, the frequencies of the $\Sigma 1$ boundaries in C/C group of Al–0.3 mass% Mg and Al–2.7 mass% Mg alloys were 57 and 75%, respectively. The frequency of the $\Sigma 1$ boundary was high in the C/C group, while the frequencies of random boundaries in the C/N and N/N groups were higher than in the C/C group for both alloys. The frequency of $\Sigma 1$ boundary between cube grains in the cube cluster was much higher than that in the other regions such as outside the clusters and the interface between cube and non-cube grains. The grain growth is caused by the successive occurrence of grain boundary migration which is controlled by the size effect of grains and the each balance of interfacial energy at each triple junction of grain boundaries. The balance of grain boundary energy at the triple junction is one of the driving forces for the grain boundary migration. Since the triple junction in materials exists under various unstable energy conditions, the triple junction moves approaching the energy balance, resulting in the grain growth. The spatial distribution of the character of the triple junction and local GBCD is an important controlling factor for the grain growth process. The high frequency of Σ1 boundary in cube clusters implied that the triple junctions composed of C/C, C/N and C/N boundaries moved toward the outside of cube grain clusters to adjust the energy balance of grain boundaries and the cube grain clusters grew consuming non-cube grains. Thus, at the initial stage of grain growth in Al–0.3 mass% Mg alloy, the volume fraction of the cube component remarkably increased. After initial grain growth, non-cube grains were isolated and surrounded by cube grains. A few very large non-cube grains which were much longer than the neighbouring cube grains appeared. Since the direction of movement of triple junctions was controlled by the size effect of grains to reduce the total grain boundary at a local region, non-cube grains turned to grow into cube clusters. Therefore, the volume fraction of cube grains decreased in the second stage in Al–0.3 mass% Mg alloy.

The increase of R-cube grains near cube clusters in Al–2.7 mass% Mg alloy changed the character and the energy balance of the triple junctions around cube clusters, resulting in suppression of the growth of cube clusters. The increase of Mg concentration may reduce the grain boundary energy. The energy of random boundaries is more affected than that of the Σ1 boundary by solute atoms since more Mg atoms segregate in the larger space of random boundaries [9]. Since the difference in grain boundary energy between the Σ1 and random boundaries becomes smaller, the driving force for the growth of cube clusters is reduced, resulting in monotonic decrease in volume fraction of cube component during the grain growth in Al–2.7 mass% Mg and Al–5 mass% Mg alloys.

Since the growth of grains with various orientation is related to the size of grains, two kinds of specimens A and B of Al–5 mass% Mg alloy with the same average grain size of about 70 μm were prepared. Although there was no significant difference in the initial volume fraction of cube component between both specimens, a notable change in the volume fraction was observed during grain growth process as shown in *Figure 4*. However, the specimens have different spatial distribution of cube clusters which induces different change in microstructure as the grain growth proceeds. Specimens A and B were primarily recrystallized at 773 K for 20 s and at 673 K for 1.8 ks, respectively. The (l/d) value represents the spatial distribution and stability of cube clusters, where d is the average diameter of cube grains and l is the mean distance between the centre positions of first nearest neighboured cube grains. In the specimens with the same volume fraction of cube grains, a small value of (l/d) approaching 1 means inhomogeneous distribution of cube grains, (formation of cube clusters): there are a large number of stable cube clusters composed of several grains and the stable clusters can get growth during annealing. Therefore the number of stable cube clusters increases with decreasing (l/d) value. In fact the volume fraction of cube component (frequency of cube clusters) for specimen B with lower (l/d) increased in the grain growth process, while the volume fraction in specimen A with higher (l/d) decreased as shown in *Figure 4*. Thus, the spatial distribution of cube clusters also controls the microstructure during the grain growth.

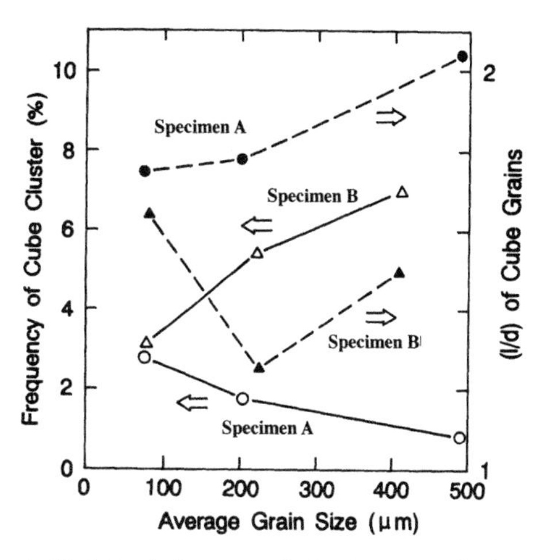

Figure 4 Variation in frequency of cube clusters and (l/d) value in Al–5 mass% Mg alloy annealed at 773 K with the average grain size. d and l represent the average diameter of cube grains and the mean distance between the centre positions of first nearest neighboured cube grains. Specimens A and B were primarily recrystallized at 773 K for 20 s and at 673 K for 1.8 ks, respectively

Conclusions

The controlling factors for recrystallization texture in Al–Mg alloys were investigated focusing on the distribution of grain orientation and grain boundary character and the following conclusions were reached:

(1) The frequency of cube component in Al–0.3 mass% Mg alloy increased in the initial stage of grain growth and then decreased accompanied by the abnormal growth of non-cube component. High Mg concentration induced monotonic decrease of cube component during grain growth.

(2) Small clusters composed of several cube grains containing Σ°1 boundaries and their spatial distribution played an important role in the change of recrystallizing texture during grain growth.

(3) The each balance of interfacial energy at triple junctions of grain boundaries affected the stability of grains, during annealing, namely the triple junctions moved approaching the energy balance.

References

1 Smith, D.A., In *Grain Boundary Structure and Properties*, ed. G.A. Chadwick. Academic Press, New York, 1976
2 Watanabe, T., *Metallurgical Transactions*, 1983, **14A**, 47
3 Watanabe, T., Fujii, H., Oikawa, H. and Arai, K.Z., *Acta Metallurgical Materials*, 1989, **37**, 941
4 Watanabe, T., *Textures and Microstructures*, 1993, **20**, 195
5 Matsumoto, K., Shibayanagi, T. and Umakoshi, Y., *Scripta Metallugical Materials*, 1995, **33**, 1321
6 Matsumoto, K., Shibayanagi, T. and Umakoshi, Y., *Materials and Science Forum*, 1996, **204–206**, 473
7 Matsumoto, K., Shibayanagi, T. and Umakoshi, Y., *Acta Metallurgical Materials*, 1997, **45**, 439
8 Brandon, D.G., *Acta Metallurgical*, 1966, **14**, 1479
9 Gleiter, H., *Acta Metallurgical*, 1970, **18**, 117

ELSEVIER

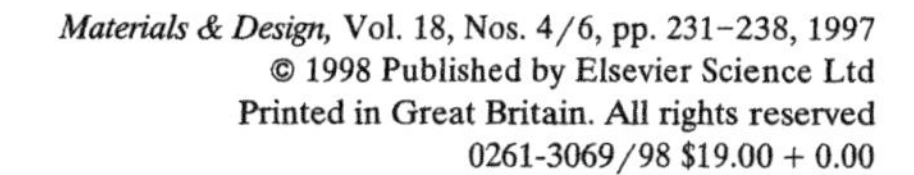
Materials & Design, Vol. 18, Nos. 4/6, pp. 231–238, 1997
© 1998 Published by Elsevier Science Ltd
Printed in Great Britain. All rights reserved
0261-3069/98 $19.00 + 0.00

PII: S0261-3069(97)00056-3

Lightweight carbon fibre rods and truss structures

R. Schütze*

DLR, Institut für Strukturmechanik, P.O. Box 3267, 38022 Braunschweig, Germany

Received 25 July 1997; accepted 8 August 1997

Lightweight carbon fibre rods and truss structures are of growing importance for modern transportation technologies. The struts of such frameworks are commonly designed as fibre-wound CFRP tubes. Here CFRP sandwich rods are an advantageous alternative. They have a lightweight foam core covered by a relative thin layer of composite material. In many real applications, however, the superior mechanical properties of such struts can only be utilized with appropriate load transfer elements. Two types of load transfer elements designed for high tensile and compressive loads with a simple screw connection will be described. The framework structures discussed in this paper refer to framework beams which have, in their simple version, cross-sections of an equilateral triangle. To realise a single point support at the ends of the beam-like truss, the rail struts must converge into a conical structure. The presented structures are connection designs without any metallic elements. An outstanding application of CFRP-trusses, rod connectors, and the mentioned formlocking load transfer elements can be found in the advanced lightweight structure of a recently developed semi-rigid airship, the Zeppelin NT. © 1998 Published by Elsevier Science Ltd. All rights reserved.

Keywords: lightweight carbon fibre rods; truss structures; transportation technologies

Introduction

Lightweight carbon fibre rods and truss structures with high stiffness, high strength and sufficient durability are more and more important for advanced transportation technologies. Of course, such structures are preferably used for aircraft and space applications. But also for many other applications, carbon fibre struts and truss structures are superior to comparable metallic components. Especially if high mass forces are limiting the functional possibilities, carbon fibre composites are very effective materials. In these cases the price, which is commonly higher for CFRP-components than for conventional materials, turns out to be no serious economical factor. However, in many composite components the superior properties of CFRP, such as high specific strength and high specific stiffness, are not sufficiently utilized due to multidirectional lay-ups with local instabilities and damages like matrix cracks and delaminations[1]. For this reason multi-purpose lightweight structural struts have been developed.

CFRP sandwich struts

Commonly, the struts which constitute a CFRP framework are designed as fibre-wound CFRP-tubes. However, there are advantageous alternatives. Most promising are the so-called 'CFRP sandwich rods'. They consist of a lightweight foam core covered by a relatively thin layer of composite material. Due to the inner support given by the foam core, the struts have a high load carrying capability although the fibre-reinforced coating is very thin[2]. There are four types of these rods available (*Figure 1*).

Most frequently used are rods of type A, which have a fibre-reinforced coating consisting of unidirectional carbon fibres oriented in longitudinal direction. This coating is again surrounded by a thin glass fibre braided hose. As a result of their construction, these rods have a high specific stiffness, particularly if high-modulus-fibres are being used. In this case the bending stiffness of the rods is significantly higher than that of steel tubes with the same dimensions. Due to the unidirectional design the torsional stiffness of type A is relatively low. CFRP sandwich rods of type B have load carrying layers consisting of one or more carbon fibre braided hoses, the lay-up angle of which can be widely adapted to the structural mechanics requirements. The mechanical properties of this rod type can be optionally adjusted between high bending and high torsional stiffness. Types BA and BAB are combinations of the two basic rods (*Figure 1*). These rod combinations can be tailored with respect to the intended application. They are particularly useful if high shear stiffness is needed as well as high bending stiffness and strength.

Figure 2 shows a solution to connect two parts of equal sandwich struts by a composite sleeve[3]. This

*Tel.: +49 531 2952303; fax: +49 531 2952875; e-mail: rainer.schütze@dlr.de

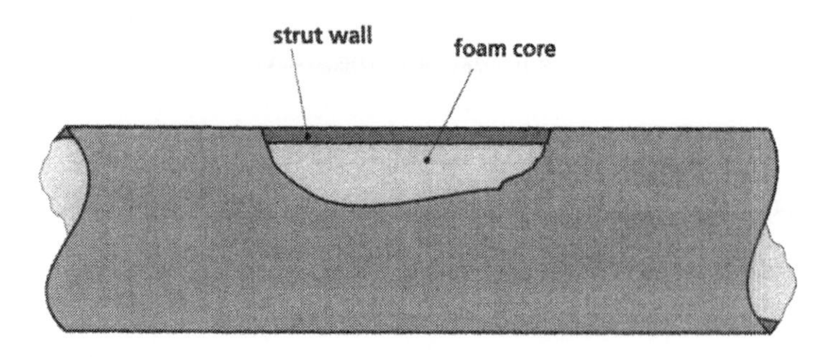

Figure 1 Lay-up of the CFRP sandwich struts

sleeve is a tube-like rod of rod type A, which will be bonded by an epoxy adhesive over a definite length to each of the semi-struts. To ensure a reliable and reproducible bonding between the sleeve and the strut parts, the sleeve must be prepared in a special way before bonding. The preparation consists of, (1) slitting the sleeve in longitudinal direction, and (2) breaking the unidirectional wall of the tube at the opposite side of the slit in two parts. The unbroken thin glass fibre braided hose will then serve as a film hinge. The amount of unidirectional fibres embedded in this slitted sleeve is equal to that contained in the two combined semi-struts. This method to connect two strut parts can be used as a repair method for damaged struts too. As can be seen in *Figure 3*, the thin glass fibre braided hose on the surface of the struts serves also as an indicator for possible damages, which might occur after strut fabrication. In particular, impact damages are indicated by white colours, so that an optical quality control of struts is quite easy. The three pictures in *Figure 3* show the damage progression under tension-compression fatigue loading at a load level of ± 7.0 KN observed on struts predamaged by impact. The predamage depicted above (at $N=0$) was performed by a special device, where a type A strut provided with load transfer elements is being dropped from a height of 1 m onto a sharp blade. The results of damage tolerance investigations on such CFRP-sandwich struts showed that defects extending over a quarter of the rod circumference lead to a drop in compression strength of more

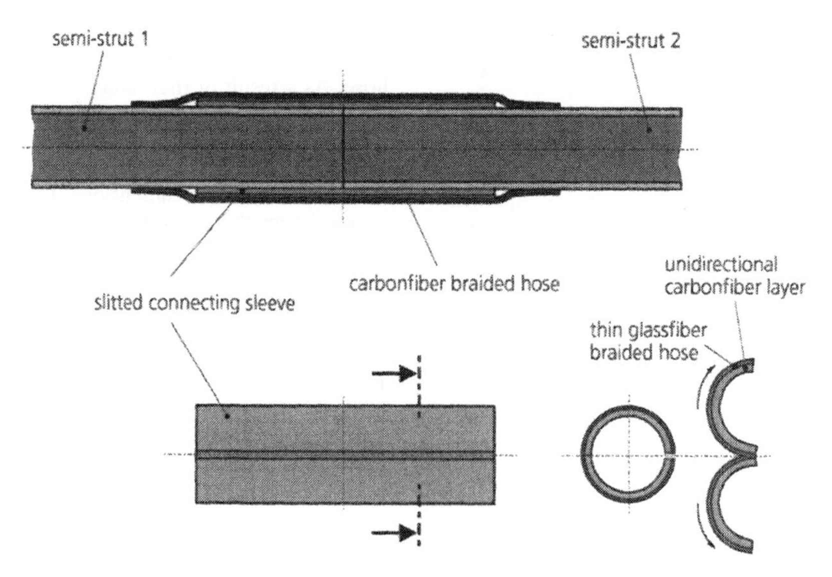

Figure 2 Connection of two equal struts

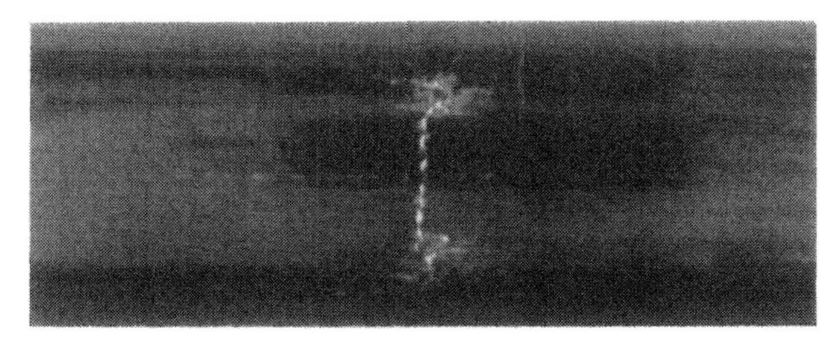

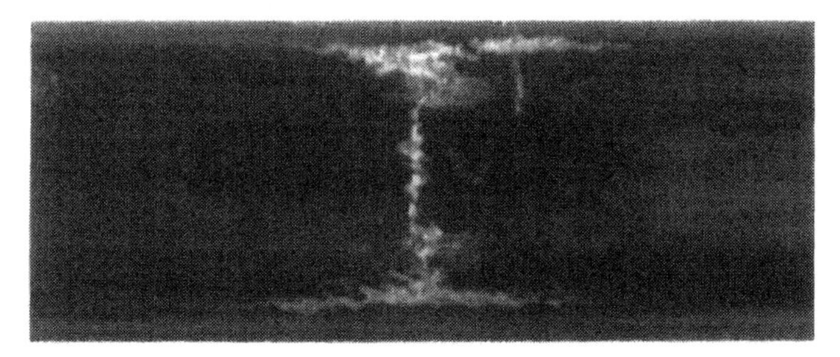

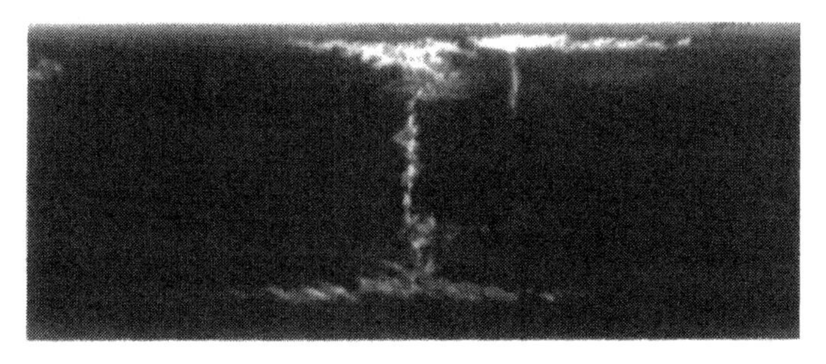

Figure 3 Damage detection on the CFRP sandwich struts

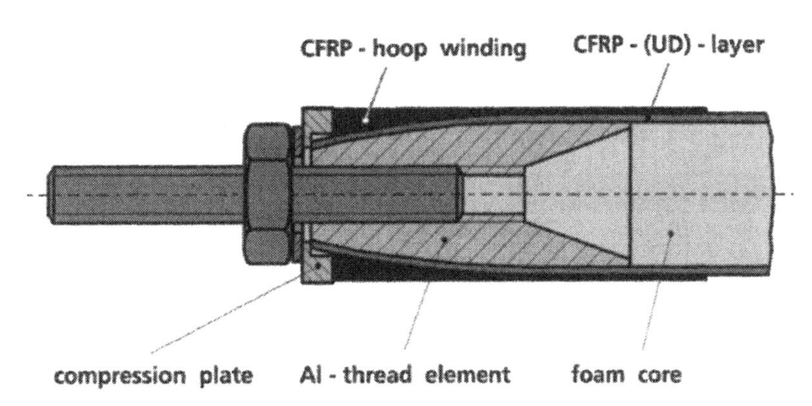

Figure 4 Design of the integrated form locking load transfer in the CFRP sandwich struts

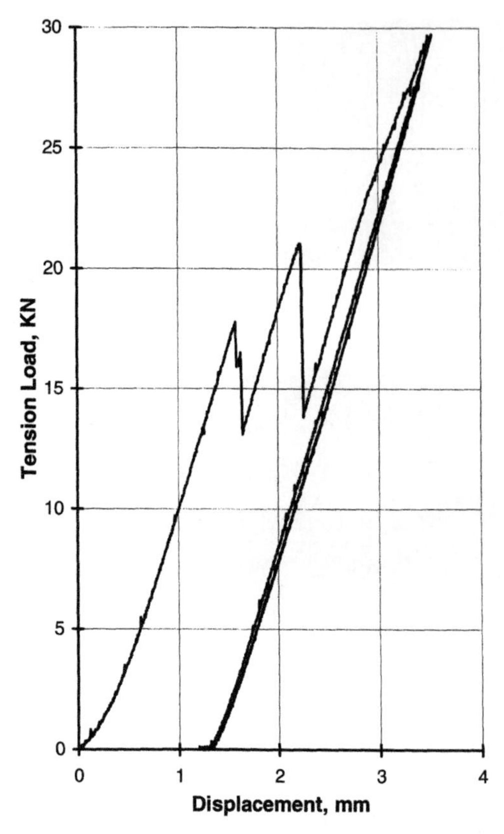

Figure 5 Load-displacement curves of struts with load transfer elements

than 50%. The reason for this large reduction is the asymmetry of the residual cross-section of the strut. But it is rather interesting that, as far as the load level remains below the static compression strength of the damaged strut, the maximum number of fatigue load cycles remains high in spite of possible crack growth in longitudinal direction.

Load transfer elements in CFRP sandwich struts

In many applications, however, the superior mechanical properties of the struts described above can only be utilized with appropriate load transfer elements, because the introduction of tensile as well as compressive longitudinal forces requires careful design which avoids any joining of incompatible partners. A design of a special form locking load transfer element for high tensile and compressive loads with a simple screw connection is described in *Figure 4*[4]. It comprises a conical metallic element with an axial threadhole, a carbon fibre hoop winding provided in a second production process, and a compression plate at the strut end. This design leads under tension loading to a load transfer via the conical thread element. In the compression regime the load will go via the compression plate directly into the unidirectional CFRP layer. It has to be mentioned that struts with such load transfer elements have to be preloaded by a tension load higher than the maximum service load. Under first high tension loading the bonding of the conical elements breaks; the element itself is tightened into the conical strut end. Thus the load transfer is performed only by form-locking elements and not by any kind of bonding. For struts of 25 mm diameter and a CFRP-tube wall thickness of only 0.6 mm the quasi-static load-displacement behaviour is displayed in *Figure 5*[2]. The two rapid drops of the tension load are typical and indicate the slipping of both conical elements into their final position. Once prestressed to a load of 30 KN, the struts established under subsequent quasi-static tension-compression loading a stress-strain behaviour with very small hysteresis loops up to the required operating load level of 30 KN. The high compressive strength is achieved by the precisely aligned longitudinal carbon fibres, which are supported on the inner side by the foam core and on the outer side by the glass fibre braid. As a consequence, even a very thin carbon fibre layer can transmit high compression loads without local buckling. More than the static behaviour, the tension-compression fatigue performance of such struts with integrated load transfer elements is of particular interest. Cyclic fa-

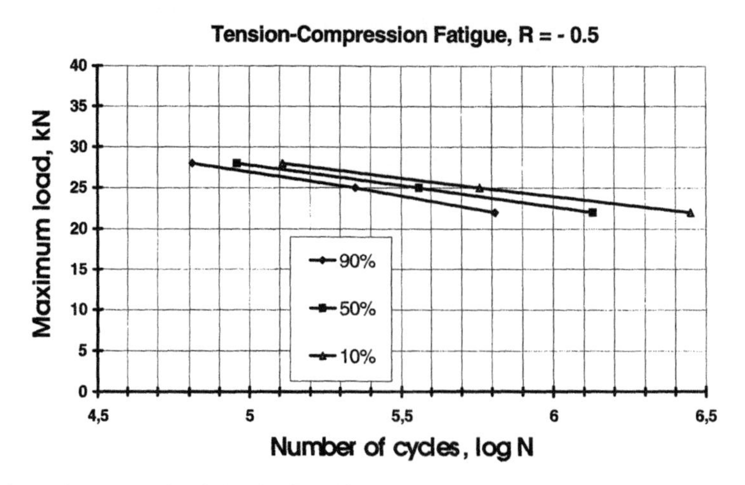

Figure 6 S–N curves under tension-compression fatigue loading with $R = -0.5$

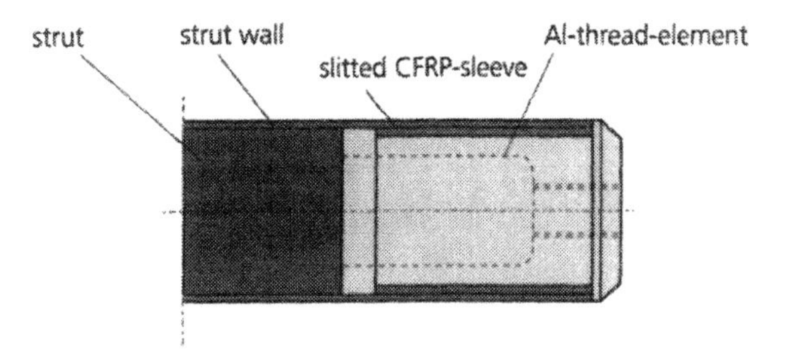

Figure 7 Design of a load transfer element assembled after fabrication

tigue tests on struts with 16 mm diameter and a CFRP-tube wall thickness of 0.6 mm were conducted with a stress ratio of $R = -0.5$ and maximum loads of $+28$ KN and -14 KN, respectively. The S–N-curve displayed in *Figure 6* shows the excellent fatigue properties of the struts. Fatigue failure occurred preferably in the sections between the load transfer elements, thus establishing the good mechanical design. Therefore, the described CFRP sandwich struts with load transfer elements can serve as general purpose structural elements for a great variety of applications.

Apart from the form-locking clamp connectors, which have to be inserted during the production process, there are also load transfer elements which can be bonded later in a strut as a semiproduct[5]. *Figure 7* illustrates a version consisting of a lathe machined aluminum part having a large back stop, a recess, a small back stop, and a slitted CFRP-sleeve fitted into the recess. Under compression loading the large back stop will press directly against the unidirectional fibres of the strut. Tension loads are resisted by the adhesive bond between the inserted CFRP-sleeve and the inner strut wall. In the case of compression loading the load transfer is effected by form-locking. In the tension regime there is a form-locking connection between the two incompatible partners (Al-element and CFRP-sleeve), while there is a force-locking connection between the two compatible parts (CFRP-strut wall and CFRP-sleeve).

Figure 8 shows an adapter, which concentrically connects two struts with different diameters[6]. This adapter contains a number of foam parts (optionally four) covered by carbon fibre braided hoses, which are symmetrically distributed around the strut with the smaller diameter. The braided hoses create shear webs corresponding to the number of foam pieces. At one side the adapter is grinded conically while on the other side it is bonded into the strut with the larger diameter. The open cut in the conical strut region is covered with a carbon fibre braided hose.

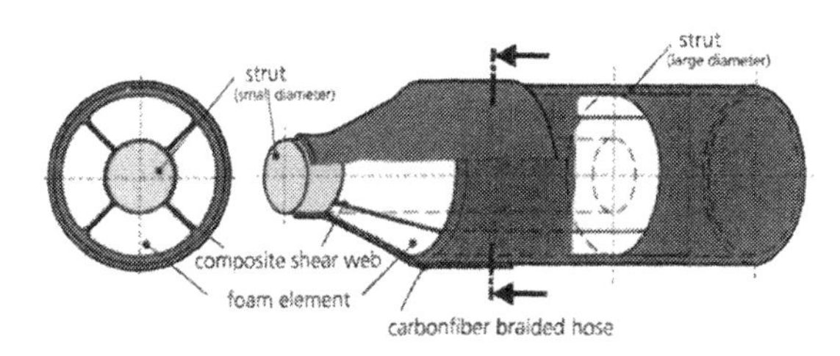

Figure 8 Load transfer of two struts with different diameter

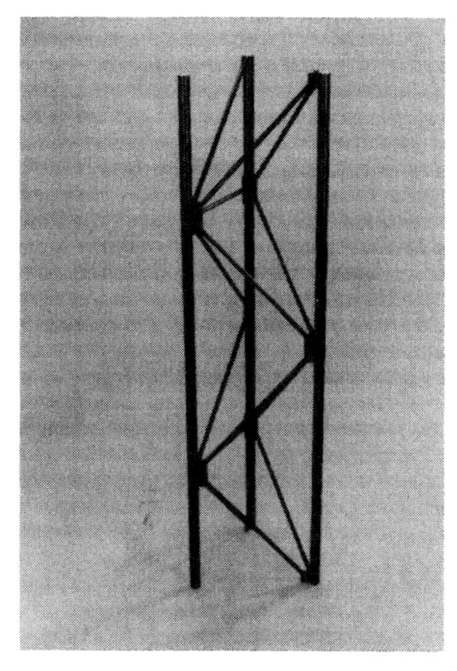

Figure 9 Truss structure of CFRP-struts connected with bonded nodal plates

Truss structures

The framework structure discussed in this paper refers to beam-like trusses which have, in their basic version, cross sections of an equilateral triangle (*Figure 9*)[7]. This framework consists of three rail struts interconnected with cross and diagonal struts. The framework is based on the idea of combining stiffening rods and rail struts with nodal plates (*Figure 10*). In this case, the nodal plate is an integral part of a CFRP-connection element (*Figure 11*), which is bonded on the rail strut. The fixation of cross- and diagonal struts on the nodal points is achieved by bonding. For that purpose the

Figure 10 Nodal point of the truss structure shown in *Figure 9*

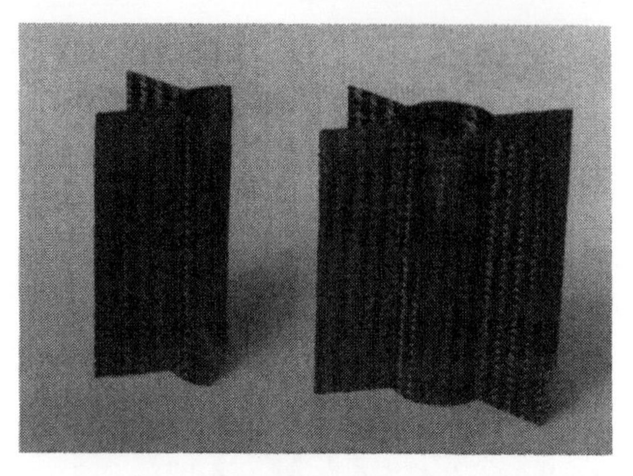

Figure 11 CFRP-connection element with nodal plates. Left: finished connection element, right: as a semiproduct

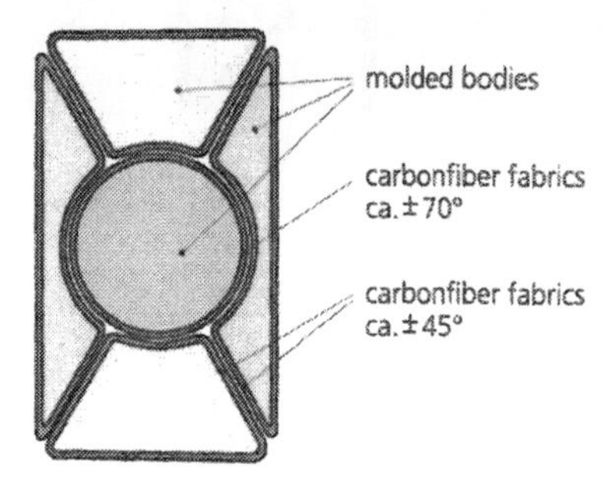

Figure 12 Cross-section of the cylindrical molded bodies for manufacturing of the semiproduct shown in *Figure 11*

strut ends are (1) filled with a shear-stiff filler (2), cured (3), slitted, and (4) afterwards provided with an adhesive glue pushed on the nodal plate. It is very important, that this rod connection avoids any weakening of the cross section of the rail struts. *Figures 10* and *11* show the element in detail. It consists of a half tube piece equipped with two nodal plates. In case of the triangular truss both nodal plates form an angle of 60°. Considering the manufacturing process the design of the CFRP-connection elements becomes quite clear. *Figure 12* shows cylindrically molded bodies with trapecoidal cross sections assembled around a rod-like form with a circular cross section. Between them there are resin-wetted carbon fibre fabrics with ±45° orientation. These textile fabrics may be replaced also by appropriate carbon fibre-braided hoses. After curing and removing the molded bodies a semiproduct with longitudinal extension arises which can be cut in several pieces of definite lengths. This semiproduct consists of only one tube with four nodal plates, if the nodal plates have previously been shortened to a definite width (*Figure 11*, right). Cutting the tube pieces along their middle plane will result in two finished CFRP-connection elements (*Figure 11*, left). This truss design offers considerable advantages in production.

The framework mentioned above is a beam-like truss structure, which is used in many cases with conical structures leading to one point of support. Such a conical CFRP-truss structure is shown in *Figure 13*. Special attention has to be given to the connection between the three rail struts and the central one. A solution for a design fulfilling the strength requirements can be seen in *Figure 14*. The main part of the connection is a prefabricated element consisting of four carbon fibre tubes made out of braided hoses and a carbon fibre fabric with ±45° fibre direction wrapped around the central and the adjacent tubes. After curing, the carbon fibre fabrics create shear webs which lead the longitudinal forces from the central to the adjacent rods. The whole element will be surrounded with a carbon fibre hoop winding, thus becoming a semiproduct with longitudinal extension which can be cut in pieces of the required lengths. Other important

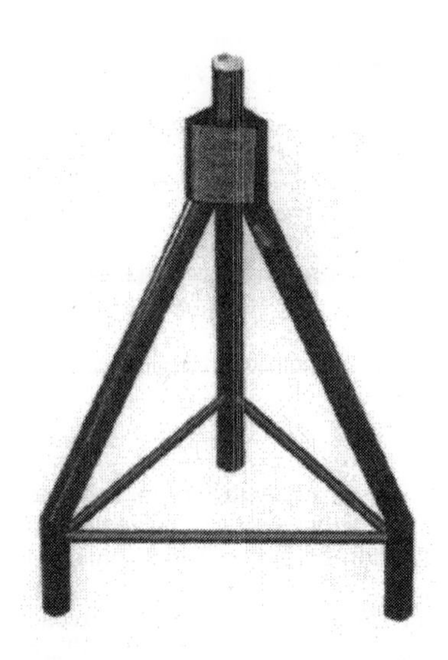

Figure 13 Conical CFRP-structure as an end part of a beam-like triangular truss structure

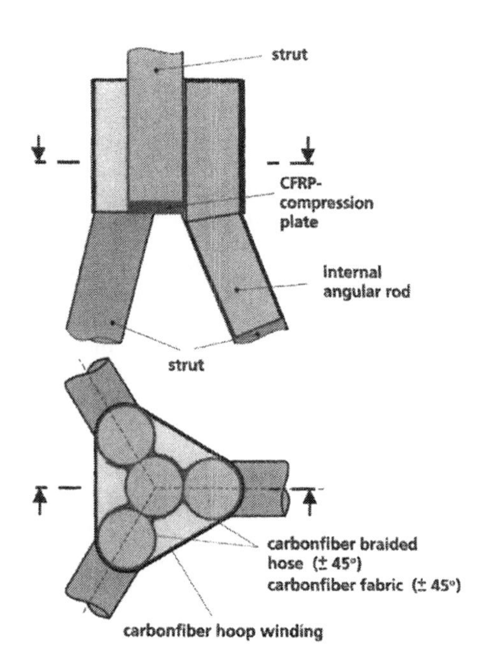

Figure 14 Assembling of three rail struts of a triangular truss structure on a central strut

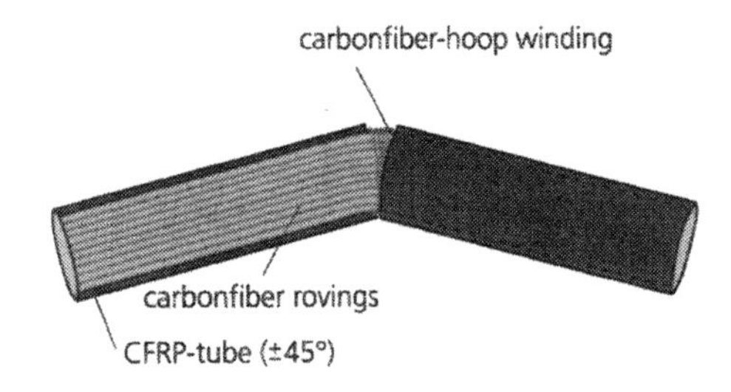

Figure 15 Angular rod element for connecting CFRP-struts in a definite angle

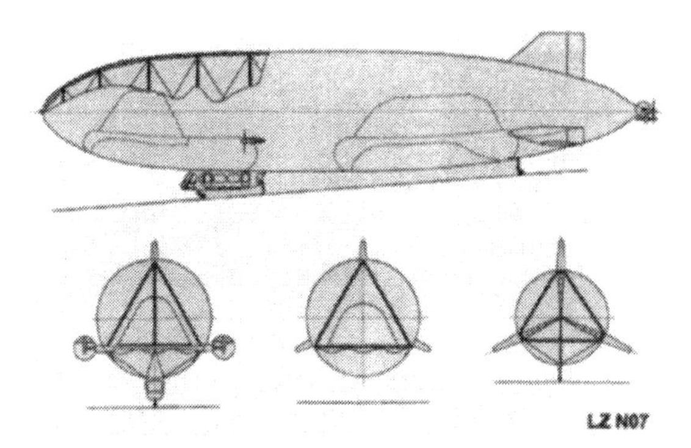

Figure 17 Structure of the Zeppelin NT. Data: length: 75 m; diameter: 14.2 m; volume: 8200 m^3; take off mass: 7900 kg; pay load: 1850 kg (2 + 12 seats)

Figure 16 The new Zeppelin NT, (courtesy of Zeppelin Luftschifftechnik GmbH)

Figure 18 Beam-like truss structures of the new Zeppelin (courtesy of Zeppelin Luftschifftechnik GmbH)

parts of the connection are the three internal angular rod elements which are bonded in the three adjacent tubes and in the three hollow struts (*Figure 15*). The same internal angular rods are used to connect the conical with the cylindrical truss. The central strut with an appropriate load transfer element will be bonded in the central tube of the connection element. Under compression load the three adjacent struts show a tendency to separate from the central one. This is prevented by the mentioned fibre hoop winding. Under tension loading the opposite tendency is observed. In order to prevent in this case a compression and even a crushing of the thin-walled central tube, a round CFRP-compression plate is inserted at the end of the central tube.

Application of the lightweight truss structures in the new Zeppelin

The development of the lightweight struts lead to an application in a new rigid airship, the Zeppelin NT (*Figure 16*), which is planned to fly in summer 1997. A sketch of the Zeppelin with its most important data is given in *Figure 17*[8]y. The structure of the Zeppelin consists of three longitudinal aluminum beams divided into sections of 6 m length, and of triangular spar elements. These spar elements are again three beam-like triangular truss structures built up by the above-mentioned lightweight carbon fibre struts. *Figure 18* shows a section, where two spar elements lead to a connection of two longitudinal beams. While the longitudinal beams are aluminum structures the both spar elements consist, except of their conical structures, of the CFRP-sandwich struts. The weight of the CFRP-trusses is about half of the comparable aluminum trusses.

Conclusion

Different options of lightweight CFRP sandwich rods and struts have been presented. A great problem associated with advanced CFRP-struts is usually the load transfer with high loadbearing capacity. One form-locking load transfer element with high static and fatigue strength is presented. It does not rely on any kind of bonding, but purely on clamping. Another load transfer element for applications with lower loads can be assembled after fabrication and will be bonded in the rod as a semiproduct. In this case bonding acts between two compatible parts while there is a form-locking connection between incompatible materials (aluminum and CFRP). Also a solution to connect struts with different diameters is described. Load transfer elements are needed for beam-like trusses consisting of

CFRP-sandwich struts. These trusses which serve, e.g. in the new rigid airship Zeppelin NT as spar elements, have a weight being half of that of the comparable aluminum truss.

References

1 Bergmann, H.W. and Block, J., *Fracture / Damage Mechanics of Composites: Static and Fatigue Properties*, DLR-Mitt. 1992, pp. 92–03

2 Schütze, R. and Goetting, H.C., Development of Adaptive CFRP struts for vibration suppression in high-loaded lightweight structures. In *Conference on Spacecraft Structures, Materials and Mechanical Testing*, vol. 1, Noordwijk 27.-29.3. ESA-SP-386, 1996, pp. 199–205

3 Schütze, R., Verbindungshülse, Patent P 44 19 691.1-12, 1995

4 Schütze, R., Stab mit integrierten Krafteinleitungen, Patent P 41 35 695. 0-09, 1993

5 Schütze, R., Nachträglich montierbare Krafteinleitung für Faserverbundstäbe, Patentanmeldung P 196 45 467. 0-12, 1996

6 Schütze, R., Verbindungselement zur Verbindung von Stäben mit unterschiedlichen Durchmessern, Patentanmeldung P 197 01 445. 3-12, 1997

7 Schütze, R., Lightweight structures based on CFRP sandwich struts and CFRP connections. In *Proceedings of the ESA-ESTEC International Symposium on Advanced Materials for Lightweight Structures*, Noordwijk, 1994

8 Prospekt der Zeppelin Luftschifftechnik GmbH, Friedrichshafen, Germany

ELSEVIER

Materials & Design, Vol. 18, Nos. 4/6, pp. 239–242, 1997
© 1998 Published by Elsevier Science Ltd
Printed in Great Britain. All rights reserved
0261-3069/98 $19.00 + 0.00

PII: S0261–3069(97)00057–5

Chemical vapor deposition of titanium nitride on carbon fibres as a protective layer in metal matrix composites

N. Popovska[a,*], H. Gerhard[a], D. Wurm[b], S. Poscher[c], G. Emig[a], R. F. Singer[b]

[a]*Lehrstuhl für Technische Chemie I, Universität Erlangen-Nürnberg, Egerlandstr. 3, 91058 Erlangen, Germany*
[b]*Lehrstuhl Werkstoffkunde und Technologie der Metalle, Universität Erlangen-Nürnberg, Martensstr. 5, 91058 Erlangen, Germany*
[c]*Fraunhofer Institut für Integrierte Schaltungen, Schottkystr. 10, 91058 Erlangen, Germany*

Received 11 August 1997; accepted 20 August 1997

The significance of carbon fibres for reinforcing metals has increased in the last years, because of their excellent mechanical properties. However, to avoid the weakening reaction during MMC fabrication between the fibre and the liquid metal, a protective coating has to be applied. Continuous carbon fibre roving with 6000 filaments were coated with TiN by thermal induced chemical vapour deposition (CVD) using a gas mixture of $TiCl_4$, N_2 and H_2 as a precursor. The deposition process in the reactor was simulated by a modified Phoenics-CVD software program using a 2D-axisymmetric model. Carbon fibres reinforced magnesium matrix composites are fabricated by a pressure infiltration casting process. The mechanical properties of the MMCs can be used to demonstrate the efficacy of the coated fibre approach. The rule of mixture is realized to 98/ for the coated fibre, and only 48% for the uncoated system. The infiltration pressure during the processing of composites was lowered from 10 to 1 MPa for the TiN coated system. © 1998 Published by Elsevier Science Ltd. All rights reserved.

Keywords: carbon fibres; reinforcing metals; metal matrix composites; titanium nitride

Introduction

Unidirectional carbon fibres reinforced metal matrix composites (MMCs) based on magnesium are promising candidates for many aerospace and transport applications because of their high specific mechanical properties, low thermal expansion coefficient in fibre direction and good electrical and thermal conductivity[1]. High tenacity carbon fibres have the advantage of high strength at low price in comparison to other advanced fibres. However, a significant disadvantage of carbon fibres as reinforcing materials lies in their poor wettability as well as in their reactivity and strength loss when in contact with molten metals[2]. Therefore, carbon fibres should be coated with metallic or ceramic layers as a protective barrier.

Chemical vapour deposition (CVD) is the most common technique for applying thin films on individual fibre surfaces in a bundle consisting of thousands of filaments. The aim of this work is to investigate the effect of thin (15–40 nm) TiN layers on carbon fibres on the properties of magnesium metal matrix composites.

*Correspondence to N. Popovska. Tel.: +49 9131 857428; fax: +49 9131 857421; e-mail: n.popovska@rzmail.uni-erlangen.de

Experimental

The chemical vapour deposition of TiN onto carbon fibres was performed continuously in a vertical laboratory scale equipment (*Figure 1*) under atmospheric pressure using a gas mixture of $TiCl_4$, N_2 and H_2. according to the following reaction:

$$2TiCl_4 + N_2 + 4H_2 \rightarrow 2TiN + 8HCl \quad (1)$$

The CVD reactor consists of a ceramic tube of 250 mm heated length and 40 mm diameter heated by a IR-oven. As a substrate, commercially available continuous carbon fibres Tenax 5331 HTA (from Akzo) delivered as roving with 6000 filaments were used. The roving velocity through the reactor was varied between 60 and 130 m h^{-1}. The sizing on the fibre was removed thermally prior to the coating.

A 4:1 molar ratio of H_2 to N_2 was found to be optimal for a high deposition rate of TiN. The concentration of $TiCl_4$ in the gas stream was controlled by changing the temperature of the saturators (28–40°C) assuming the $TiCl_4$ vapour pressure to be in equilibrium with the $TiCl_4$ liquid. The reaction temperature was varied between 750°C and 1200°C. The deposition rate of the TiN and the mean layer thickness were calculated gravimetrically from the increase in weight of the substrate per surface area and time. The me-

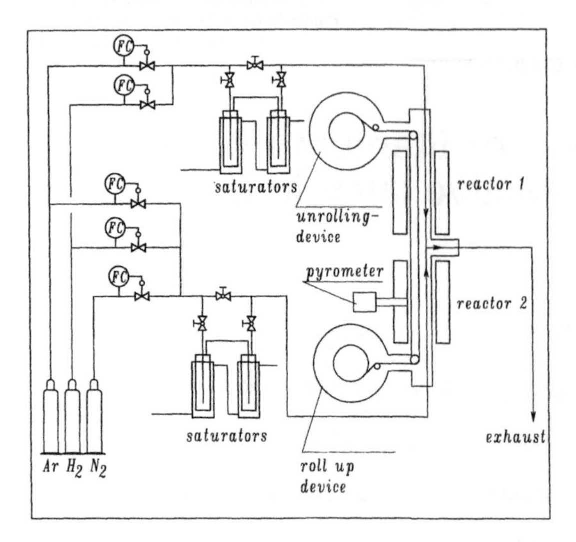

Figure 1 CVD equipment for coating of continuous fibres (schematically)

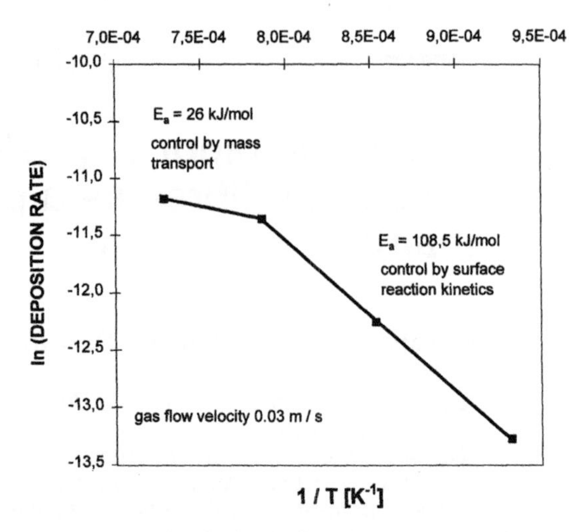

Figure 2 Arrhenius plot for the deposition of TiN

chanical properties of both uncoated and TiN coated carbon fibres were measured by the single filament test according to ASTM 3379.

Unidirectional reinforced samples were prepared by a pressure infiltration casting process[3] at 720°C and 10 MPa pressure using commercially pure magnesium (cp-Mg) as a matrix. The metal matrix composites with uncoated and TiN coated carbon fibres were characterized by 3-point bending strength according to DIN 29 971. The element mappings were investigated by microprobe (EPMA).

Results and discussion

Kinetics of the chemical vapour deposition of TiN

The chemical vapour deposition is a typical heterogeneous process involving the following steps: (a) mass transfer of the reactants from the bulk gas phase to the substrate surface; (b) adsorption on the substrate and chemical reaction, and (c) desorption of the adsorbed species and back diffusion of the by-products into the bulk. The slowest step determines the rate of the overall process. Therefore, the growth rate of the deposit can be either determined by: (a) the surface reaction kinetics, or (b) the mass transport.

An uniform coating of the filaments within the carbon fibre bundle can only be achieved when the surface reaction is rate controlling. It is possible to switch from one limiting step to the other by changing the temperature and/or gas flow velocity. This is illustrated in *Figure 2* where the Arrhenius plot (logarithm of the deposition rate vs reciprocal temperature) is shown. The apparent activation energy for the deposition of TiN using 2 vol.%$TiCl_4$ is 108 kJ mol^{-1} in the temperature range between 800° and 1000°C. From this high activation energy in combination with the low input concentration of the educt in the gas phase it can be concluded that the overall process is controlled by surface kinetics as the rate determining step[4]. The transition point to surface reaction kinetics control lies at 1000°C, but it can be shifted to higher temperatures by increasing the gas flow velocity. Performing the deposition process at higher temperature leads to increased growth rates, which make the CVD-equipment more effective.

Simulation of the TiN deposition process

The deposition process of TiN by CVD was simulated using the 'PHOENICS-CVD' computing program which is the standard Computational-Fluid-Dynamics-Program PHOENICS extended by special models for chemical vapour deposition[5]. In order to consider the convective mass transport, the continuity equation and the Navier-Stokes equation without time dependence were solved. Multicomponent diffusion and the chemical reaction were considered by Stefan-Maxwell and species balance equations. Solving these equations results in local values of velocities, pressure and concentrations, from which parameters like viscosity, density, binary diffusion coefficients, and reaction rates are calculated. The CVD reactor could be described in good approximation by a 2D-axisymmetric model in which the coupled partial differential equations mentioned above were solved on a grid with 15 cells in radial and 24 cells in axial direction. For each cell inside the fibre or having contact to the reactor wall the deposition rate of TiN (r) was calculated according to Eq. (2) which leads to a consumption of $TiCl_4$, H_2 and N_2 and to formation of HCl in the gas phase. The activation energy in Eq. (2) was experimentally determined, constant A was fitted.

$$r = A\,\frac{P_{TiCl_4}}{RT}\,\exp\left(\frac{-108\ \mathrm{kJ\ mol^{-1}}}{RT}\right) \tag{2}$$

The best fit for the mean growth rate over the fibre length between simulation and experiment was achieved for $A = 5.925$ kg m^2 mol^{-1} s^{-1}

As a result of the deposition reaction, $TiCl_4$ depletion in the gas phase and accordingly TiN growth profiles in axial and radial direction were calculated by simulation. The educt consumption along the reactor was calculated to be about 50%. Fibre roving and gas

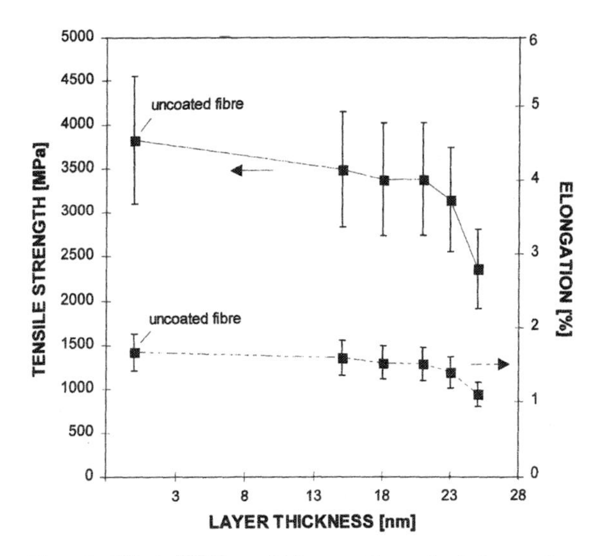

Figure 3 Effect of TiN layer thickness on the mechanical properties of carbon fibres

mixture passed in opposite direction through the reactor, so that at the fibre entrance the educt concentration is lower, which is favourable because of promoting the nucleation. The difference in the growth rate in the radial direction was found to be about 13%, which pointed to a relatively uniform coating of the single filaments within the fibre bundle.

Characterization of the TiN-coated carbon fibres

All deposits are of yellow gold colour which is typical for stoichiometric TiN[4]. Depending on the fibre velocity in the reactor (60–120 m h^{-1}), a mean TiN layer thickness in the range between 15 and 40 nm was achieved. The temperature and the gas flow velocity were kept constant, 1100°C and 0.05 m s^{-1} respectively.

The investigation in scanning electron microscope (SEM) shows that the filaments within the carbon fibre bundle are uniformly coated with closed and non-porous TiN layers even at a layer thickness of about 15–20 nm.

The single filament test on TiN coated fibres (*Figure 3*) shows that the tensile strength and the elongation to fracture of the uncoated carbon fibre was only slightly affected up to a layer thickness of 20 nm. However, a strong reduction of the mechanical properties of the fibre was observed with thicker TiN layers. Therefore, carbon fibres coated with a TiN layer in a range of 15–22 nm were used for reinforcing of magnesium.

Reinforcing of magnesium matrix with TiN-coated carbon fibres

The main problem of the magnesium matrix composites when using high tenacity carbon fibres is the degradation of the fibres during the infiltration process, caused probably by the penetration of the magnesium into the fibre[2,6]. The TiN coating on the fibre acts as a diffusion barrier even at a layer thickness of 15–20 nm and prevents the penetration as can be seen in the element mappings. The TiN coating remains unaffected after the infiltration process. No traces of magnesium were found inside the carbon fibre.

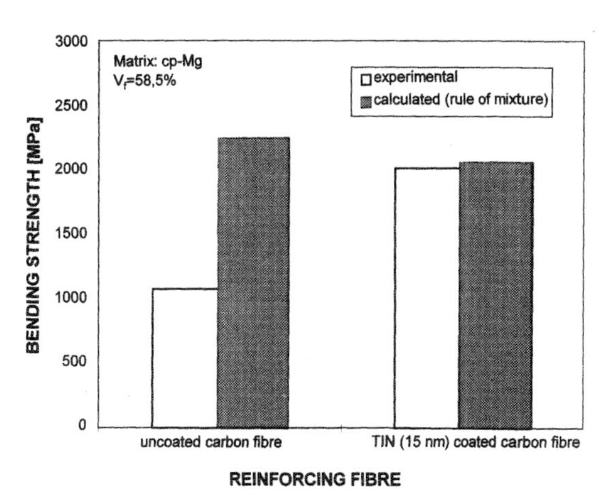

Figure 4 Bending strength of MMCs reinforced with uncoated and TiN-coated carbon fibres

The mechanical properties of the MMCs can be used to demonstrate the efficacy of the coated fibre approach (*Figure 4*). In the case of a pure magnesium matrix the bending strength was increased from 1180 MPa for the uncoated carbon fibre up to 2130 MPa for the TiN (15 nm)-coated carbon fibre. The rule of mixture is realized to 98% for the coated fibre, and only 48% for the uncoated system. As a result of the much better wettability of the coated fibre surface by molten magnesium, the infiltration pressure during the processing to composites could be lowered from 10 to 1 MPa.

Conclusions

(1) Thermal induced chemical vapour deposition (CVD) technique was applied to coat continuous carbon fibre roving with thin (15–20 nm) TiN layer without significant degradation of its mechanical properties.

(2) The study of the kinetics and the simulation of the deposition process allow to optimise the reaction conditions resulting in dense, well adherent TiN layers on the carbon fibre surface.

(3) A thin (15–20 nm) TiN layer acts as diffusion barrier preventing the diffusion of magnesium into the carbon fibres during the processing of MMCs. As a result, the rule of mixture is realized to 98% for the coated fibre, and only 48% for the uncoated system

(4) The excellent wettability of the TiN-coated carbon fibres by molten magnesium allowed to reduce the infiltration pressure significantly.

Acknowledgements

The authors would like to thank the Bavarian Research Foundation, Munich for the financial support of this study as a part of the FOROB program.

References

1 Hall, I.W., *Scripta Metallurgica*, 1987, **21**, 1717
2 Öttinger, O., Grau, C., Winter, R., Singer, R.F., Feldhoff, A., Pippel, E. and Woltersdorf, J., In *Proc. 10th International Conference on Composite Materials*, vol. VI, eds. A. Pourtsartip and K. Street. Woodhead Publishing, Cambridge, 1995, pp. VI-447/VI-454
3 Gruber, M., Grau, C., Öttinger, O. and Singer, R.F., In *Jahrbuch der Deutschen Gesellschaft für Luft und Raumfahrt*, ed. G. Bürgener. Bonn, Deutsche Gesellschaft für Luft und Raumfahrt, pp. 235–242
4 Dekker, J.P., van der Put, P.J., Veringa, H.J. and Schoonman, J., *Journal of Electrochemical Society*, 1994, **141**, 787
5 Spalding. B. ed., *The PHOENICS Journal of Computational Fluid Dynamics and its Applications*, vol 8, CHAM, Wimbledon, UK, 1995, 4, p. 402
6 Viala, J.C.,Fortier, P., Glaveyrolas, G., Vincent, H. and Bioux, J., *Journal of Materials and Science*, 1991, **26**, 4977.

ELSEVIER

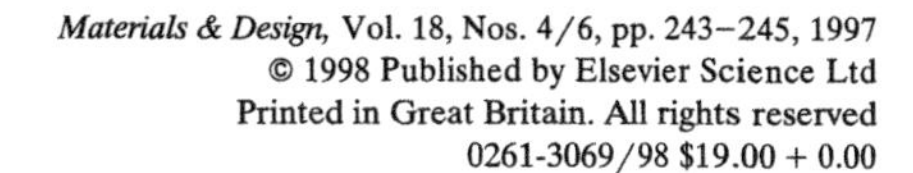
Materials & Design, Vol. 18, Nos. 4/6, pp. 243–245, 1997
© 1998 Published by Elsevier Science Ltd
Printed in Great Britain. All rights reserved
0261-3069/98 $19.00 + 0.00

PII: S0261-3069(97)00058-7

Hot isostatic processing of metal matrix composites

H. V. Atkinson[a,*], A. Zulfia[a,1], A. Lima Filho[a], H. Jones[a], S. King[b]

[a]*Department of Engineering Materials, Hadfield Building, University of Sheffield, Mappin St., Sheffield S1 3JD, UK*
[b]*Bodycote HIP Ltd., Carlisle Cl., Sheffield Rd., Sheepbridge, Chesterfield S41 9ED, UK*

Received 13 August 1997; accepted 20 August 1997

The effect of four different Hot Isostatic Pressing (HIPping) treatments on porosity in the aluminium casting alloy A357 and stir-cast A357/15 vol% SiC particulate MMC has been investigated and the optimum treatment identified. The bend strength increased after HIPping relative to as-received. Such ceramic particle reinforced MMCs, however, do not have adequate toughness for many commercial applications. Metal reinforced MMCs made by combining an Al alloy matrix with stainless steel wire can provide a compromise between weight saving and toughness. The production, involving HIPping, of such a composite is discussed. © 1998 Published by Elsevier Science Ltd. All rights reserved.

Keywords: hot isostatic processing; aluminum; metal matrix composites

Introduction

Ceramic particle reinforced MMCs are already in use for some transportation applications because of their improved specific strength, specific strength, wear resistance and tailorable thermal expansion characteristics[1]. They can be made by both solid-phase (powder metallurgy) and liquid-phase (casting) processes. Casting methods are cheap but give poorer ductility and toughness than solid state routes, mainly because of porosity and particle clusters[2,3].

Hot Isostatic Pressing (HIPping) involves the simultaneous application of a high-pressure (usually inert) gas and an elevated temperature in a specially constructed vessel[4]. The pressure applied is isostatic because it is developed with a gas, so that, at least as a first approximation, no alteration in component geometry occurs. Under suitable such conditions of heat and pressure, internal pores or defects within a solid body collapse and weld up. The effects of HIPping on the microstructure and properties of cast MMCs are not yet fully understood.

Several previous studies have been carried out. Loh *et al.*[5] HIPped two as-cast SiC A359-based composites at pressures in the range 100–150 MPa and temperatures in the range 450–550°C. HIPping increased the ductility but reduced the yield strength drastically. This is in sharp contrast with the results obtained for Al-Li based MMCs[6] and alumina reinforced A356 MMCs[7]. The decrease in strength obtained by Loh et al was thought to be largely due to matrix softening with the high HIP temperatures they used, and could be restored by age hardening. Pagounis *et al.*[8] found that rigid-rigid contacts and the formation of ceramic particle networks increased the pressure required for densification. There is a critical volume fraction of reinforcement above which continuous networks start to form. Percolation theory suggest this is around 16%[9].

The toughness of ceramic particle reinforced MMCs is inadequate for many commercial applications. Aluminium alloys reinforced with continuous stainless steel wires do not give enhancements in specific properties as large as for continuous fibre ceramic reinforced alloys, but nevertheless there are worthwhile gains and the toughness is expected to be higher than with the ceramic reinforced materials. Considerable work was carried out on aluminium alloy/stainless steel combinations in the 1960s and 1970s[10]. If these are processed by liquid metal routes the reactions at the interfaces between the molten aluminium and the stainless steel are generally extensive and detrimental. There has been a renewal of interest in the 1990s. Colin *et al.*[11] made squeeze cast Al-MMCs reinforced with 12 μm stainless steel fibres. The extent of interfacial reaction could be controlled via the infiltration parameters and the volume fraction of fibres. Barbier and Ambroise[12] found that squeeze cast stainless steel short fibre Al MMCs had a tensile strength 3–4 times that of the Al matrix.

In this paper, we describe the effect of HIPping on stir-cast MMCs and preliminary work on the HIPping of stainless steel reinforced aluminium alloys.

*Correspondence to H. V. Atkinson. Tel.: +44 114 2225512; fax: +44 114 2225943; e-mail: h.v.atkinson@sheffield.ac.uk
[1]Present address: University of Jakarta, Indonesia

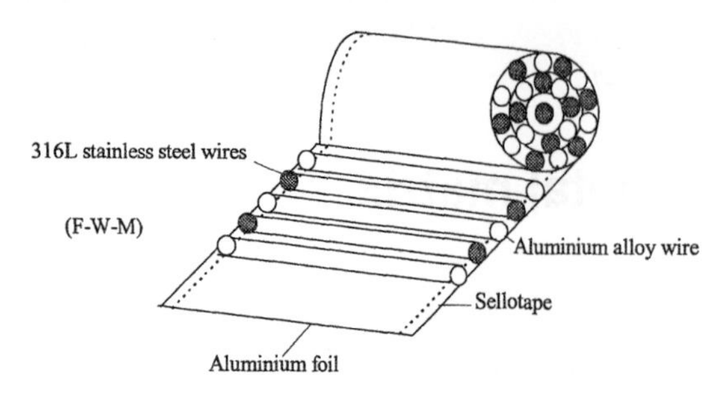

Figure 1 Schematic diagram of the lay-up method for arranging stainless steel wires alternately with Al alloy wires on an Al foil backing strip

Experimental method

Ceramic reinforced MMCs

The materials used (supplied by Norsk Hydro) were based on the aluminium foundry alloy A357 (Al6.7Si0.3Mg0.2Ti0.1Fe unreinforced, Al5.7Si0.5Mg0.3Ti0.1Fe reinforced). Unreinforced material was compared with stir-cast MMC with A357 as the matrix and 15 vol%SiC particle size fraction as reinforcement. The average SiC particle size is 30 μm.

Four cycles of HIPping were carried out in an Autoclave Engineers laboratory unit at Bodycote HIP Ltd., Chesterfield, UK. Firstly, HIPping was attempted at 550°C/103 MPa/2 h, i.e. below the solidus which is at 555°C. The results were not satisfactory, as porosity was only slightly reduced relative to as-received. Secondly, HIPping was attempted at 575°C/103 MPa/2 h, ie. between the solidus and the liquidus which is at 615°C. This was too severe a treatment in the semi-solid region, in that the unreinforced specimen distorted and porosity was significantly increased relative to as-received. In the third cycle, HIPping was carried out at 565°C/103 MPa/15 min, followed by 535°C/103 MPa/2 h. This short burst in the semi-solid region followed by a sustain below the solidus gave significantly reduced porosity relative to as-received. In the final cycle, which was less effective, there was a longer period in the semi-solid region of 570°C/103 MPa/40 min, followed by a sustain of 535°C/103 MPa/ 2 h below the solidus.

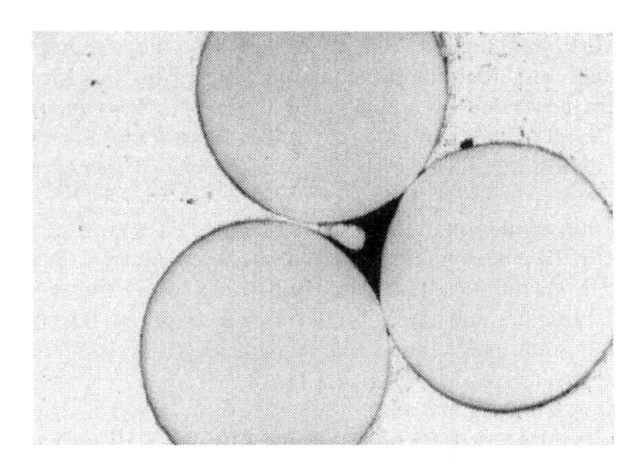

Figure 2 Al extruded into the void between 0.3 mm diameter 316L stainless steel wires during HIPping

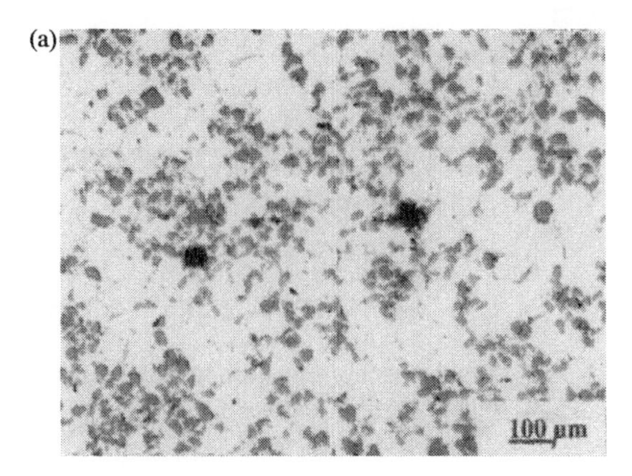

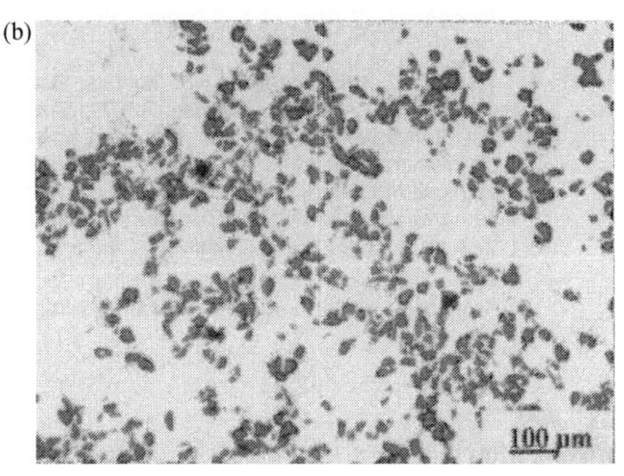

Figure 3 (a) Stir-cast A357/15 vol%SiC as-received; 2.2% porosity from densitometry. (b)Porosity is reduced in stir-cast A357/15 vol%SiC after HIP cycle 3 (565°C/103 MPa/15 min followed by 535°C/103 MPa/2 h); 1.6% porsity from densitometry

The percentage porosity was found by densitometry using Archimedes Principle according to British Standard 5600. After HIPping, the specimens were heat treated. The T6 condition was found to be: solution treatment at 530°C for 17 h, quench in hot water, ageing at 170°C for 9 h. Mechanical testing was carried out by four point bending on specimens from HIP cycle

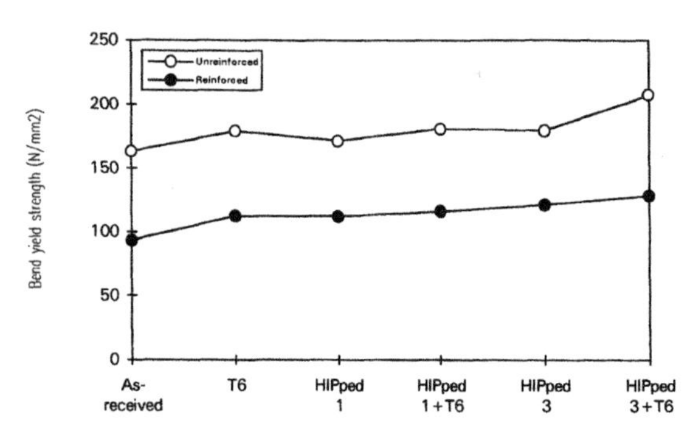

Figure 4 Bend yield strength for A357/15 vol%SiC before and after HIPping and T6 heat-treatment

1 and HIP cycle 3. The specimen geometry followed that in[13]; there are no standards as yet defined for four point bending of these materials. The two side faces of each bend bar were polished to 6 μm diamond finish prior to testing on a Universal Screw Driven Mayes instrument.

Metal reinforced MMCs

The aim here was to make reinforced aluminium alloys with stainless steel wires arranged parallel to each other in a longitudinal array. A technique had to be found of laying-up the wires and the matrix by hand prior to consolidation. The optimum method found was to alternate stainless steel wires and Al-alloy matrix wires on an Al-foil backing strip (*Figure 1*). Sellotape was used on either edge of the strip during lay-up, to hold the wires in place. After the strip had been rolled up, the edges were cut off to remove the sellotape and avoid organic contamination during HIPping. Alternating the wires, and the use of the backing foil, helped to prevent any two stainless steel wires directly touching. Otherwise, phenomena such as that shown in *Figure 2* occurred where aluminium matrix did extrude between two wires in direct contact but did not fill the pore. The wire bundle was encapsulated in a mild steel can of slightly wider diameter than the bundle. The can was evacuated, sealed and HIPped (525°C/100 MPa/120 min).

Results and discussion

Ceramic reinforced MMCs

Figure 3 shows a comparison between the unreinforced and the reinforced material, before and after HIP cycle 3, which was found to be the optimum in terms of reduction in porosity, although pores were not completely removed. This may be because of the presence of some particle networks resisting deformation, as the volume fraction of reinforcement was close to the percolation threshold. Mechanical properties after HIPping and T6 heat treatment are shown in *Figure 4*. This shows a slight improvement in bend yield strength after HIPping and T6 heat treatment gave a further slight increase. The results for the reinforced material are consistently lower than for the unreinforced. This could be due to brittle phases occurring in conjunction with the reinforcement or to the higher porosity content in the reinforced material.

Metal reinforced MMCs

Full consolidation was achieved by HIPping. A small amount of interfacial reaction product was present but this is minimal in comparison with liquid metal routes. The can has collapsed giving non-symmetric bundles after HIPping and, near the base, some stainless steel wires have become elliptical in section suggesting anisotropic deformation. The wires are still cylindrical in section in the central area. Some reinforcement wires are touching after HIPping. This could be due to defects in the lay-up or indeed to the hard wires forcing the soft matrix out from between them during HIPping.

Conclusions

For A357 reinforced with 15 vol%SiC particle volume fraction, the optimum HIPping treatment, of those investigated, was 565°C/103 MPa/15 min, followed by 535°C/103 MPa/2 h. This gave significantly reduced porosity relative to the as-received but the bend strength was still significantly lower than for the unreinforced material. For aluminium reinforced with stainless steel wires, a technique has been established for laying up the wires prior to consolidation and full densification has been achieved by HIPping at 525°C/100 MPa/120 min.

Acknowledgements

We are grateful to Bodycote HIP for carrying out the HIPping, to Norsk Hydro for the supply of material, and to Dr Eduardo de los Rios for helpful discussions. We would also like to thank the Indonesian Government for support for A. Zulfia and CNPq and UNESP of Brazil for support for A. Lima Filho.

References

1 Lloyd, D.J., *International Materials Review*, 1994, **39(1)**, 1
2 Ray, S., *Journal of Materials Science*, 1993, **28**, 5403
3 Wei, L. and Huang, J. C., *Materials Science and Technology*, 1993, **9**, 841
4 Atkinson, H.V. and Rickinson, B.A., *Hot Isostatic Processing*, Adam Hilger, London, 1991
5 Loh, N.L., Wei, Z. and Hu, Z., In *Proc. Int. Conf. on Hot Isostatic Pressing*, Andover, MA, 1996, pp. 187–192
6 Hanada, K., Tan, M.J., Murakoshi, T., Negishi, T. and Sano, T., *Journal of Materials Processing Technology*, 1995, **48**, 399
7 Li, Q.F., Loh, N.L. and Hung, N.P., *Journal of Materials Processing Technology*, 1995, **48**, 373
8 Pagounis, E., Talvitie, M. and Lindroos, V.K., *Materials Research Bulletin*, 1996, **31(10)**, 1277
9 Zallen, R., *Physics of Amorphous Solids*, ch.4. Wiley, New York, 1983
10 Pinnel, M.R. and Lawley, A., *Metallurgical Transactions*, 1971, **2**, 1415
11 Colin, C., Marchal, Y., Boland, F. and Delannay, F., *Journal de Physique IV*, C7, 1993, **3**, 1749
12 Barbier, F. and Ambroise, M.H., In *Composites: Design, Manufacture and Application*, ed. S.W. Tsai and G.S. Springer. Honolulu, 1991, p. 18, J1
13 Lewandowski, J.J., Liu, C. and Hunt, W.H., Jr., *Materials Science and Engineering A*, 1989, **A107**, 241

Materials & Design, Vol. 18, Nos. 4/6, pp. 247–252, 1997

0261-3069/98 $19.00 + 0.00

PII: S0261-3069(97)00059-9

Structure, texture and mechanical properties of AlZnMgCuZr alloy rolled after heat treatments

J. Dutkiewicz*, J. Bonarski

Institute of Metallurgy and Materials Science of the Polish Academy of Sciences, 25 Reymonta St., 30-059 Kraków, Poland

Received 15 July 1997; accepted 30 July 1997

The aluminium alloy containing 6.7 wt.% Zn, 2.6 wt.% Mg, 1.6 wt.% Cu and 0.1 wt.% Zr was continuously cast and either quenched from 465°C, or furnace cooled down to 100°C to find the best ductility for further cold plastic deformation. The alloys were then cold rolled down to the highest possible degree of deformation. The initial texture in both alloys can be described by (211)[111], (321)[346] and (110)[112] ideal orientations. With increasing deformation other orientations like {110}⟨001⟩ and cubic {100}⟨001⟩ appear after both types of treatments. TEM studies revealed increase of subgrain misorientation up to approx. 9° after 75% of deformation by rolling. On ageing at 120°C for 24 h the maximum hardness of 210 HV was reached. The alloys deformed prior to ageing at 120°C attained 230 HV. Very small GP zones, up to a few nanometers in size, grow after several days of ageing giving diffused diffraction effects. After ageing for 1 day at 120°C, precipitates grow and were identified as η'. © 1998 Published by Elsevier Science Ltd. All rights reserved.

Introduction

Crystallographic texture has a strong effect on the anisotropy and fracture toughness of hot-rolled 7XXX aluminium alloy plates[1–4]. As stated by Engler et al.[2], even if a significant part of the microstructure is recrystallized the rolling texture is not altered. This behaviour was explained by the randomisation effect during recrystallization and texture sharpening within the recovered matrix. The presence of particles was an important factor influencing the deformation and recrystallization texture. Depending on the type of precipitates (shearable, non-shearable) various types of deformation texture in Al-alloys were reported[5]. Copper orientation existed in all alloys investigated (homogeneous and with particles), while the presence of small particles additionally caused appearing of brass type and 'S' orientations {321}⟨364⟩[5]. The other problem which is important in the 7000 series alloys is the effect of deformation on age-hardening. It compensates the loss of strength that normally occurs during the overaging part of T73 ageing treatment. Deformation should be carried out above the GP solvus temperature prior to ageing what leads to a microstructure containing high densities of dislocations together with the η' and η precipitates, characteristic for a high strength[6]. The influence of post-quench deformation in AlZnMg alloys depends strongly on the heating rate to the ageing temperature and the holding time at room temperature[7] due to competition between the GP dissolution and η' precipitation. The kinetics of ageing increases after prior deformation and in addition planar anisotropy is affected by a room-temperature processing[8]. The aim of the present article is to determine the effect of the initial structure on the texture development and the effect of prior deformation on ageing behaviour.

Experimental procedure

Aluminium alloy of composition close to 7278A, given in the *Table 1*, was cast as a 200-kg ingot in a semi-continuous way. The alloy was homogenised at 48.0°C for 12 h and then extruded at 400°C at a flow rate of 1 m/min. Two variants of heat treatment were applied: (A) quenching from 465°C to room temperature (RT) in water; and (B) slow cooling with a furnace at rate 30 K/h. Then the alloy was cold rolled up to maximum possible deformation without crack formation, i.e. up to 75% in case of variant (A) and up to 90% in the case of variant (B).

Texture was analysed using the orientation distribution function (ODF). It was calculated by the discrete

Table 1 The chemical composition of the investigated alloy (in wt.%)

Alloy	Al	Zn	Mg	Cu	Mn	Zr	Fe	Cr	Ti	Si
7278A	Bal.	6.7	2.65	1.6	0.03	0.12	0.01	0.01	0.01	0.05

*Correspondence to J. Dutkiewicz

ADC method[9] from the incomplete pole figures measured by means of Philips X'Pert diffractometer system equipped with texture goniometer ATC-3 using CoKα filtered radiation. Structure of alloys was studied by means of the optical, the scanning Philips XL30 and the transmission electron microscopes CM20 with Oxford Instruments energy-dispersive spectrometers. Samples for scanning and optical microscopy were polished using diamond paste, then etched using reagent containing 10 ml HF, 10 ml HNO_3 and 30 ml of glycerine, while for TEM studies using jet electropolishing in perchloric acid/methyl alcohol electrolyte at subzero temperatures.

Results and discussion

Figure 1 shows two hardness/ageing curves of the alloy aged at 120°C, i.e. just after quenching and after prior deformation to 75%. It can be seen that after deformation the highest hardness of 230 HV is obtained after approx. 8 h of ageing, while after quenching the maximum hardness of 210 HV is reached after 20 h. The earlier RT ageing does not increase the hardness maximum. Its increase caused by deformation prior to ageing is not very high, as in the earlier studies[8].

Figure 2 shows a set of optical microstructures taken from the investigated alloy after variant (A) of heat treatment and plastic deformation by rolling to 15% (a), 45% (b) and 75% (c). One can see that elongated grains existing after extrusion do not change their shape after 45% of deformation, but after 75% the grains are much more extended in the rolling direction. This is connected with a dense propagation of deformation bands and increased misorientation of subgrains formed during the deformation process. The particles present in the as quenched state at the grain boundaries and within them were identified using SEM combined with non-dispersive energy spectrometer as $Al_6MgCuZn$ and Al_6Zr with other dissolved elements. Their size after heat treatment type (B) is significantly larger.

Figure 3 shows a set of TEM micrographs taken from the alloy deformed by 75%, where a deformation band with very fine grains of a large misorientation can be seen in the central part of the micrograph. The upper part shows elongated subgrains with a black/white contrast due to misorientation above 10° as results from several tilt experiments. It increases with a degree of rolling since an average subgrain misorientation after 45% of rolling was in the range 3–5°.

Texture studies were carried out on the cold-rolled samples with deformation 0, 15%, 30%, 45%, 60%, 75% and 90% after the mentioned A and B variants of heat treatment. The basis of the texture investigation were ODFs presented in the space of Euler angles (φ_1, Φ, φ_2). Analysing the texture functions, it follows that the textures of both series of the examined samples can be described in general by a few components, such as: {110}⟨001⟩ 'Goss', {100}⟨001⟩ 'cubic', {211}⟨111⟩ 'copper' as well as close to the 'S' {321}⟨346⟩ and {411}⟨011⟩ orientations (see *Figure 4* and 5). The copper component is the minor component of the texture.

In both A and B series of the samples, the Goss component disappears at simultaneous rising of a com-

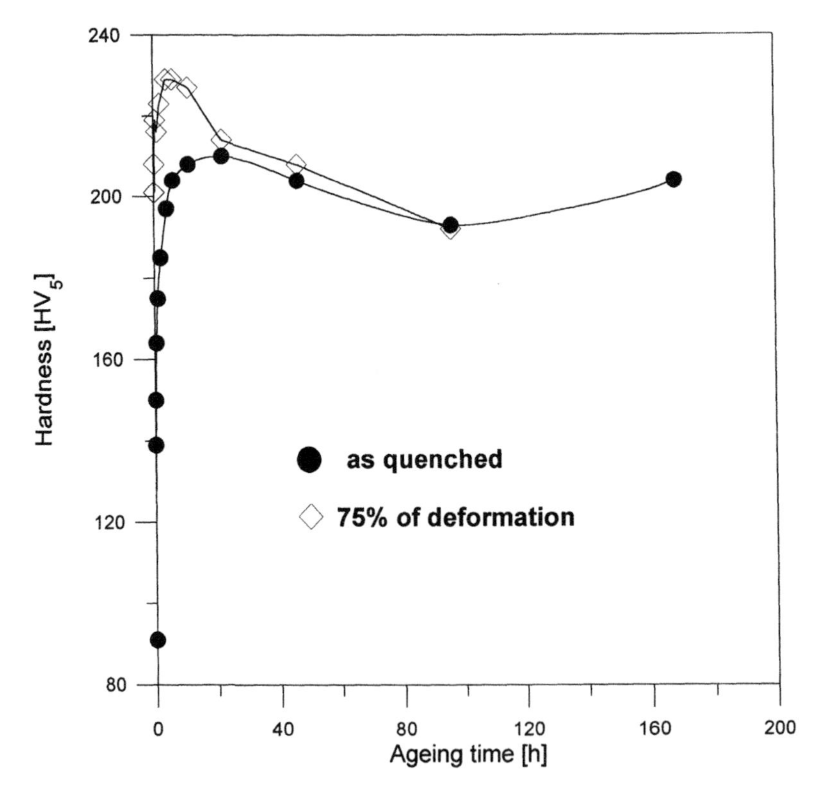

Figure 1 Hardness changes of investigated alloy aged at 120°C after quenching and after additional deformation of 75% by rolling

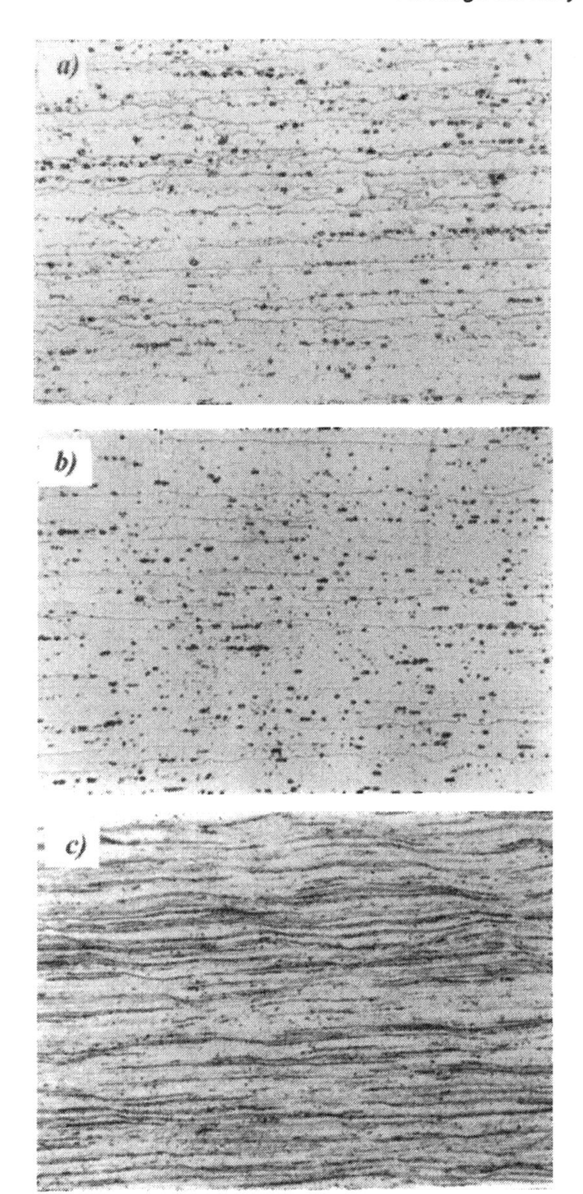

Figure 2 Set of optical micrographs of investigated alloy deformed by rolling after quenching 15% (a), 45% (b) and 75% (c)

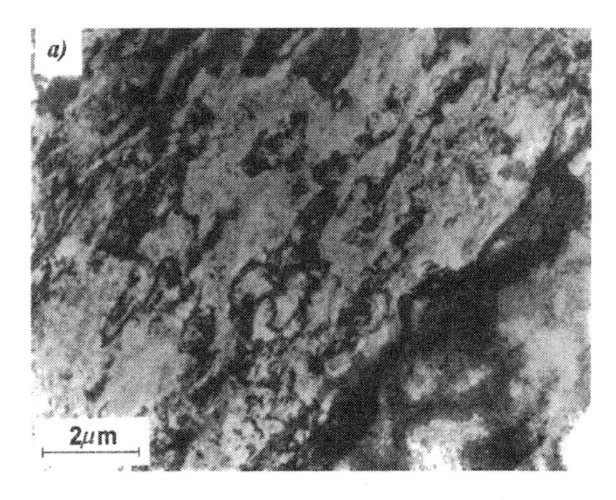

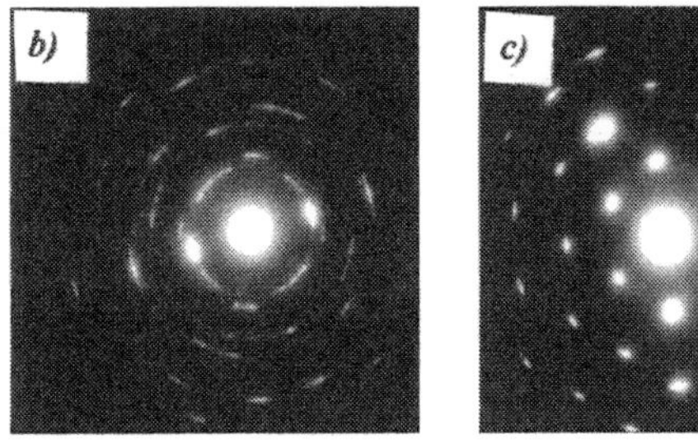

Figure 3 Alloy deformed 75% by rolling: (a) transmission electron micrograph; (b) SADP from the deformation band in the left upper part of micrograph; and (c) SADP from the right lower part

ponent close to {411}⟨122⟩ orientation (twin to the Goss' one) when deformation degree increases. However, deformation twins were not observed in the structure but twin formation can not be excluded.

The regularity is observed in both series with comparable intensity and is the most distinct at 60% deformation. With increasing deformation up to 90% the copper component (the mentioned twinned component of Goss is placed in its neighbourhood) becomes more distinct and simultaneously more spread (see *Figure 5*). Moreover, fluctuation of contribution of the cubic component can be seen. It becomes distinct already after a low deformation (up to 30%) and then disappears at 45%, to be observed again at stronger deformation of 60–75%. The fluctuations are stronger for the B variant samples. The described fluctuation corresponds to the {321}⟨346⟩ and its contribution to the texture is strongest at 45% deformation.

Conclusions

The investigated alloy of composition close to 7278A shows a good workability allowing 90% of deformation by rolling after slow cooling from 465°C. In the as-quenched state, the plastic properties are worse.

The alloy hardens during RT ageing from 85 up to 150 HV. Ageing at 120°C allows to attain 210 HV after 20 h of ageing. A combined treatment, i.e. RT + 120°C ageing gives a small additional hardening. After the deformation applied prior to the ageing treatment, a slightly higher maximum hardness of 230 HV can be reached.

The recrystallization texture becomes stronger after 15% of deformation by rolling, particularly after variant A of the heat treatment. After both heat treatments development of texture was similar with the β-type copper fibre texture with (112)[111], (213)[364] and (101)[121] ideal orientations. Depending on the degree of deformation, some fractions of the Goss and cube textures were observed in both cases.

Acknowledgements

This work was supported by the Research Grant No.: PBZ012-06/4 from the Polish Committee for Scientific Research (KBN).

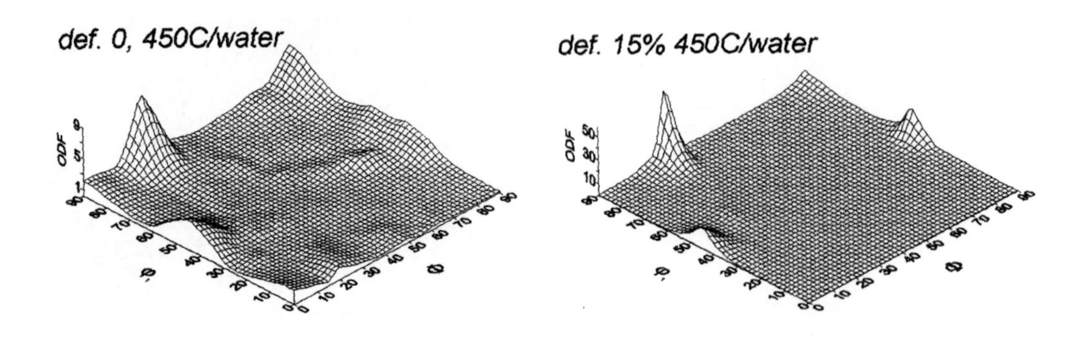

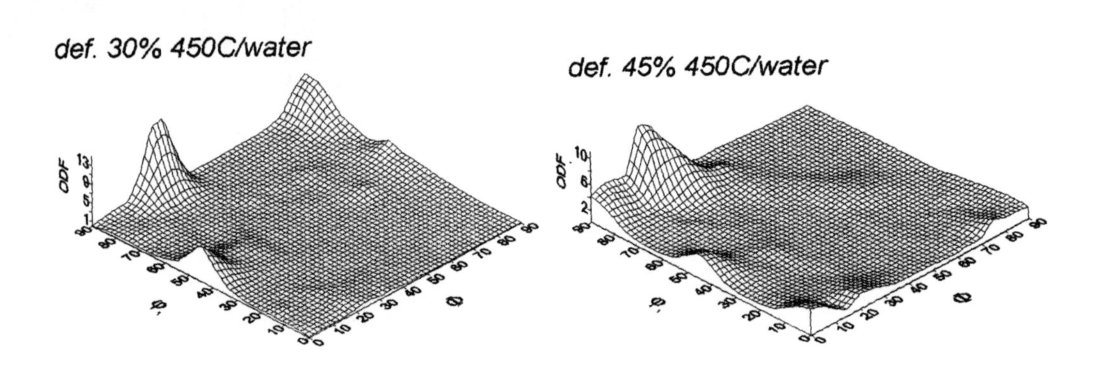

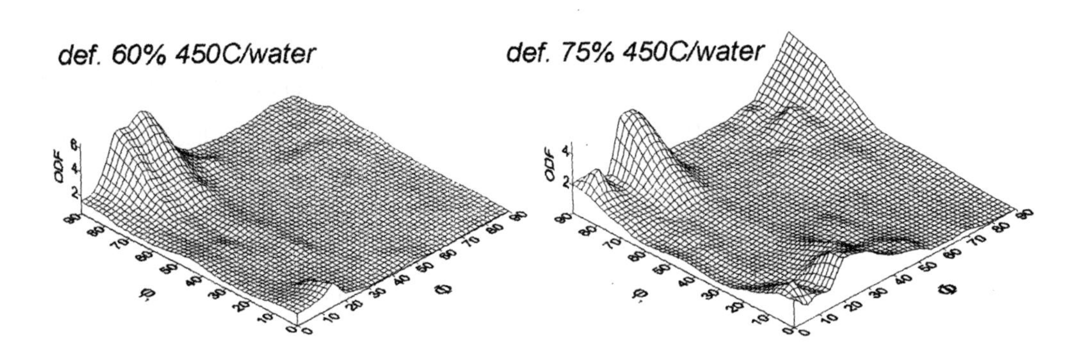

Figure 4 ODF profiles ($\varphi_2 = 45°$ for deformed samples after variant A of heat treatment

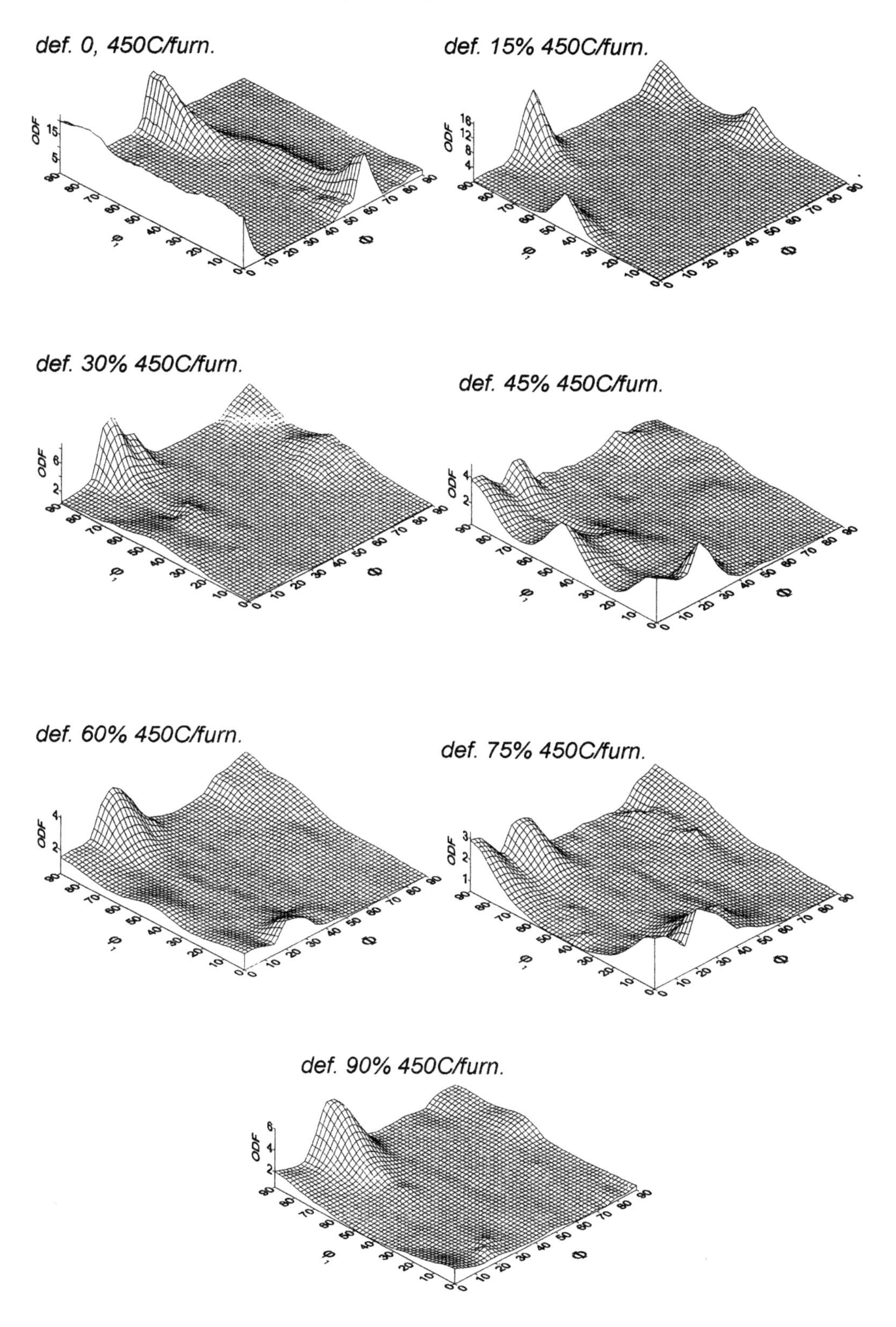

Figure 5 ODF profiles ($\varphi_2 = 45°$ for deformed samples after variant B of heat treatment

References

1 Staley, J. T. In *Properties Related to Fracture Toughness* STP 605, ASTM, Philadelphia, PA, 1976, pp. 71–103

2 Engler, O., Sachot, E., Ehrstroem, J. C., Reeves, A. and Shahani, R. *Materials Science and Technology*, 1996, **12**, 717–729

3 Ehrstroem, J. C., Shahani, R., Reeves, A. and Sainfort, P. In *Proceedings of the Fourth International Conference on Aluminium Alloys*, ed. T. H. Sanders and E. A. Starke, Georgia Institute of Technology, Atlanta, GA, 1994, pp. 32–39

4 Liu, J. In *ASM-TMS Fall Meeting*, ASM International, Cincinnati, OH, USA, 1991

5 Luecke, K., Engler, O. In *Proceedings of the Third International Conference on Aluminium Alloys*, Trondheim, Norway, ed. L. Arnberg, O. Lohne, E. Ness and N. Ryum, Trondheim SINTEFF 1992, p. 439, (1996), pp. 465–470

6 Polmear, I. *Materials Trans. JIM*, 1996, **37**, 12–31

7 Deschamps, A., Brechet, Y., Livet, F. and Gomiero, P., *Materials Science Forum*, 1996, **217 / 222**, 1281–1286

8 Poole, W. J. and Shercliff, H. R., *Materials Science Forum*, 1996, **217 / 222**, 1287–1292

9 Pawlik, K., *Physica Status Solidi*, 1986, **134(b)**, 477

ELSEVIER

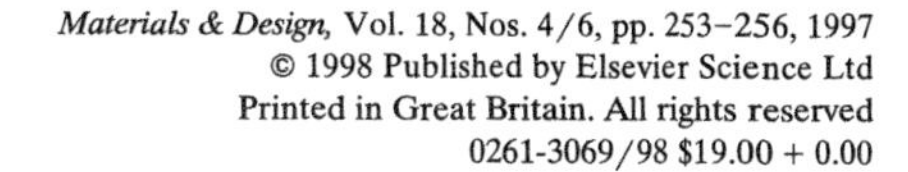
Materials & Design, Vol. 18, Nos. 4/6, pp. 253–256, 1997

0261-3069/98 $19.00 + 0.00

PII: S0261-3069(97)00060-5

Pre-ageing of AlSiCuMg alloys in relation to structure and mechanical properties

W. Reif[a,*], S. Yu[a], J. Dutkiewicz[b], R. Ciach[b], J. Król[b]

[a]*Institute for Metallic Materials, Technical University of Berlin, Strase des 17 Juni 135, D-10623 Berlin, Germany*
[b]*Institute of Metallurgy and Materials Science, Polish Academy of Sciences, Reymonta 25, Kracow 30 059, Poland*

Received 30 June 1997; accepted 15 July 1997

The effect of magnesium addition to the AlSi9Cu3.5 alloy on the hardening and precipitates morphology during ageing at RT, 160°C or two stage ageing (TSA) was studied using TEM and XSAS methods. It was found that only alloys with Mg addition harden during RT ageing and they also attain the highest hardness maximum at 160°C or during TSA. Two types of precipitates (starting from 0.4 and 1.2 nm) were identified during ageing at RT using XSAS method. They cause streaks in the electron diffraction patterns. In alloys aged at 160°C with Mg addition the S′ phase was identified using lattice imaging technique in addition to the θ' plates formed during ageing of the ternary AlSiCu alloy. © 1998 Published by Elsevier Science Ltd. All rights reserved.

Introduction

In AlSiCuMg alloys following phases (in addition to the Al solid solution) were found during solidification: Mg_2Si, Q, Si and θ (Al_2Cu)[1]. In alloys of industrial purity formation of $Al_{15}(FeMn)_3Si_2$, Al_5FeSi and $Al_5Mg_8Cu_2Si_6$ was additionally observed[2]. The addition of copper and magnesium increase considerably the strength of AlSi alloy[3,4] after ageing treatment. The mechanism of strengthening is not clear due to several alloying additions and resulting complexity of phase transitions. The precipitation of Θ'', Θ' due to the presence of copper and S′ and β'' (Mg_2Si) due to addition of magnesium and silicon has been reported[3–5]. The identification of precipitates is very complicated since Θ', S′ and β' are all needle like particles forming along the $\langle 100 \rangle$ directions of the matrix[7]. From the investigations of Eskin[6], it shows that the best combination of mechanical properties corresponds to the presence of all hardening phases, i.e. Θ', S′ and β''. This statement was not supported, however, by the electron diffraction identification of phases, but was based only on the precipitates morphology. On the other hand Sagalowicz *et al.*[5] suggest the precipitation of β'' and Θ', a quaternary phase, while Sakurai and Eto[7] proposed the θ' formation in the presence of copper. The precipitation process in AlSiCuMg alloys requires then further studies to explain the mechanism of hardness increase due to magnesium addition to AlSiCu alloys. Another problem, which has not been reported is the room temperature (RT) hardening of these alloys and their effect on the final properties after ageing at elevated temperatures.

*Correspondence to W. Reif. Tel.: +49 30 21421127; fax: +49 30 31423222

Experimental procedure

Three alloys of composition: AlSi9Cu3.5 (1), AlSi9Cu3.5Mg0.5 (2) and AlSi4Cu3.5Mg1.5 (3), 0.8 Fe, 0.8 Zn, 0.2 Mn and Ti, Ni, Cr below 0.05 (in wt.%) were cast under salt, then homogenised in 520°C and quenched in RT water and aged at RT and/or at 160°C. Hardness tests were performed using Vickers method under load of 5 kg. Structure studies were carried out using Philips CM20 microscope, Philips PW 1830 diffractometer and X-ray small angle scattering using Rigaku-Denki diffractometer with XSAS camera with perpendicular slits standardised by M1811 Lupolen sample. Thin foils for TEM studies were obtained by dimpling, then jet electropolishing in 1/3 HNO_3 2/3 CH_3OH electrolyte.

Results and discussion

Results of hardness measurements

The hardness changes during RT ageing are presented in *Figure 1a* and indicate a very strong effect of a magnesium addition. Alloy 1 shows a very slow hardness increase $\Delta H = 8$ HV in 100 h, while the growth of hardness in alloy 2 is equal to 40 HV after 30 h of ageing. The similar hardness increase was observed in alloy 3 with 1.5% Mg addition (50 HV) during the same time interval. The highest rate of the hardness change was observed from 1 up to 20 h of ageing. During ageing at 160°C (*Figure 1(b)* the alloys with Mg addition show similar hardness changes like at RT, but attaining higher maximum after 20 h. Alloy 1 show a lower hardness maximum after similar ageing time like for the other two alloys.

The TSA two-stage ageing (preageing 4 h at RT and

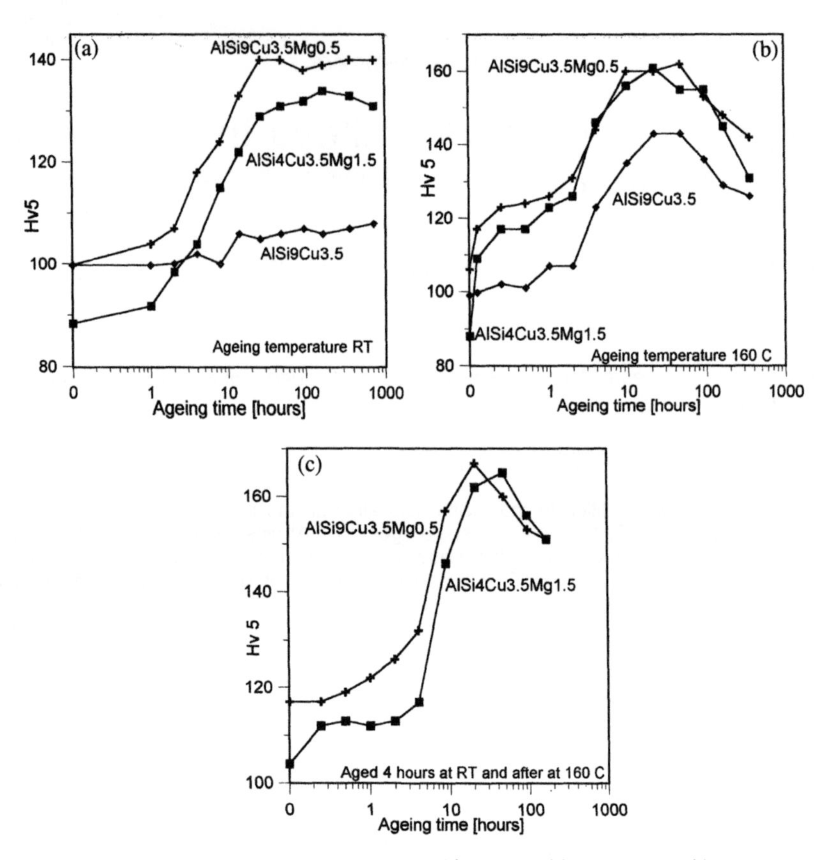

Figure 1 Hardness changes during ageing of alloy 1, 2 and 3, aged at RT (a), at 160°C (b) and two stage (c) 4 h RT and then at 160°C

then ageing at 160°C) has a similar character like that at 160°C (*Figure 1(c)*). The hardness maximum after about 20 h of TSA is only slightly higher than that at 160°C.

X-Ray phase analysis and small-angle-scattering studies

In both investigated alloys with magnesium additions Al_2Cu and Mg_2Si phases were identified in the as quenched state. In alloy 3, Al_2CuMg phase was additionally found. In the overaged alloys, the precipitates of the both previously reported phases Θ' and β''[1–5] and the ternary $Al_7Cu_3Mg_6$ were identified. In the alloy with 1.5% Mg, addition beyond the above mentioned phases the Al_3Mg_2 and the ternary $Cu_{16}Mg_6Si_7$ were found.

Guinier radius (R_g) related to precipitate size (thickness for the plate-like precipitates) and integral intensity (Q)—related to amount of the precipitates were calculated in a manner presented in earlier paper[8]. They are shown in *Figure 2(a),* and (*b*) as a function of ageing time at RT for alloy 2. Two kinds of precipitates can be distinguished in both alloys. The Guinier radius of the small particles grows from about 0.4 nm at the beginning to about 0.7 (alloy 2) and 0.8 nm (for the alloy 3) after about 40 h of ageing. The large precipitates grow almost uniformly from about 1.2 nm after 2 h of ageing to about 2 nm after 300 h of ageing. The precipitates are generally smaller in size in the alloy with higher Mg content. The amount of precipitates (integral intensity) grow fast up to 10 h of ageing for the small particles and for a longer time for the large ones. The amount of small particles was much higher than the large ones (*Figure 2(b)*). The change of size and the amount of precipitates is in a good accordance with the hardness changes.

Three types of precipitates were formed during two stage ageing as presented for alloy 3 (*Figure 3*). Two kinds of a small precipitates 1–2 nm in size were observed in a small quantities (according to integral intensity values), while much more was found of the of large ones. One can see that the growth of size and of the amount of largest precipitates proceeds at highest rate between 4 and 20 h of ageing. The changes of size and amount of a small precipitates during ageing are less pronounced. Similar types of R_g and Q changes of three types of precipitates shows alloy 2, however, largest precipitates grow at a lower rate than in the case of alloy 3.

Transmission electron microscope studies

In alloy 3 aged 72 h at RT, no precipitates were found even at higher magnification. Dislocations were created during dimpling which disturbed observation. At the electron diffraction pattern one can distinguish very weak streaks in ⟨100⟩ directions through 100 positions. They may result from either the S′, β'or Θ' precipitates since the streaks match the positions of reflections from these precipitates at this orientation[5,9,10].

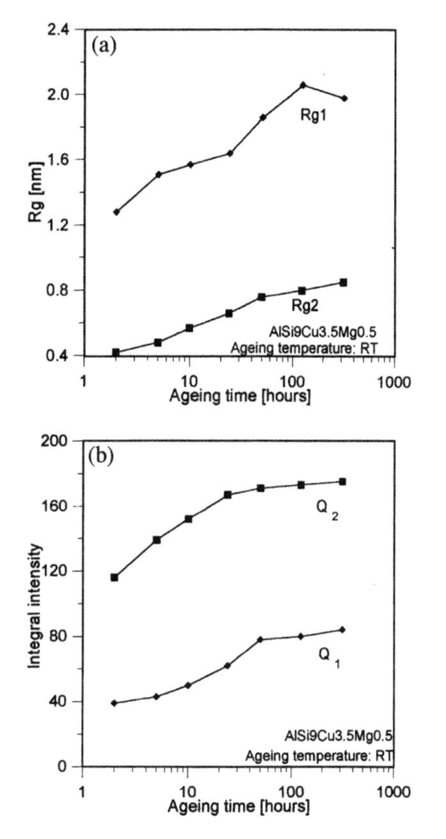

Figure 2 Changes of Guinier Radius R_g (a) and integral intensity Q (b) during RT ageing of alloy 2

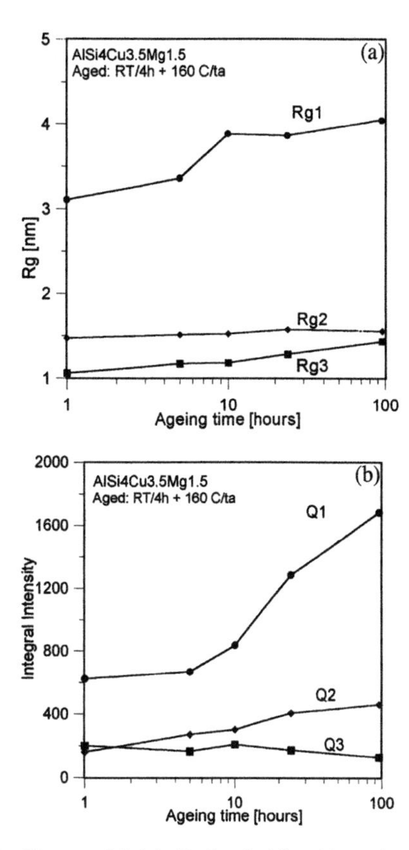

Figure 3 Changes of Guinier Radius R_g (a) and integral intensity Q (b) during two stage ageing of alloy 3

Figure 4 shows TEM microstructure of alloy 2 aged 24 h at RT and then 1 h at 160°C. At relatively high magnification one can see small elongated precipitates giving streaks in ⟨100⟩ directions as visible in the electron diffraction pattern (*Figure 4(b)*). One can distinguish reflections within streaks corresponding to a large lattice spacing, i.e. 010 S′. The precipitates without preageing grow larger and possess an elongated shape. *Figure 5* shows a lattice image of elongated precipitates in alloy 2 aged 24 h at 160°C, where fringes within elongated precipitates parallel to 200 Al plane can be seen. Their distance measured from the micrograph is equal to 0.1 nm and corresponds to 010 S′ lattice spacing. It does not match the largest spacing of θ'[10] or the quaternary Q-phase reported in[5]. It agrees with the suggestion of Eskin[6] that S′ forms during ageing additionally to θ' and β' with increasing magnesium addition.

Conclusions

(1) AlSiCu alloy shows insignificant hardness increase during RT ageing, while alloys with 0.5% and 1.5% magnesium addition show hardness increase of 40 and 50 HV respectively during 100 h. Two stage ageing (RT/160°C) causes only slightly higher hardness maximum of 165 HV than in alloys aged at 160°C (160 HV).

(2) Small angle X-ray scattering revealed two kinds of precipitates (of size close to 0.4 and 1.2 nm) in both alloys with Mg addition. They grow in size during 100 h at RT what corresponds to hardness measurements. After two stage ageing three kinds of precipitates can be distinguished in both alloys.

(3) Transmission electron microscopy allowed to identify the θ' precipitates in the AlSiCu alloys and additionally the S′ precipitates formed in the alloys with Mg addition during ageing at 160°C.

Acknowledgements

This work was supported by the Polish-German Cooperation Project No. X082.1 supported by Kernforschungszentrum Karlsruhe and by Committee of Scientific Research (KBN) Project No. 7S20107406.

References

1 Chakrabarti, D.J. and Murray, J.L., *Materials Science Forum*,1996, **177**, 217–222

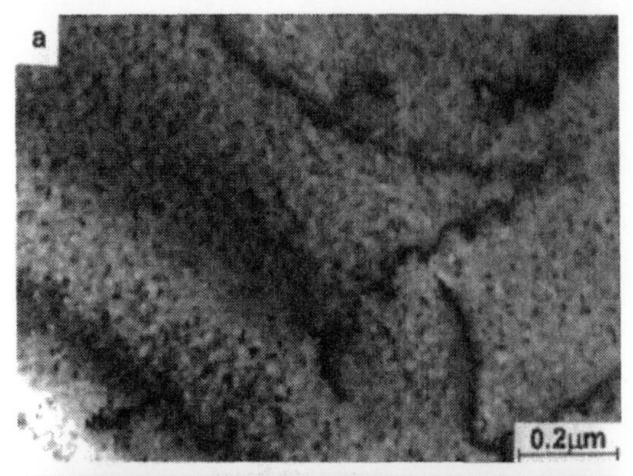

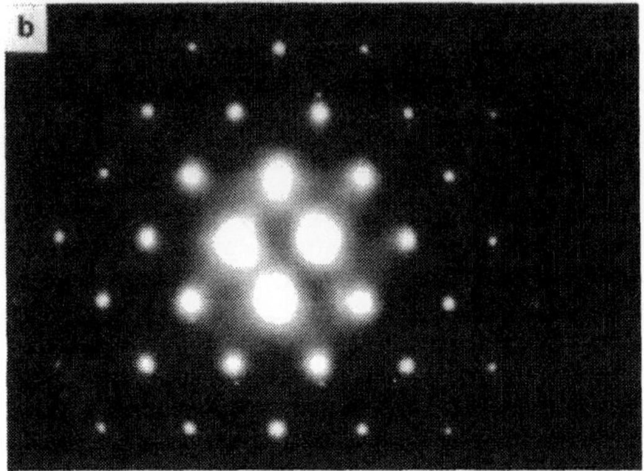

Figure 4 Transmission electron micrograph (a) and corresponding SADP (b) of alloy 3 after 24 h ageing at RT and then 1 h at 160°C

2 Yao, J.-Y., Edwards, G.A., Zheng, L.H. and Graham, D.A., *Materials Science Forum*, 1996, **183**, 217–222
3 Gowri, S. and Samuel, F.H., *Metallurgical Transactions, [Section] A*, 1994, **25A**, 437
4 Das Gupta, R., Brown, C.C. and Marek, S., *AFS Transactions*, 1989, **9**, 245.
5 Sagalowicz, L., Hug, G., Bechet, D., Sainfort, P. and Lapasset, G., 4th Int. Conf on Al Alloys, ed. T.H. Sanders, E.A. Starke, Georgia Institute of Technology, Atlanta, GA, 1994
6 Eskin, D.G., *Zeitschrift für Metallkunde*, 1995, **86**, 60
7 Sakurai, T. and Eto, T., 3rd Int. Conf. on Al-Alloys, Trondheim, Norway, ed. L. Arnberg, O. Lohne, E. Ness and N. Ryum, Trondheim SINTEFF, 1992, p. 208
8 Dutkiewicz, J. and Król, J., *Physica Status Solidi (a)* 1994, **141**, 317–327
9 Matsuda, K., Uetani, Y., Anada, H., Tada, S. Ikeno, S., 3rd Int. Conf. on Al-Alloys, Trondheim, Norway, ed. L. Arnberg, O. Lohne, E. Ness, N. Ryum, Trondheim SINTEFF 1992, p. 272
10 Papazian, J.M., *Metallurgical Transactions [Section] A*, 1981, **12A**, 269.

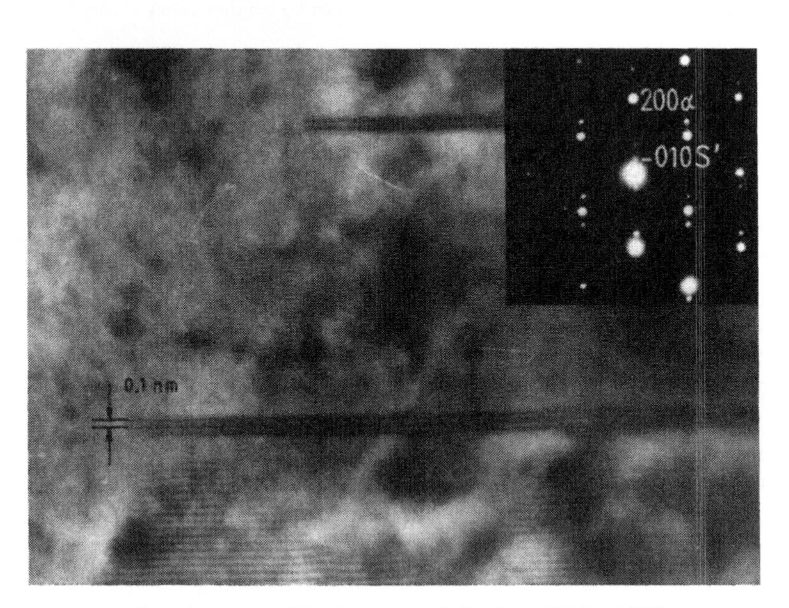

Figure 5 Lattice image and corresponding selected area diffraction pattern of alloy 2 aged 24 h at 160°C

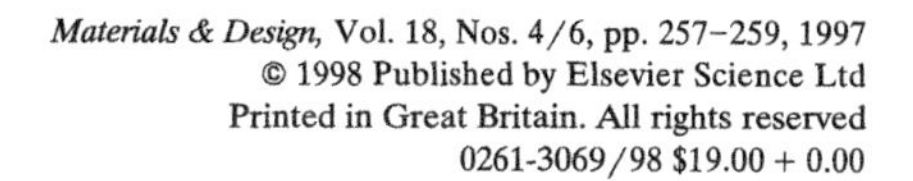

Materials & Design, Vol. 18, Nos. 4/6, pp. 257–259, 1997
© 1998 Published by Elsevier Science Ltd
Printed in Great Britain. All rights reserved
0261-3069/98 $19.00 + 0.00

PII: S0261-3069(97)00061-7

Structure and mechanical properties of Al–B composite powder

F. Muktepavela[a,*], I. Manika[a], V. Mironovs[b]

[a]*Institute of Solid State Physics, University of Latvia, 8 Kengaraga Str., LV-1063 Riga, Latvia*
[b]*Technical University, 1 Kalku Str., LV-1658 Riga, Latvia*

Received 30 July 1997; accepted 12 August 1997

Al–B composite powder has been obtained by crushing pieces of composite material presenting industrial waste. Structural peculiarities and microhardness of separate powder particles (d ~ 1 mm) have been investigated. Original design of high precision microhardness tester made it possible to detect the properties of powder both in near-surface layer and below it. The powder represents a new structurally non-homogenous material with the increased microhardness (1.5 GPa) which grows up to 4 GPa in near-surface layers. Stable oxide compounds are formed on internal surfaces and defects of the aluminium alloy. Powder compacts were obtained. Adhesion on Al–B and Al–Al interfaces at various temperatures and pressures were investigated. The applications of the powder compacts were considered.

Keywords: Al-B powder; disintegration; microhardness

Introduction

Mechanical disintegration of materials is widely used in various powder technologies. A large number of indirect data of processes, such as mechanical alloying, mechanoactivation of adhesion, etc., show that through crushing, the material acquires special properties[1,2]. At the same time, the changes in the material are insufficiently investigated. It is connected with the small size of particles and the highly deformed state of the materials. The nature of highly deformed states in bulk materials has already been investigated[3] yielding new data concerning formation and rearrangement of dislocation structures. Correlation of relaxation processes with the dynamics of dislocations and point defects was established.

Powder production is accompanied by numerous acts of fracturing. High density of dislocations in the subgrains and network of microcracks has been obtained in this case. When the process of milling takes place in the active medium, like in air, oxides and other chemical compounds may be formed both on external and internal interfaces. As a result, powders obtained through crushing may serve as a way of producing a new heterogeneous material. Some of the recently obtained results confirm such a possibility[4].

The subsequent compacting of the powder, depending on interface adhesion attracts special attention. As is well known from thermodynamics, these processes depend on adhesion work W_a. Adhesion work W_a is defined as $W_a = g_0^A + g_0^B - g_{pb}^{AB}$, where g_0^A, g_0^B are surface and g_{pb}^{AB} phase boundary energies. In the case of heterogeneous materials, g_{pb}^{AB} is an important parameter which determines the kinetics of diffusion and structure formation. The structure and mechanical properties of Al–B composite powders obtained in the air were investigated in the present paper. The processes of adhesion interaction on Al/Al and Al/B interfaces have been studied on bimetallic cold-welded joints as model objects.

Experimental procedure

The waste of commercially produced AlB composite was used for crushing. Fritch's disintegrator was used for that purpose. The samples, showing lack of bonding between fibers and matrix, were specially chosen. During crushing of such material, the free boron fibers served as additional deforming rigid elements.

The microhardness method was used for estimation of the mechanical properties of powder particles. The depth dependence of microhardness characterizes the mechanical properties of particles both in near-surface layers and in the bulk. The Vickers microhardness tests were performed by the original device described by Upit and Varchenya[5], which is insensitive to vibration and is suitable for accurate hardness measurements over a load range from 3 mN to 2 N. Under the load range from 2 to 10 N, measurements were made by the microhardness tester PMT-3. The indentation depth was calculated as 1/7 the diagonal of the marks left by the indenter. Structure investigations were carried out using Neophot-30 microscope, SEM and X-ray diffractometer.

* Correspondence to Dr F. Muktepavela

Table 1 Density, size and microhardness of Al-B powder particles

No.	Density $\times 10^3$ (kg m^3)	Size of particles (mm)		Microhardness of particles (GPa)	
		Al	B	Al	B
Al-B (1)	0.60	1.5–2.5	2.5–5.0	2.3	70
Al-B (2)	0.75	1.0–1.5	2.0	4.3	70
Al-B (3)	1.0	0.2–0.5	1.0	4.3	70
Al-B (4)	1.35	0.2–0.5	2.0	2.5	70

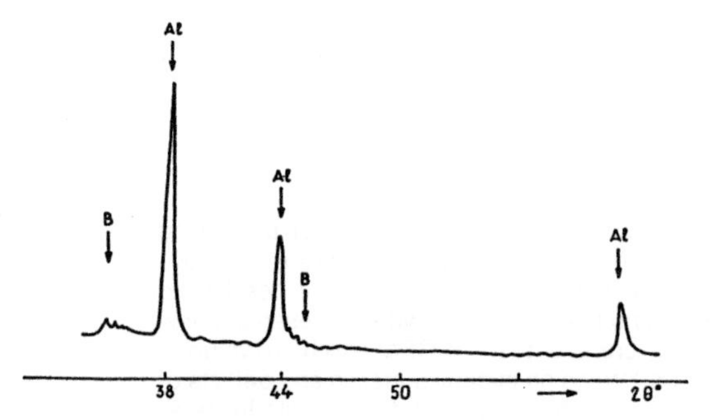

Figure 1 X-ray diffraction spectra for Al–B powder

Results and discussion

After disintegration powder particles had diameter from 0.5 to 1.0 mm. The length of boron fibers was 2 mm (*Table 1*). X-ray analysis of the powders showed that no new phases such as AlB_2 and AlB_{12} appear after disintegration. Boron and aluminum lines were broadened (*Figure 1*). For boron it is connected with its fine dispersed quasi-amorphous state which is caused by fibers production method[6]. Elastic stresses of deformation nature caused the broadening of the aluminium lines. Influence of oxide complexes, mixed with metal during disintegration is also possible.

Structural observations have shown that aluminium powders are strongly deformed by crushing. In the overstressed regions, the large cracks could be seen. Each Al powder particle is structurally inhomogeneous. Boron fibers were frequently broken but showed no evidence of interaction with aluminium. The above corresponds to the results of X-ray analysis. The conclusion is that plastic deformation of aluminium is not an activating factor for aluminium borides formation at 293 K.

The results of Al powder particle microhardness measurements are presented in *Figure 2*. For comparison, in the same figure values of microhardness for aluminium matrix in composite (curve 3) and for pure Al in the annealed state (dotted line 4) are shown. As it is seen the powder particles after disintegration have much higher microhardness than aluminium matrix in the composite. In the near-surface layer, microhardness is even higher and reaches the value of 4.5 GPa. It can be assumed that we deal with strongly deformed state of aluminium. However, the results of microhardness measurements for powder annealed at 800 K did not confirm this conclusion. As it is seen from *Figure 2*, annealing caused the reduction of strengthening only in the surface layer, while microhardness in the bulk did not change, being 2–3 times higher than that of aluminium in the composite matrix before disintegration.

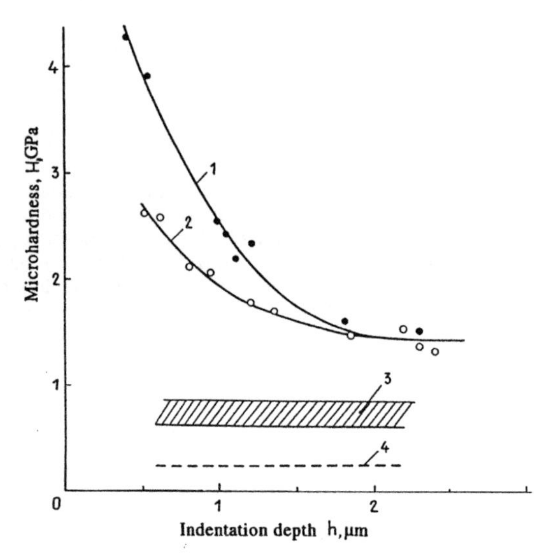

Figure 2 The dependence of microhardness on the indentation depth for aluminium powder after disintegration (1) and after annealing (2), Al matrix of the as-prepared fiber composite (3) and after its annealing (4)

From these data it could be inferred that oxide compounds occur in the Al powder particles hardening them simultaneously. This is in agreement with the results of X-ray analysis and structural observations. Deep oxidation of aluminium is possible due to oxygen penetration through developed networks of dislocations, grain boundaries and microcracks, which appear in the regions of stress concentrating during material disintegration. Every Al powder particle presents itself as a 'composite'. It could be the reason for so large microhardness and brittleness of aluminium powder particles after Al-B composite disintegration. As it concerns boron fibers, their microhardness after disintegration corresponded to the initial value and was approx. 70 GPa.

AlB powder compound after compacting at 700–800 K showed a strength $\sigma = 1.8$ GPa on a bending test. These values are coincident with the initial strength of Al–B fiber composite. Stable oxides determine also the strength and adhesion on the Al/B and Al/Al interfaces. Investigations of the adhesion were carried out on the bimetallic joints obtained by the special cold welding method[7]. It has been shown that the adhesion strength of the Al/Al joints increases under annealing at 400 K while their softening after annealing at 800 K was observed (*Figure 3*). For comparison, the data for Al–Cu–Mg alloy, as well as data for atomically-clean surfaces, are shown. The strength of Al/B joints in-

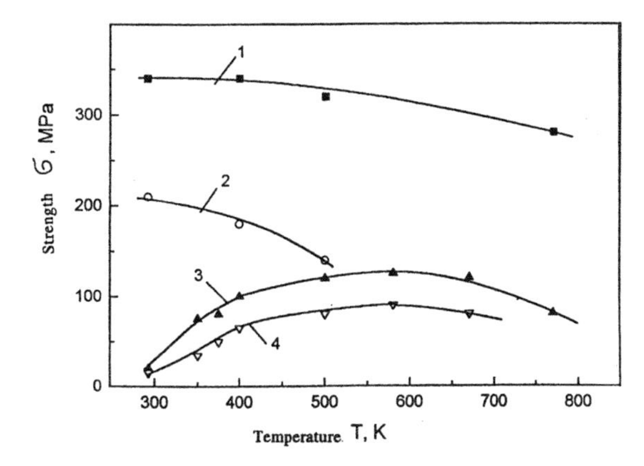

Figure 3 Effect of annealing temperature on the strength of cold-welded joints of Al/Al (2,4) and Al–Cu–Mg/Al–Cu–Mg alloy (1,3). Contact of atomically-clean (1,2) and real (3,4) surfaces

creased only under annealing at 800 K. *Figure 3* shows that the effect of annealing on the deformation behavior of atomically-clean (curves 1,2) and oxidized interfaces (curves 3,4) is quite different. These data together with the results of our earlier fractographic investigations[7] allow to conclude that annealing leads to increase of the adhesion and plasticity of the oxidized interface layer.

We note, in conclusion, that Al–B powder possessing high hardness properties can be successfully applied as an abrasive material for doping grinding pastes. Another example of applying the Al–B powder is fabricating the powder-based grinding tools. For that purpose, Al–B powder was mixed with polymer filler and then pressed and polymerized. The obtained samples containing 30–60% Al–B were used for fabricating the abrasive wheel which could be applied to the treatment of glass, metals and ceramics.

References

1 Benghalem, A. and Morris, D.G., *Acta Metals et Materials*, 1994, **42**, 4071
2 Oehring, M., Yan, Z.H., Klassen, T. and Borman, R., *Phys. Stat. Sol. A*, 1992, **131**, 671
3 Panin, V. E., Likhatchev, V. E. and Grinyaev, Yu. V., *Structural Levels of Solid State Strain*. Nauka, Novosibirsk, 1985 (in Russian)
4 Corrias, A., Ennas, G., Morongiu, G., *Abst. 6th Int. Conf. on the Structure of Non-crystalline Materials*. NCM 6, Praha, 1994, p. 80
5 Upit, G.P. and Varchenya, S.A., In *Scientific Instruments*, ed. I.G. Matis. Zinatne, Riga, 1986, p. 12 (in Russian)
6 Shorshorow, M.Kh., *Strength Physics of the Metallic Matrix Based Fiber Composite Materials*. Metallurgia, Moscow, 1989, (in Russian)
7 Muktepavela, F., Maniks, J., Astanin, V., Simanovskis, A. and Manika, I., *Izvestiya Latv. Akademii Nauk., Seriya Fiz*, 1990, **6**, 44, (in Russian)

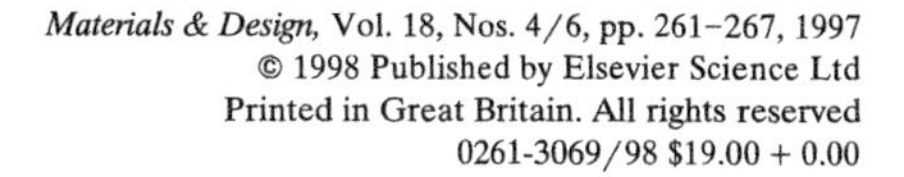
Materials & Design, Vol. 18, Nos. 4/6, pp. 261–267, 1997

0261-3069/98 $19.00 + 0.00

PII: S0261-3069(97)00085-X

Laser beam welding of low weight materials and structures

K. Behler, J. Berkmanns, A. Ehrhardt*, W. Frohn

Fraunhofer Inst Lasertechnik, Steinbachstrasse 15, D-52704 Aachen, Germany

Received 10 September 1997; accepted 15 September 1997

In this presentation an overview will be given about laser beam welding of aluminium. Different aspects regarding process parameters, metallurgical aspects, weld seam properties and possible applications will be discussed.

Keywords: laser beam welding; low weight materials; structure

Introduction

In the transportation industry there is increasing use of aluminium alloys because of greater requirements concerning exhaust emissions, energy consumption and recycling of material. Aluminium seems to be a possibility with regard to these requirements because of it's small specific weight, high recycle potential, relatively good mechanical properties and positive weldability[1]. Beside the use of extruded aluminium profile structures in railroad construction which led to a 20–30% reduction in mass[2], it was in the car industry the first time a complete aluminium car was produced by the Audi company[3].Laser beam welding is used widely in industrial production[4,5]. This is because of the low heat input, high welding speed and productivity as well as it's flexibility. However, these applications are mainly oriented to the material group of different steels. In this presentation an overview will be given about laser beam welding of aluminium. Different aspects regarding process parameters, metallurgical aspects, weld seam properties and possible applications will be discussed.

Laser beam welding process

Two types of lasers can be used: the CO_2-laser (beam power up to 40 kW, wavelength 10.6 μm) or the Nd:YAG-laser (beam power up to 4 kW, wavelength 1.06 μm). Whereas the first type of laser gives the potential to weld material with thickness up to 15 mm or to use high welding speed in the case of the Nd:YAG-laser optical fibers can be used to guide the laser beam to the focusing optic. The principle of the welding process is schematically shown in *Figure 1*. The laser beam is focused onto the surface of the workpiece. Above a certain intensity the material starts to evaporate, a keyhole is formed which leads to a strong increase in beam absorption. The threshold intensity to ignite this process depends on the material and the normal absorptivity. As a consequence of these influences in *Figure 2* it is shown that the threshold intensity and the weld depth which can be reached in dependence on beam power or intensity are effected by the type of laser, i.e. the wavelength.

Influence of the welding process on the material

The laser beam welding process is similar to other welding techniques, i.e. a thermal effecting melting process. This causes metallurgical changes in the weld metal as well as in the heat affected area beside the weld seam. A microscopic cross-section of the transition area weld metal/heat affected zone is shown in *Figure 3*. In this case an alloy out of the 6000-series has been welded applying a 5% Si-contenting additional wire. Because of the rapid (and in the transition area mainly unidirectional) cooling it can form a fine grained solidification structure with a nearly linear orientation to the center of the weld seam. In the central area of the weld seam typically a fine disperse globular structure of the solidified material occurs as has been published by Berkmanns et al.[6]. In the picture in *Figure 3* it is not possible to show this region because of the magnification. The center of the photograph (transition weld metal/heat affected zone) shows the typical change in the structure of hot cracking sensitive materials. Because of the large temperature region between melting and solidification (solidus–liquidus interval) in the heat affected zone the material is partially molten in the solid aluminum matrix. On the weld seam side there is a region of some microns where small defects (solidification cracks) occur during the solidification process. Typical dimensions of these defects are: length

*Correspondence to A. Ehrhardt

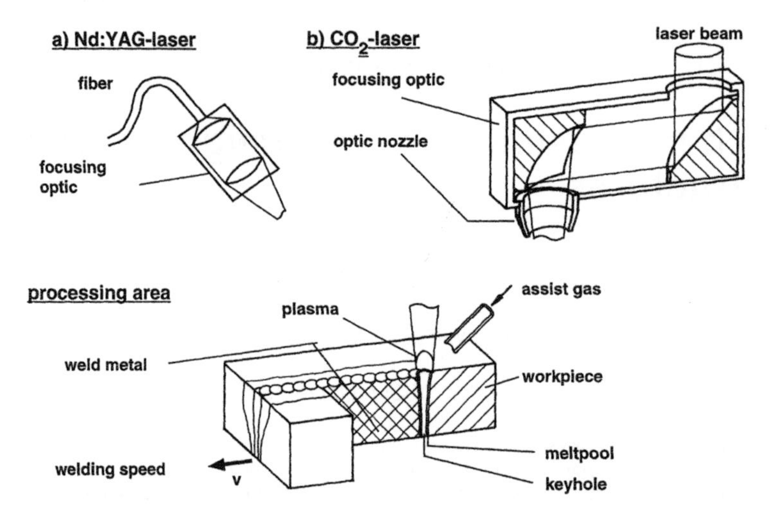

Figure 1 Principle of laser beam welding

< 3 μm and width < 1 μm. These kind of defects are typical in the transition area of weld seams on hot cracking sensitive material. However, because of its size it is not possible to detect them by using non-destructive methods such as X-ray transmission photography. Because of the 'molecular' size of these defects there is no special negative influence on the mechanical properties of the weld seam. On the right hand side of the picture no small cracks occur because there is a dilution between base material and additional wire. This leads to a change in the chemical composition resulting in crack free areas.

Beside the internal structure of the weld seam the rapid process and short cycle time also influence the surface structure of the weld seam and heat affected zone. *Figure 4* shows a typical micrograph of the surface of a laser beam welded seam ($AlMg_{4.5}Mn$). The structure and the direction of the solidification lines are an effect of the relative high welding speed. Especially in the center line of the seam a slight oxidation occurs. At the side of the weld seam small pores as well as a recondensed layer occur. This area is located in the transfer area between weld seam and heat affected zone. So it can be assumed that this is caused by melting or evaporating material with low transfer temperatures similar to the solidification defects which are shown in *Figure 3*. Additionally some small solidification cracks occur at the surface. This is also an effect of the rapid welding process in combination with the solidification characteristics of the used alloy. In general there is no negative influence of the pores as well as of the small cracks on the mechanical properties.

Another principle aspect in welding is the possible influence of the process energy on the chemical compo-

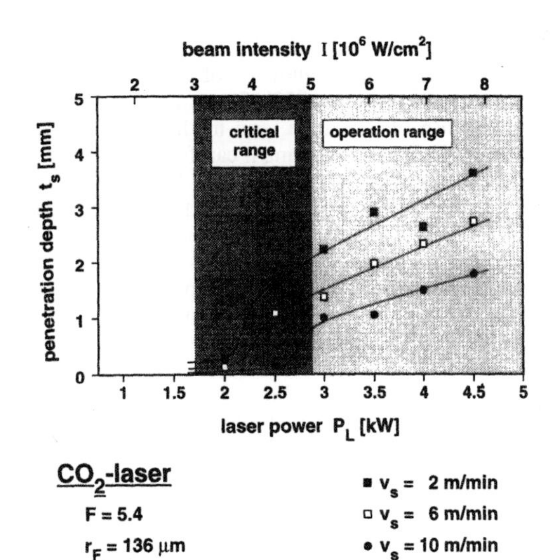

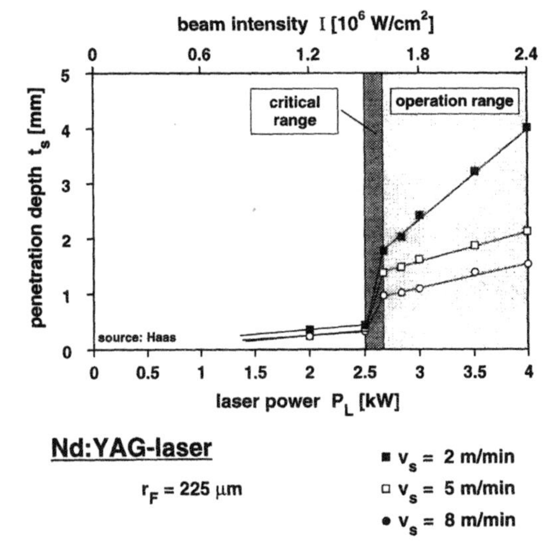

Figure 2 Intensity threshold for laser beam welding ($AlMgSi_1$)

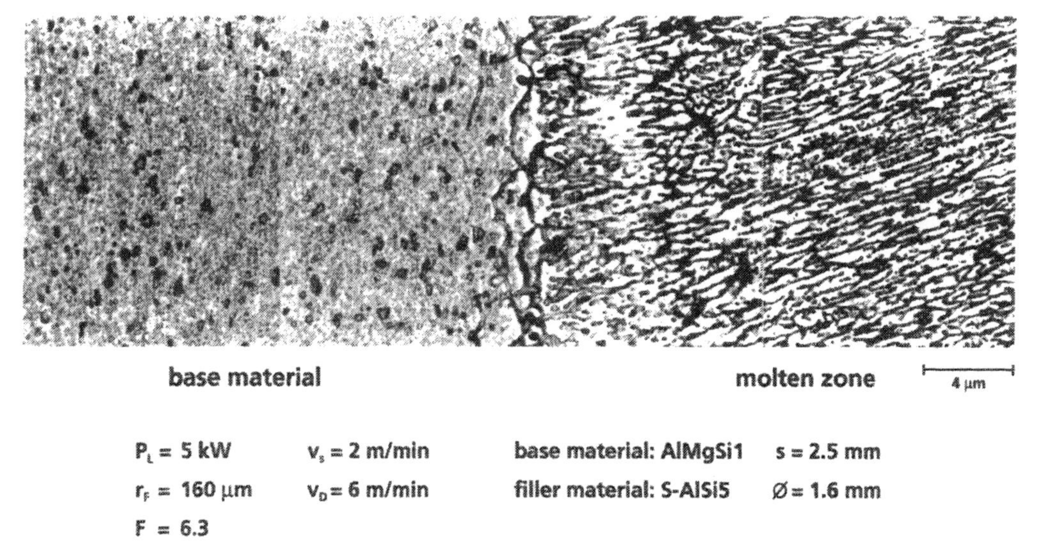

Figure 3 Laser beam welding of aluminium with filler material

sition and the grain structure in the weld zone. Conventional welding processes which generate a meltpool as arc welding methods partially require additional material to reach a chemical composition in the weld area which is similar to the base material. In addition to that, the heat input of the welding process often lead to a coarse grain structure in the weld zone. Refining elements in additional materials are used to homogenize the solidification structure and to reduce the typical dimensions of the grains. In *Figure 5* line scans of the distribution of the alloy elements Mn, Fe, Mg and Si in a laser beam welded seam on $AlMg_{4.5}Mn$ are shown. Because of precipitation effects (AlFeSi- and AlFeSi(Mn)-) in the base material there are large distribution fluctuations of these elements except the element Manganese which is not included in precipitation processes. In opposition to that in the weld metal there is a very fine and homogeneous distribution of all the different elements. In addition to that, the level of the element content is comparably high as in the base material. As a consequence it can be interpreted that there is no change in the chemical composition. In this example of welding the alloy $AlMg_{4.5}Mn$, the grain structure is refined by the welding process itself. The main reasons for these positive results are:

- the high welding speed;
- the short cycle time of the melting and solidification process;
- the deep penetration effect of the keyhole with a

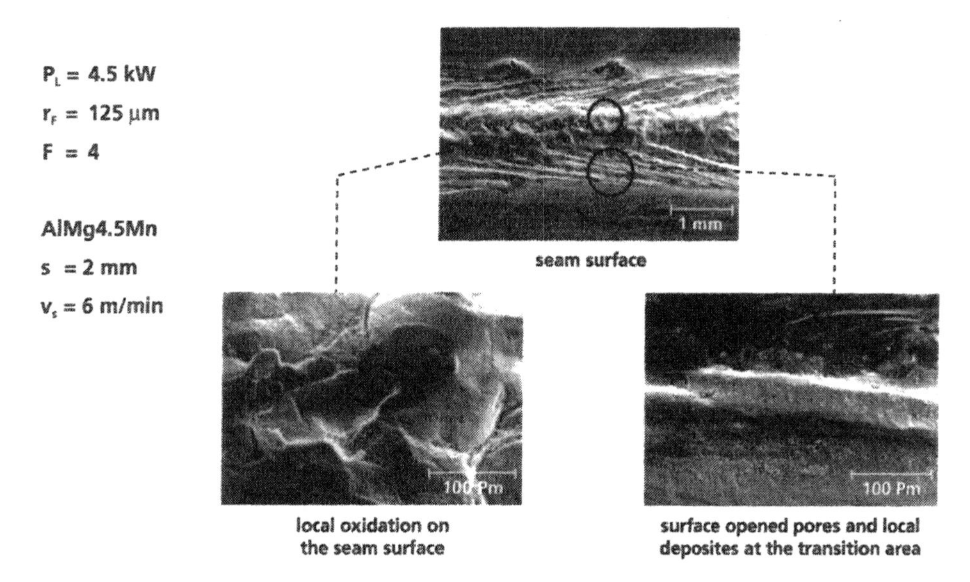

Figure 4 Surface of laser beam welded seams on aluminium

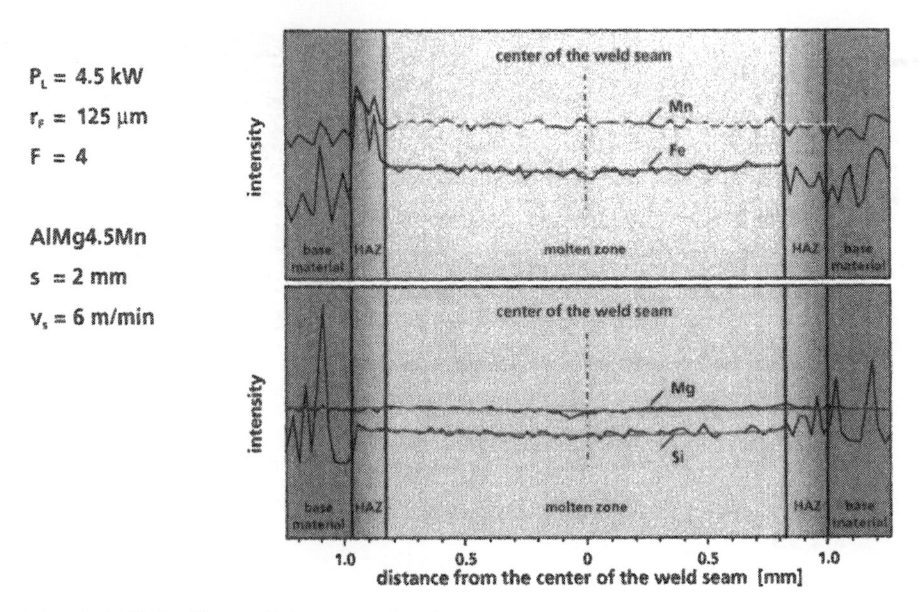

Figure 5 Chemical analysis (EMS) of laser welded seams on aluminium

very large ratio of depth to width; and

- the turbulent melt pool dynamic.

There are different advantages in laser beam welding which not only give positive results in production and manufacturing but also concerning material and solidification effects.

The thermal influence on the material structure beside the weld seam itself is very small. If an alloy is welded which is sensitive to hot cracking additional wire can be used to overcome this problem as in conventional welding. In the transition area between the weld seam and heat effected zone some defects occur but with 'near molecular' dimensions. So there is no negative influence on the properties of the welded component.

The seam surfaces are characterized by a structure which is defined by the high welding speed and rapid solidification. Effects such as oxidation reactions are comparable to other thermal treatments in the case of aluminum; but because of the short cycle time oxidation of the weld seam is very low.

Welding an alloy which does not require additional wire because of metallurgical effects shows that there is no change in chemical composition compared to the base material. If the base material is characterized by relatively strong fluctuations of the local chemical composition the laser beam welding process homogenizes and refines the composition and the structure of the material. This gives positive effects on mechanical and corrosion properties of the welded component.

Mechanical properties

Figure 6 exemplifies the results of tensile strength tests

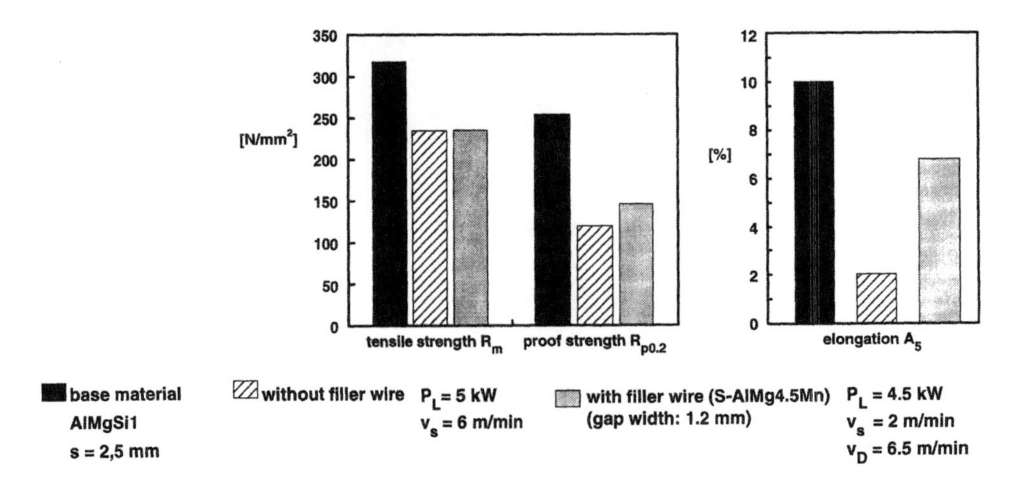

Figure 6 Comparison of mechanical properties of base material and weld seam (with and without filler wire)

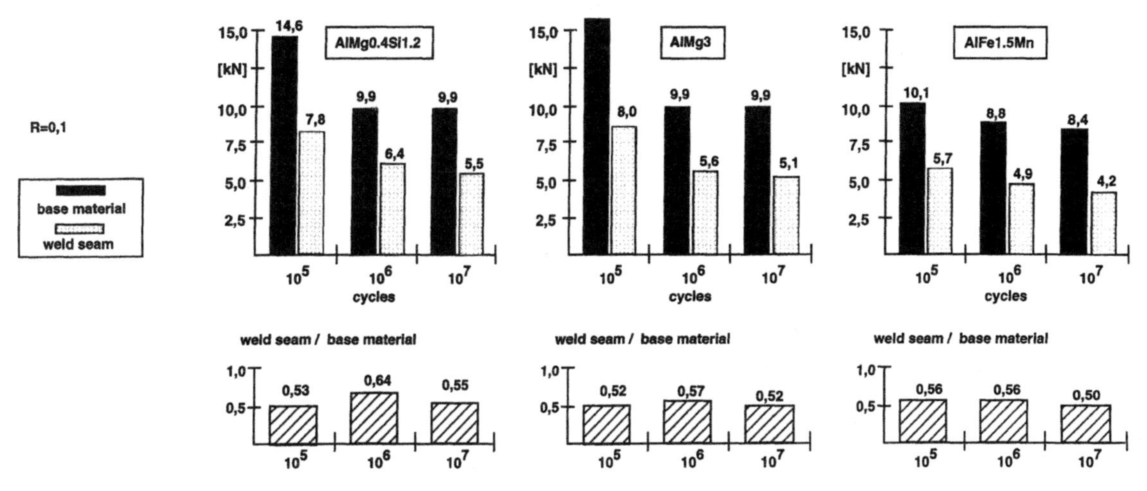

Figure 7 Fatigue strength of laser beam welded butt joints on aluminium

on base material as well as on laser beam welded seams on $AlMgSi_1$. The weld seams were carried out with and without additional material. From the tensile strength of the base material it can be estimated that the material has been delivered in the strength level F32. Because of the heat input during welding the tensile strength in the weld seam area is reduced because of the annealing effect of the precipitated material. This occurs in both cases, welding with or without additional material. The proof strength is also reduced if one compares the results of the weld seams to the base material. The reason for this is also the annealing in the weld zone area. It can be seen in the elongation results that welding without additional wire there is only 2% elongation of the whole specimen. By using additional wire the elongation can be increased to 7% which is in the typical level of arc welding methods[7]. The low level of the elongation in the case of welding without additional wire is caused by the annealing effect in the weld zone as well as by the small width of laser beam welded seams. During the strength test the complete elongation is concentrated in the weld seam area because of its reduced strength. Because of this concentration the local elongation of the weld seam area is very much more than the elongation level of the complete specimen. If additional material is used the seam width is enlarged so the local elongation potential will increase. This results in an increase of the specimen elongation.

Results of fatigue tests on laser beam welded specimens, butt weld, on different materials ($AlMg_{0.4}Si_{1.2}$, $AlMg_3$ and $AlFe_{1.5}Mn$), are shown in *Figure 7*. As a reference the fatigue properties for each base material are included. This allows a better comparison and gives an indication of the influence of the weld seam on the mechanical properties which for example have to be taken into account in designing welded constructions out of aluminum alloys. It can be seen in *Figure 7* that

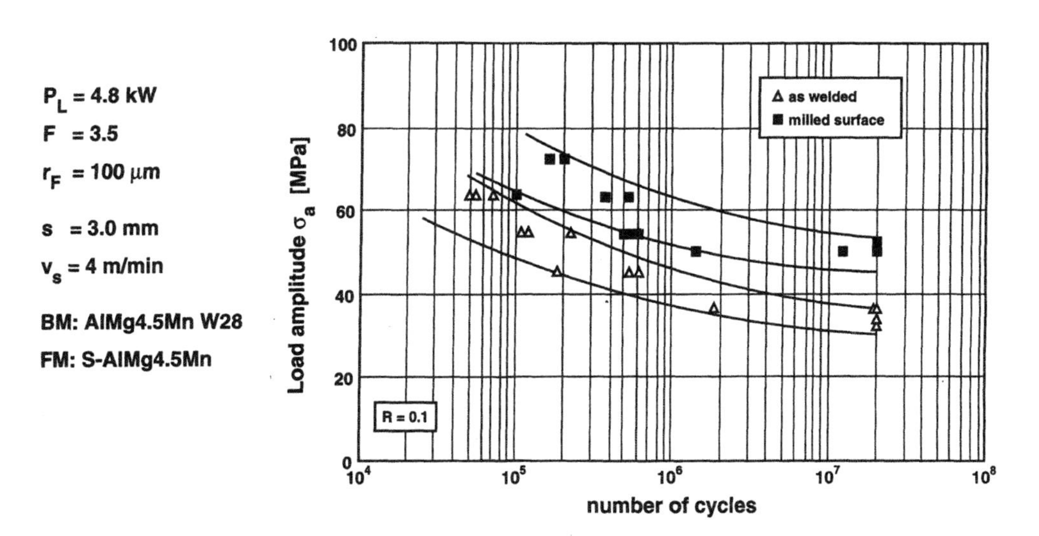

Figure 8 Fatigue properties of laser beam welded seams on aluminium

P_L = 3.5 kW
f = 150 mm
F = 4.7

v_s = 8 m/min
Δz = +1mm
AlMg4.5Mn
Ar/He 12/8 l/min

Figure 9 Light weight panel out of aluminium

there is a typical drop down of the fatigue strength to approximately 60% of the base material. The reasons for this change in the properties are the reduced strength capabilities in the weld metal and heat affected zone as it has been shown in the static strength tests. There is also a strong influence of geometrical defects (especially in fatigue).

This can be seen from the results in *Figure 8*. Here the results of fatigue tests on welded specimens with two different surface conditions are shown. The results in the lower level curves are examined with specimen in the 'as welded' condition. Because the welds were carried out using additional wire there was a seam overfill of 0.9 mm at the top and the bottom of the weld seam. FEM-calculations show that the geometrical function of the transition curve from the weld seam to the original surface of the base material can be interpreted as a geometrical defect which because of stress concentration enhances the initiation of a first crack during the running fatigue load. If the weld seam surfaces were post-machined so that there is a plane surface at top and bottom of the specimen the fatigue load level will increase by nearly 50% compared to the condition 'as welded' (see *Figure 8*).

Applications

The results discussed in the chapters before were examined in research and development projects. Knowledge at this level is required to enable the possibility of adapting for industrial applications.

In *Figures 9* and *10* typical examples are shown of the potential of laser beam welding of aluminium in industrial production. *Figure 9* shows a light weight

floor panel structure:
AlMg5Mn (EN AW-5182)
space frame:
AlMg0.7Si (EN AW-6063)

P_L = 2.8 kW
f = 100 mm
r_F = 150 μm

v_s = 2 m/min (inside)
v_s = 3 m/min (outside)

Figure 10 Laser beam welding of an aluminium–space–frame (floor panel structure)

panel out of a flat sheet in connection with u-formed profiles. The profiles have been welded on the outer skin in overlap configuration. The aim of this construction is to get a sandwich panel with

- low weight;
- high stiffness; and
- high pressure capabilities.

Laser beam welding is the right technology to manufacture light weight panels like this one because of the low energy input and the high welding speed and productivity. Shipbuilders as well as the aerospace and railroad industry are interested in elements like this one because of its advantageous functional properties.

In *Figure 10* welding and space frame construction of a light weight car is shown. Here a light weight aluminium sandwich has been welded as a floor panel into the space frame. A Nd:YAG-laser has been used because the beam can be guided via an optical fibre and a robot can be used as it is shown in the right photograph.

Some other applications as transmission parts, filter boxes or pressure vessels are established in industrial production[8,9] mainly in the car industry.

Conclusion

Starting from some short process oriented remarks within this presentation metallurgical aspects have been discussed in laser beam welding of aluminium.

It has been shown that there is typically no influence on the chemical composition of the material if the process is well adapted. Comparable to other welding techniques, especially solidification phenomena, have to be taken into account during welding. For example hot cracking can occur but this problem can be solved by using additional wire.

To show the usability of laser beam welding in industrial manufacturing the mechanical properties of the weld seam have to be considered mainly. Here as well as in other welding applications the influence of the heat input on the mechanical properties of the material have to be taken into account. Typically the strength of the aluminium material will be reduced be annealing effects or it will be kept on a nearly constant level if the base material is delivered and welded in the weakest condition. In fatigue one has to consider the aluminium-specific sensitivity against geometrical edges which lead in general to stress concentration but in the case of aluminium reduce the fatigue load capability.

Taking into account all the different aspects regarding material, metallurgy, construction, processing phenomena, etc., it is shown in some examples that laser beam welding of aluminium will be a very useful technology in industrial production.

References

1 N. N. *Aluminium-Taschenbuch*, 15th edn, 1, 2 Aluminium-Zentrale, Düsseldorf, Germany, 1995, 1996
2 Schnaas, J., Gebogene Aluminium-Strangpreßprofile im Fahrzeugbau. *Aluminium*, 1995, p. 1
3 Timm, H., Konstruktion und Entwicklung des Audi Space Frame. Auditorium, 30.09. 08.10.1993
4 Hanicke, L. and Strandberg, Ö., *Roof Laser Welding in Series Production*. Int. SAE-Conf., Detroit, Michigan, SAE Technical Paper 930028, 1993
5 Roessler, D. M., Jenuwine, W. C., Koons, J. N. and Speranza, J. J., *Laser Material Processing in General Motors Corporation*, 25th ISATA Conference, Florence, June 1992
6 Berkmanns, J., Imhoff, R., Behler, K. and Beyer, E., *Laser Welding of Aluminium*, Int. Workshop: Automotive Laser Application Dearborn, Michigan, March 1995
7 N. N.EN 288, Part 4 Beuth Verlag, Berlin, 1992
8 Radaj, D., Dausinger, F. and Rapp, J., *Laserstrahlschweißen von Aluminium*, Merkblatt der Daimler Benz AG, 1996
9 Information of Bayrische Druckgußwerke, 1996

Materials & Design, Vol. 18, Nos. 4/6, pp. 269–273, 1997
© 1998 Published by Elsevier Science Ltd
Printed in Great Britain. All rights reserved
0261-3069/98 $19.00 + 0.00

PII: S0261-3069(97)00062-9

Friction stir welding for the transportation industries

W. M. Thomas*, E. D. Nicholas

TWI, Abington Hall, Abington, Cambridge CB1 6AL, UK

Received 16 June 1997; accepted 17 June 1997

This paper will focus on the relatively new joining technology—friction stir welding (FSW). Like all friction welding variants, the FSW process is carried out in the solid-phase. Generically solid-phase welding is one of the oldest forms of metallurgical joining processes known to man. Friction stir welding is a continuous hot shear autogenous process involving a non-consumable rotating probe of harder material than the substrate itself. In addition, FSW produces solid-phase, low distortion, good appearance welds at relatively low cost. Essentially, a portion of a specially shaped rotating tool is plunged between the abutting faces of the joint. Once entered into the weld, relative motion between the rotating tool and the substrate generates frictional heat that creates a plasticised region around the immersed portion of the tool. The contacting surface of the shouldered region of the tool and the workpiece top contacting surface also generates frictional heat. The shouldered region provides additional friction treatment to the weld region as well as preventing plasticised material being expelled. The tool is then translated with respect to the workpiece along the joint line, with the plasticised material coalescing behind the tool to form a solid-phase joint as the tool moves forward. Although the workpiece does heat up during FSW, the temperature does not reach the melting point. Friction stir welding can be used to join most aluminium alloys, and surface oxide presents no difficulty to the process. Trials undertaken up to the present time show that a number of light weight materials suitable for the automotive, rail, marine, and aerospace transportation industries can be fabricated by FSW. © 1998 Published by Elsevier Science Ltd. All rights reserved.

Keywords: friction stir welding; transportation; solid phase

Introduction

Recently, a novel friction welding process for non-ferrous materials has captured the attention of the fabrication industry. This relatively new process called Friction Stir Welding (FSW) is a solid-phase process giving good quality butt and lap joints[1–4]. The FSW process has proved to be ideal for creating high quality welds in a number of materials, including those which are extremely difficult to weld by conventional fusion processes[5].

The basic principle of the process is illustrated in *Figure 1*. The process operates by generating frictional heat between a rotating tool of harder material than the workpiece being welded, in such a manner as to thermally condition the abutting weld region in the softer material. The tool is shaped with a larger diameter shoulder and a smaller diameter, specially profiled probe. The probe first makes contact as it is plunged into the joint region. This initial plunging friction contact heats a cylindrical column of metal around the probe as well as a small region of material underneath the probe. The depth of penetration is controlled by the length of the probe below the shoulder of the tool. The contacting shoulder applies additional frictional heat to the weld region and prevents highly plasticised material from being expelled during the welding operation. Once the shoulder makes contact the adjacent thermally softened region takes up a frustum shape corresponding to that of the overall tool geometry. The thermally softened region appears much wider at the top surface in contact with the shoulder, tapering down to the probe diameter. The combined frictional heat from the probe and the shoulder creates a plasticised almost hydrostatic condition around the immersed probe and the contacting surface of the shouldered region of the workpiece top surface. Material flows around the tool and coalesces behind the tool as relative traverse between substrate and the rotating tool takes place. Friction stir welding can be regarded as a autogenous keyhole joining technique. The consolidated welds are solid-phase in nature and do not show fusion welding defects. No consumable filler material, shielding gas, or edge preparation is normally necessary. The distortion is significantly less than that caused by any fusion welding technique.

Exploratory development work has encompassed aluminium materials from 1 to 75 mm thick.

FSW welding trials

Early exploratory development trials were carried out

*Correspondence to W. M. Thomas. Tel.: +44 01 223891162; fax: +44 01 223892588; e-mail: wmthomas@twi.co.uk

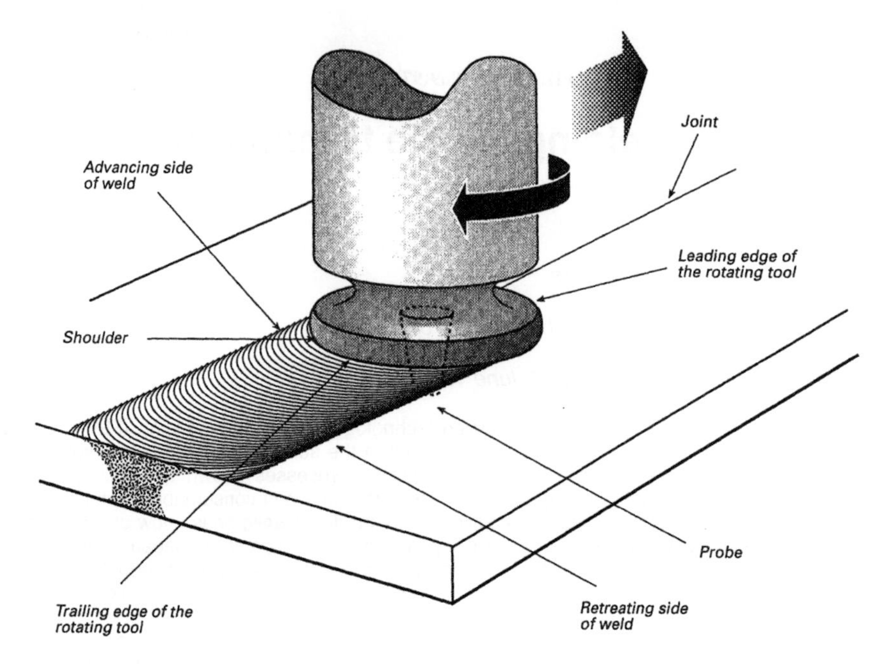

Figure 1 Friction stir welding technique

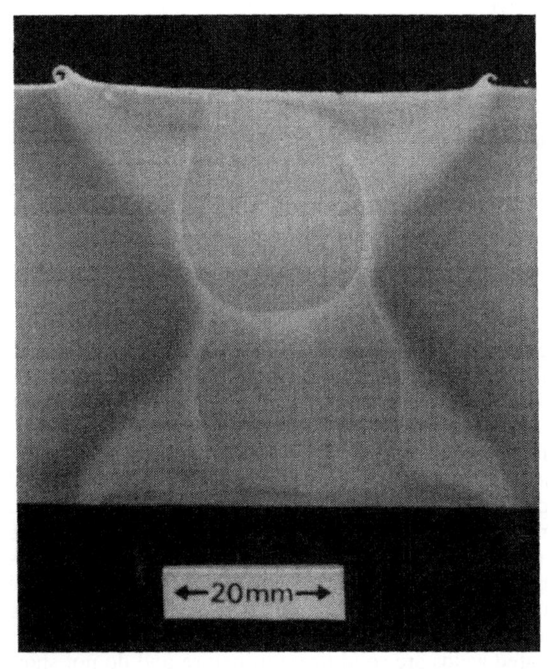

Figure 2 Transverse macrosection of 50 mm thick 6082 T6 aluminium alloy. Plate material welded from both sides, showing weld nugget profile and flow contours

with 6082 T6 aluminium alloy material in thicknesses ranging from 1.6 to 12.7 mm. Recent trials have extended the thickness range upwards to 75 mm in two passes.

Metallographic examination

A series of 50 and 75 mm thick welds in 6082 T6 condition aluminium alloy, which demonstrate the possibilities for the process for thick plate are shown in *Figures 2–4*. The macrosections from these welds are characterised by well defined weld nuggets and flow contours, almost spherical in shape, these contours are dependant on the tool design and the welding parameters and process conditions used. For heat treatable materials a well-defined heat affected zone surrounds the weld nugget region and extends to the shoulder diameter of the weld at the plate surface. The weld nugget itself is the region, where full dynamic re-crystallisation occurs that comprises of a fine equiaxed grain structure. The measured grain size is in the order of 2–4 μm in diameter. Typically the parent metal chemistry is retained, without any segregation of alloying elements. A hardness traverse taken from 50 mm thick test weld recorded the following values:

- Parent metal 100 $HV_{2.5}$
- Weld nugget 65 $HV_{2.5}$
- HAZ region 52 $HV_{2.5}$

Fractography

Samples of welded 50 mm thick 6082 T6 plate were notched in the parent material, and the weld nugget region and then fractured by bending. The fracture surfaces were examined using scanning electron microscopy. Both the weld nugget and the parent material failed in a ductile manner by the microvoid coalescence mechanism as shown in *Figure 3*. However, there was an absence of relatively large microvoids in the weld nugget sample and this may be a consequence of the break up of the primary constituent particles during stir welding.

Mechanical integrity

For plate thickness up to 50 mm thick, transverse

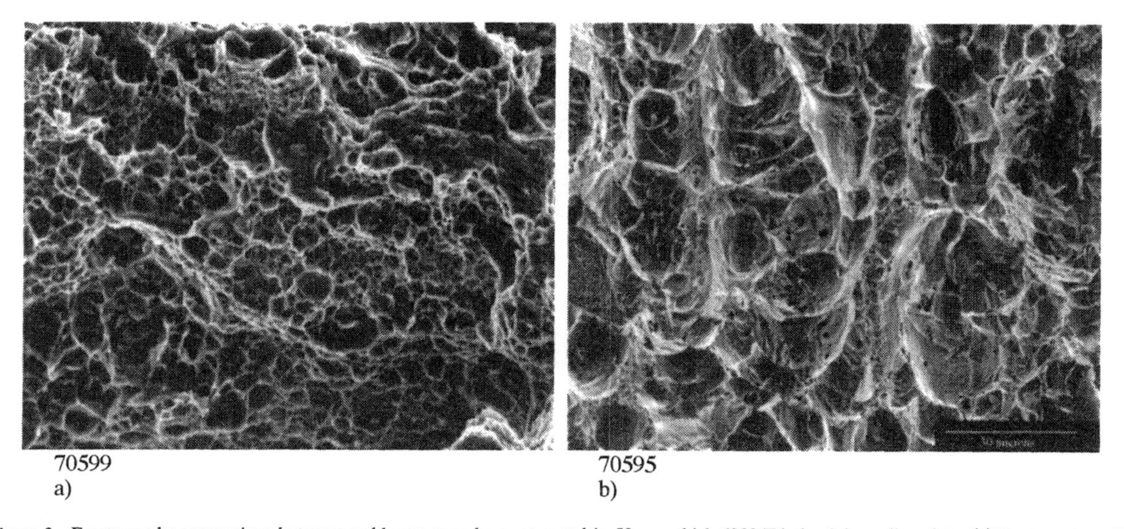

Figure 3 Fractography comparison between weld nugget and parent metal in 50 mm thick 6082 T6 aluminium alloy plate. (a) Fracture in weld nugget. Scanning microscopy of nick break bend, fracture face. (b) Fracture in parent metal. Scanning microscopy of nick break bend, fracture face

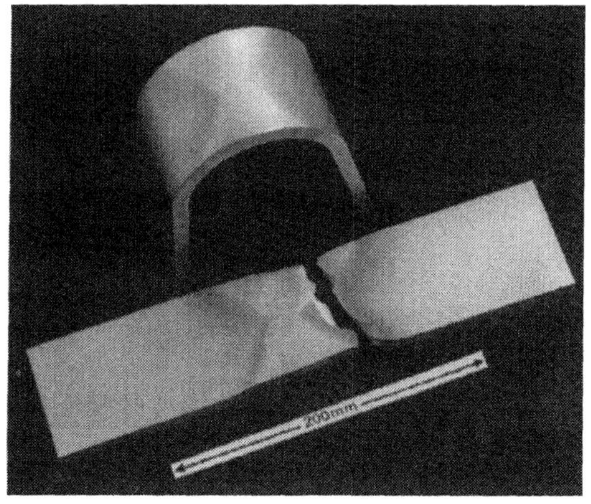

Neg No 66336/3

Figure 4 Three point bend and tensile test in 75 mm thick 6082 T6 aluminium alloy FSW plate

sections were hammer bend tested to 180°. While transverse sections taken from 75 mm thick plate were three point bend tested as shown in *Figure 4*. A number of tensile tests were carried out with failure typically occurring in the HAZ region at 175 N mm^2 as shown in *Figure 4*. The localised reduction in specimen thickness during the tensile testing corresponds with regions of reduction in hardness.

Process characteristics

Experimental research work is being carried our at TWI to evaluate a range of materials and develop other FSW tools designed to improve the flow of plasticised material around the probe itself and to enable substantially thicker plates to be joined and enable relatively high traverse rates to be achieved.

Figure 5 illustrates the natural dynamic orbit inherently associated with every type of rotary machine. This eccentricity must to a greater or lesser extent be part of the friction stir welding process characteristics. Eccentricity allows hydromechanically incompressible plasticised material to flow more easily around the probe. It follows that a nominal bias off-centre or non-circular probe will also allow plasticised material to pass around the probe. Essentially it is the relationship between the greater volume of the 'dynamic orbit' of the probe and the volume of the static displacement of the probe, that helps provide a path for the flow of plasticised material from the leading edge to the trailing edge of the rotating tool.

A number of tool geometries and tool attitude for different materials have been reported in the literature[3,6,7]. For tools positioned perpendicular to the workpiece the leading edge of the rotation tool provides a frictional preheat effect heat and subsequent thermal softening of the workpiece in front of the probe. This preheat can be of advantage when dealing with harder or difficult to weld materials. The greater the area of the shouldered region of the rotating tool making contact with the joint surface the greater the frictional heat available. Increasing the diameter of the shouldered region, however, has practical limitations and tends to produce side flash on the weld surface.

Potential for the FSW process in the transportation industry

The potential scope for FSW initially lies with joining materials like aluminium, copper, copper alloys, lead, titanium, zinc etc. The applications range from the following:

- Airframes, fuel tanks, and thin alloy skins in the aerospace

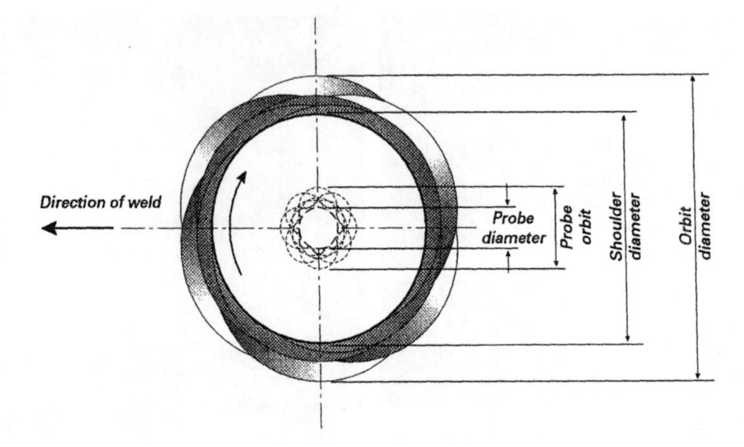

Figure 5 Dynamic orbit—plan view of rotating tool and probe

- Sheet bodywork and engine support frames for the automotive industry
- Railway wagon and coachwork, and bulk carrier tanks for the transportation industry
- Hulls, decks, and internal structures for high speed ferries and LPG storage vessels for the shipbuilding industry

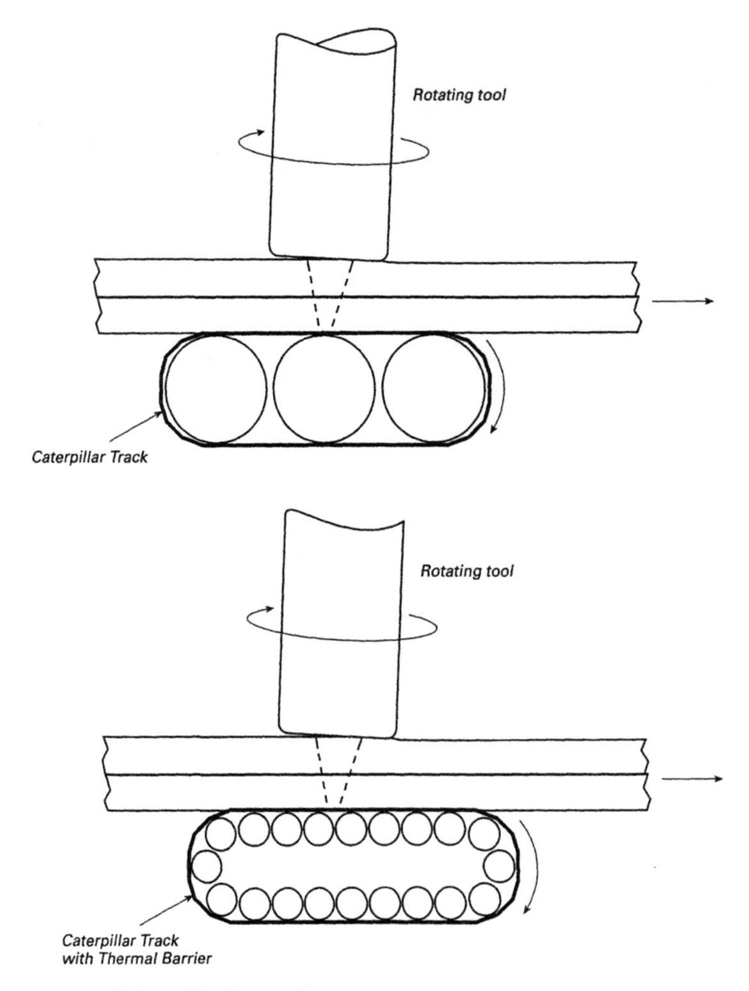

Figure 6 FSW seam welding of lapped sheet with roller\track support

The FSW technique will move ahead with thinner sheet and foam filled sections being produced at comparatively high traverse speeds. Developments that will enable the technique to be more flexible for thin sheet lap joining are being studied.

Moving reactive support-anvil

Figure 6 shows a miniature caterpillar track that could be developed to provide local support for the rotating tool, a 'moving anvil'. Such a device would be able to fit into similar joint configurations as those accessible by traditional resistance welding techniques and tackle material thickness of about 1.5 mm or less. The moving anvil could be suitably interfaced with a robot for fully automatic seam and tack welding. The 'moving anvil' approach could also be considered as an alternative for a relatively large machine design. For example, instead of a rotating head traversing along a fixed reactive support, the material could be transported continuously over carrier rollers to a fixed work station, fitted with a 'moving anvil', much in the same way as a domestic sewing machine works. The 'moving anvil' concept may be worth considering for aerospace and automotive applications.

Concluding remarks

There is no doubt that the use of FSW will open up new markets and new opportunities as the technology gets wider recognition as a welding process that can produce superior welds, of improved reliability and of increased productivity. The FSW process is already in commercial use and has been found to be a robust process tolerant, technique that has much to offer.

Acknowledgements

The authors wish to thank Mike Gittos and Phil Threadgill for metallurgical support, and Steve King, Peter Temple-Smith and Cliff Hart for their technical support.

References

1 Thomas, W.M., *Friction Stir Butt Welding*. International Patent Application No PCT/GB92 Patent Application No.9125978.8, 1991
2 Dawes, C.J., Seam welding aluminium sheet and plate, using the friction stir welding process. In *Proceedings of the 6th International Symposium*, JWS, Nagoya, 1996
3 Thomas, W.M., Nicholas, E.D., *Friction Stir Welding and Friction Extrusion of Aluminium and its Alloys*. 3rd World Congress on Aluminium, Limassol, Cyprus, 1997
4 Threadgill, P.L., Friction stir welds in aluminium alloys: Preliminary microstructural assessment. *TWI Bulletin* (to be published in 1997)
5 Mahoney, M.W., Science friction. *Welding and Joining*, 1997
6 Midling, D.T., Morley, G.J. and Sandvick, A., *Friction Stir Welding*. International Patent Application WO 95\26254. (Assigned to TWI), 1994
7 Christner, B.K. and Sylva, G.D., Friction stir weld developments for aerospace applications. In *Int. Conf. on Adv. Welding Technology Joining of High Performance Materials*. Columbus, Ohio, 1996.

Materials & Design, Vol. 18, Nos. 4/6, pp. 275–278, 1997
© 1998 Published by Elsevier Science Ltd
Printed in Great Britain. All rights reserved
0261-3069/98 $19.00 + 0.00

PII: S0261-3069(97)00063-0

Behavior of bubbles in welding for repairs in space

K. Nogi, Y. Aoki*

Joining and Welding Research Institute, Osaka University, Mihogaoka 11-1, Ibaraki 567, Japan

Received 17 July 1997; accepted 17 July 1997

In order to investigate the difference in the behavior of bubbles between in a microgravity environment and in a terrestrial environment, gas tungsten arc (GTA) welding was performed in both environments. The microgravity environment was produced for 10 seconds with less than 10^{-5} G with a drop-shaft type system at the Japan Microgravity Center (JAMIC). The materials used were an aluminum alloy and pure silver. It has become clear that more pores are left in the weld in the microgravity environment than in the terrestrial environment. The bubbles cannot easily be released from the weld pool before solidification due to the lack of buoyancy in the microgravity environment. In the microgravity environment, blowholes are distributed uniformly in the weld and are smaller than those in a terrestrial environment. In the microgravity environment, the weld bead is formed flatly though the weld shape is significantly affected by gravity in the terrestrial environment. This indicates that in the microgravity environment, a larger amount of metal can be welded at once and in any welding position. © 1998 Published by Elsevier Science Ltd. All rights reserved.

Keywords: bubbles; welding; space repairs

Introduction

A lot of space structures are travelling in the Low Earth Orbit (LEO). In the LEO, space debris and micrometeroids also are travelling at about 8 km/s and, therefore, space structures are expected to be damaged and destroyed. In order to repair the damaged parts in space, a welding technique is essential. However, because the different environments, such as microgravity and vacuum, change the welding phenomena, the welding technique in space has not been completely established yet.

The first welding experiment in space was performed in 1969, by the USSR during the 'Vulkan' experiments of the Soyuz-6 mission[1]. The first US experiment on space welding was conducted in 1973, aboard the Skylab[2]. Both the USSR and the US have used the electron beam welding process in space, which is possible to weld in vacuum[1,2]. These experiments indicated that welding can be performed in space. However, the detail information on the welding phenomena in space is not very open to the public.

In this study, the gas tungsten arc (GTA) welding was carried out using a drop-shaft type system at Japan Microgravity Center (JAMIC), which can produce 10-s microgravity with a high quality microgravity condition[3]. The difference in the behavior of bubbles, which will be a problem on welding for repairs in space, between in the microgravity environment and in a terrestrial environment, is investigated. The weld shape in both environments is also investigated.

*Correspondence to Y. Aoki. Tel.: +81 6 879 8663; fax: +81 6 8798653; e-mail: aoki@jwri.osaka-u.ac.jp

Experimental apparatus and procedure

Experimental apparatus

The JAMIC drop-shaft type facility was used to obtain a microgravity condition. The system can maintain 10-s microgravity duration with a condition of less than 10^{-5} G. *Figure 1* shows a schematic of the system. The quality of microgravity is the highest up to the present and is similar to that in space[4]. The drop capsule is composed of a double structure consisting of an inner and an outer capsule and a vacuum is maintained between them so that the free fall velocity of the inner capsule will not be affected by the air drag. A small-sized GTA welding apparatus is accommodated in the inner capsule. The size of the apparatus is $0.87\ m^w \times 0.87\ m^l \times 0.92\ m^h$. *Figure 2* shows the GTA welding apparatus used in this study. The apparatus consists of a welding chamber, a battery, a welding power source, a shielding gas supply and a welding control system.

Materials and welding condition

The material used was an aluminum alloy. It was selected as the material commonly used in aerospace. *Table 1* shows the chemical composition of the aluminum alloy. 99.99 mass% pure silver was also used as a reference material. *Table 2* shows the welding conditions. Bead-on-plate welding was performed. The welding positions were vertical up and horizontal. The polarity was the direct current electrode negative (DCEN). The flow rate of the shielding gas was op-

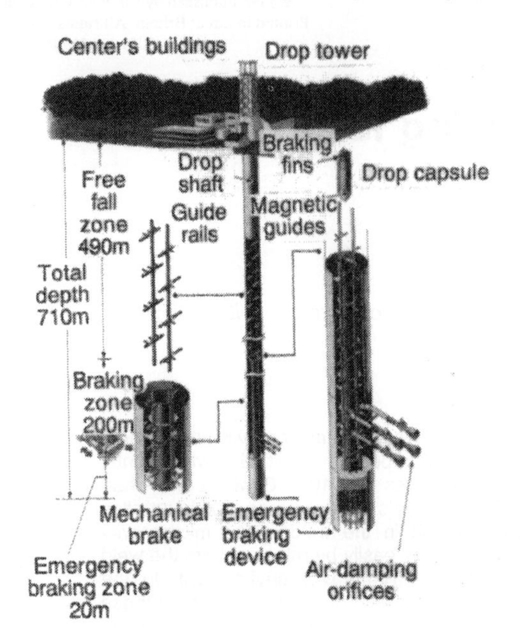

Figure 1 Schematic of the JAMIC system

timized in a terrestrial environment and all specimens were welded with the same GTA welding apparatus. An argon–1% hydrogen mixed gas was used as a shielding gas in order to investigate the behavior of bubbles. Some of them were welded in a microgravity environment and the others in a terrestrial environment. In order to investigate the weld shape and the distribution of pores, the welds were observed using a microscope after cutting, mechanical polishing and electrical etching the specimens. Pore size distribution of the samples was measured from optical micrographs by image analysis on a Power Macintosh 7100/80AV computer.

Results and discussions

Pore distribution

In order to investigate the behavior of pores, an argon–1% hydrogen mixed gas was used as a shielding gas. *Figure 3* shows transverse sections of the welds in both environments. *Figure 4* shows the pore size distribution in the welds in both environments. As shown in *Figure 3(a)*, blowholes are segregated in the upper part in a terrestrial environment, and they are not observed in the lower part. Wormholes are also observed in the transverse section of the weld. Wormholes are not distributed in the upper part, in which many large blowholes exist, but wormholes are distributed in the lower part of the weld. In the terrestrial environment, as shown in *Figure 4(a)*, pores larger than about 100 μm are mainly distributed in the upper part. Pores smaller than about 100 μm, on the other hand, are distributed uniformly in the weld without the segregation due to gravity. These results indicated that in the 1-G environment only larger bubbles are affected by buoyancy due to gravity and move to the upper part in the molten metal. In addition, in the upper part, bubbles repeatedly combine with other bubbles and thus grow. In the microgravity environment, on the other hand, as shown in *Figures 3* and *4(b)*, blowholes are distributed uniformly in the weld and are smaller than that in a terrestrial environment. Wormholes are also distributed without the segregation in the weld. It can be concluded from this result that the bubbles cannot easily be released from the weld pool due to the lack of buoyancy.

(a) Experimental apparatus

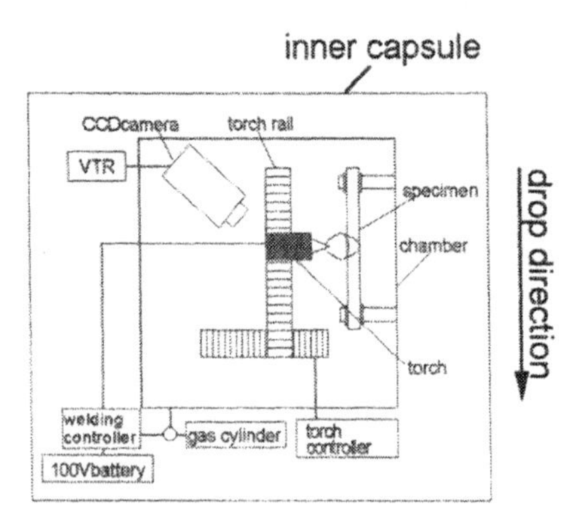

(b) Schematic of apparatus

Figure 2 GTA experimental apparatus

Table 1 Chemical composition of aluminum alloy (mass%)

Al	Mg	Mn	Fe	Cr	Si	Cu	Zn	Ti
Bal	4.41	0.62	0.21	0.12	0.09	0.02	0.02	0.02

Table 2 Welding condition

Welding current, I /A	80	81
Welding voltage, E /V	11	12
Welding velocity, v /ms^{-1}	$4.0 \cdot 10^{-3}$	$3.6 \cdot 10^{-3}$
Shielding gas	Ar	Ar–1%H_2
Shielding gas flow rate, f /m^{3}s^{-1}	$8.3 \cdot 10^{-5}$	$1.7 \cdot 10^{-4}$

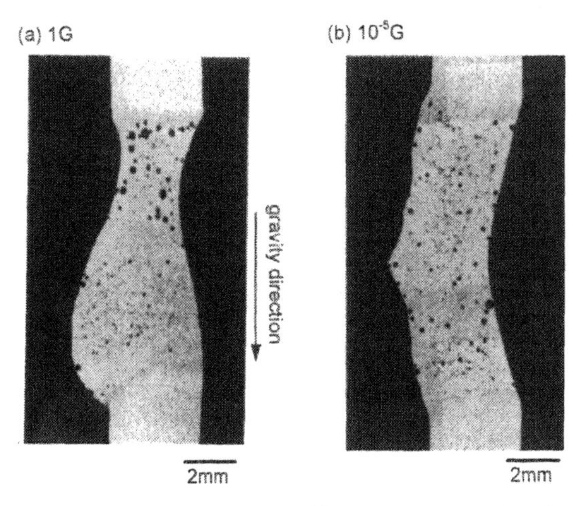

Figure 3 Transverse sections of welds in horizontal welding using Ar–1%H_2 shielding gas

Weld

In order to investigate the effect of gravity on weld shape, welding was also performed in the vertical up position. *Figure 5(a)* and *(b)* show weld bead appearances of the aluminum alloy in both environments, and *Figure 5(c)* and *(d)* show transverse sections of the welds. In the microgravity environment, the weld bead and the crater are nearly flat. Thus, it is possible to weld in any welding position in a microgravity environment. In the terrestrial environment, on the other hand, the center of the weld bead is bulged, and the upper part of the crater is deeply hollowed out due to gravity. However, the difference in the weld bead is much smaller than expected. This is probably because the flow of the molten metal is restrained by the oxide films.

In order to investigate the effect of oxide films on welding phenomena, pure silver is selected as a model material. *Figure 6* shows a weld bead of the silver in both environments. As shown in *Figure 6(a)*, in the terrestrial environment, the weld pool fell away by the effect of gravity. In the microgravity environment, on the other hand, the weld pool maintains its position on the specimen plate, even without the oxide films, and both the weld bead and the crater are flatter than those of aluminum alloy. This indicates a great advantage of

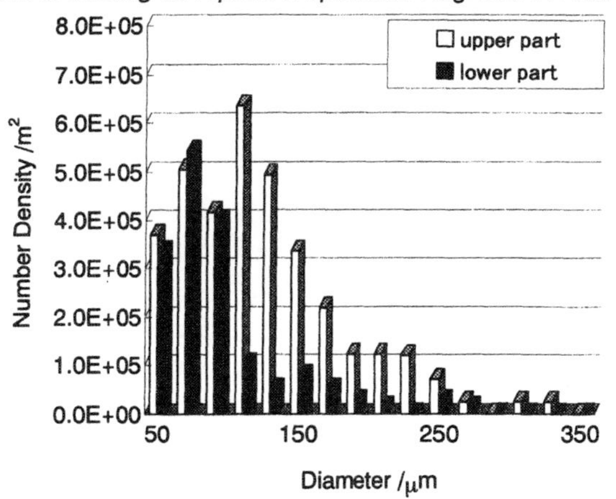

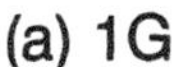

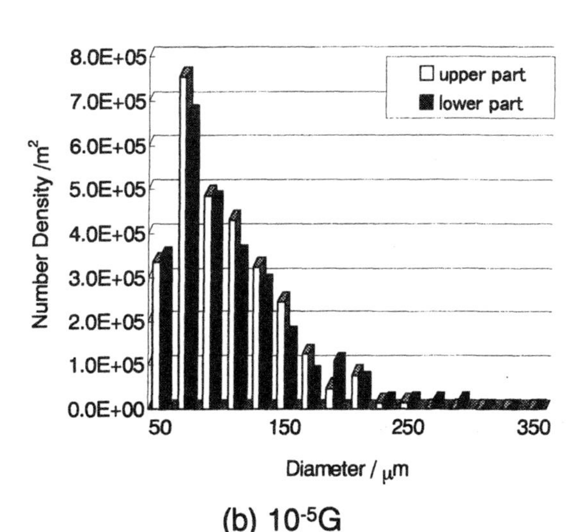

Figure 4 Pore size distribution in the welds in both environments

microgravity environment; welding can be performed with a larger energy density forming a larger weld pool in a microgravity environment.

Conclusion

By performing welding both in a microgravity environment and in a terrestrial environment, the following points were found:

1. more pores are left in the weld in the microgravity environment than in the terrestrial environment because the bubbles cannot easily be released from the weld pool before solidification due to the lack of buoyancy;
2. in the microgravity environments, blowholes are

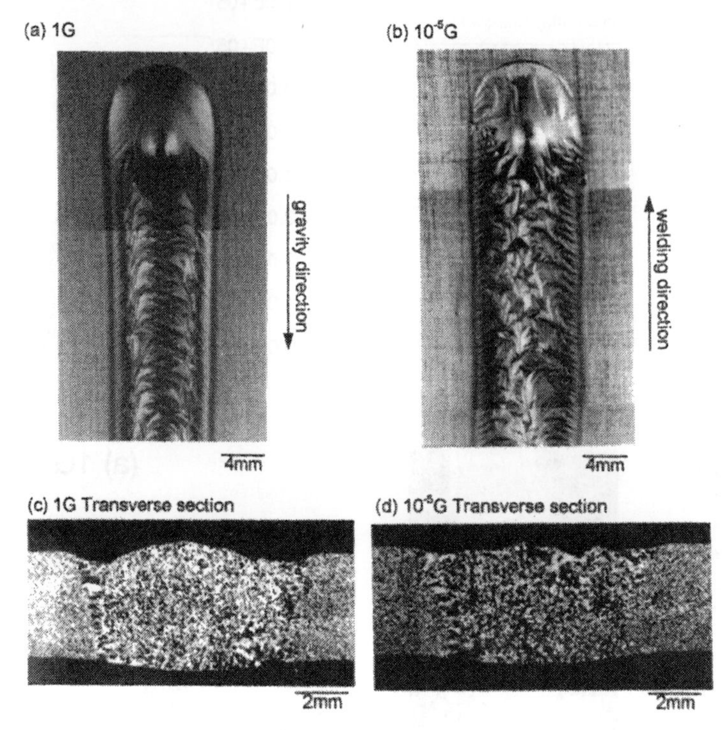

Figure 5 Weld bead appearances of weld and transverse sections in vertical up welding

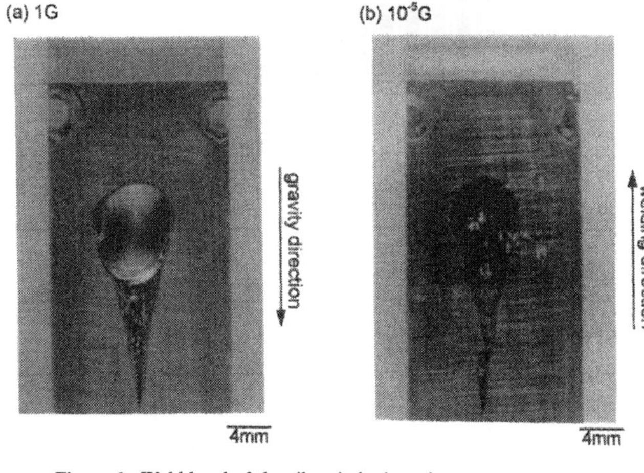

Figure 6 Weld bead of the silver in both environments

distributed uniformly in the weld and are smaller than those in a terrestrial environment;

3. the weld bead is formed flatly in the microgravity environment and is significantly affected by gravity in the terrestrial environment; and
4. in the microgravity environment, a larger amount of metal can be welded at once and in any welding position.

Acknowledgements

The authors greatly appreciate the supports of the Japan Space Utilization Promotion Center (JSUP) and the Japan Microgravity Center (JAMIC).

References

1 Paton, B.E., *Weld. Eng.*, 1972, **57**, 25
2 Nance, M. and Jones, J.E., *ASM Handbook Weld. Braz. Solder.*, 1993, **6**, 1020
3 Nogi, K., Aoki, Y., Nakata, K. and Kaihara, S., In *Proc. of In Space '96*, Tokyo, 1996, p. 307
4 JAMIC User's Guide, Japan Microgravity Center, 1995

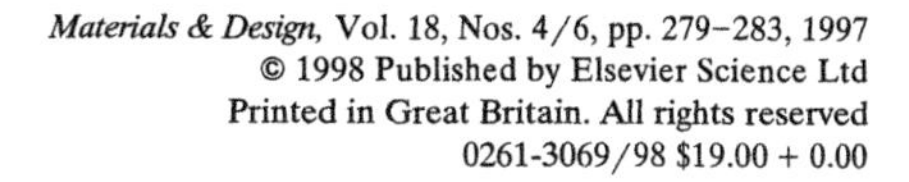
Materials & Design, Vol. 18, Nos. 4/6, pp. 279–283, 1997
© 1998 Published by Elsevier Science Ltd
Printed in Great Britain. All rights reserved
0261-3069/98 $19.00 + 0.00

PII: S0261-3069(97)00064-2

Some aspects of solidification and homogenisation of Mg–Ag alloys

A. Rakowska[a,*], M. Podosek[b], R. Ciach[b]

[a] *Technical University of Mining and Metallurgy, 30 Mickiewicza Av., 30-059 Kraków, Poland*
[b] *Institute of Metallurgy and Materials Science, Polish Academy of Sciences, 25 Reymonta Street, 30-59 Kraków, Poland*

Received 25 July 1997; accepted 1 August 1997

The model of non-equilibrium solidification based on assumption of full diffusion in liquid and its lack in the solid state has been adopted in Krupkowski's evaluation to Mg–Ag alloys to describe maximum microsegregation of components in the α solid solution resulting in appearance of maximum amounts of non-equilibrium phases. The results were then compared with experimental in MgAg 2.5 wt.% alloys with and without addition of 0.6 wt.% zirconium and 2.5 wt% neodymium (RE) by means of scanning and transmission electron microscopy equipped with an energy dispersive spectrometer. It was found that zirconium entered the solid solution, while neodymium appeared as a net of the Mg–Ag–Nd ternary eutectic.The homogenisation process has been studied based on hardness measurements and structure analysis. The times and temperatures of homogenisation to receive uniform distribution of components in the solid solution have been chosen neglecting the remaining eutectic precipitates due to economical reasons. © 1998 Published by Elsevier Science Ltd. All rights reserved.

Keywords: non-equilibrium solidification; non-equilibrium phases; homogenisation

Introduction

Magnesium–silver alloys were discovered for industrial applications by Payne and Bailey who observed the high temperature creep resistance in magnesium-rare earth metals–zirconium alloys with addition of silver[1]. The alloys showed high yield strength, good tensile strength and fatigue properties which were maintained up to 316°C. Their casting qualities were good and the alloys were to be used in the heat treated condition[2]. Due to fairly expensive components the alloys are mainly used in the aircraft industry for landing wheels, gearbox housings and rotor heads for helicopters[3].

The influence of dendritic segregation on the structure of Mg–Ag alloys was shown in detail by Podosek *et al.*[4], where the non-equilibrium solidification of binary Mg–Ag alloys was studied in a theoretical and experimental way. It was shown that in order to obtain the lowest possible amounts of non-equilibrium eutectics, the alloys should solidify at higher rates than 20 K/s, when the structure also becomes finer and promising in terms of mechanical properties. The non-equilibrium solidification affected the compound detected in the as cast binary alloy, which turned out to be rather orthorhombic ϵ' ($Ag_{17}Mg_{54}$) with lattice parameters $a = 1.4240$, $b = 1.4209$ and $c = 1.4663$ nm, than hexagonal $AgMg_4$ or $AgMg_3$ as was suggested by the phase diagram[5].

The aim of this work was to follow the influence of zirconium and rare earth metals on the solid solution of silver in magnesium taking into account microsegregation of components and other structural factors and their effect on homogenisation kinetics. Since the structure of the alloys are dependent on solidification rate, in order to make the obtained structures comparable and according to Podosek *et al.*[4], it was chosen to solidify the alloys at approximately 30 K/s.

* Correspondence to A. Rakowska

Experimental

The alloys were cast from metals of commercial purity. They solidified at a rate of 30 K/s. Samples were electrolytically polished and etched and their structure was observed either by means of an optical microscope or scanning electron microscope Philips XL30 equipped with a Link ISIS energy dispersive X-ray spectrometer (EDX) analysing system. The amounts of eutectics were established using a standard point method.

Solution treatment was carried out at 450°C for all alloys investigated, which was 20°C below the Mg–Ag eutectic temperature according to Lagowski and Meier[6] and at 530 and 550°C for the quaternary alloy, whose melting range was established to be between 550 and 640°C. The alloys heated for different times from 30 min to 16 h were then quenched in water and hardness (with Vickers tester at 5 kp) and the structure (SEM + EDX + X-ray phase analysis) were examined. X-ray diffractometer PW 1830 and transmission electron microscope (TEM) Philips CM20 were used for structure determination.

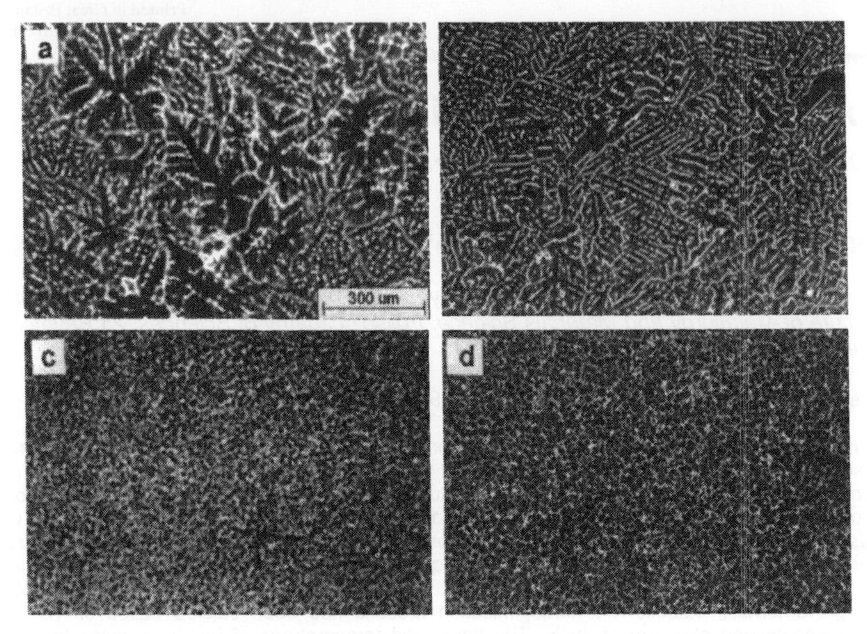

Figure 1 Microstructures of Mg–Ag alloys solidified at 30K/s: (a) MgAg2,5% alloy; (b) with 2.5 wt.% Nd addition; (c) with 0.6 wt.% Zr addition; and (d) MgAg2,5Nd2,5Zr0,6 alloy

Results and discussion

Microstructure of as cast Mg–Ag alloys

A set of microstructures of cast Mg–Ag alloys solidified at approximately 30 K/s taken at the same magnification is presented in *Figure 1*. The bright areas between the dendritic arms visible in *Figure 1a* are certainly not the eutectic precipitates but mostly the α solid solution considerably enriched in silver indistinguishable from eutectic precipitates, which are present in an amount of approximately 4% (area fraction). The accurate amount of the eutectics in the binary alloy was difficult to establish due to the redeposition of silver on the surface that blurred the image. It can be seen how the dendritic structure of the binary alloy is preserved in the alloy with neodymium together with the distance between the dendritic arms, although the amount of eutectics seems to rise up to approximately 20% [*Figure 1(b)*]. Its amount decreases to 0.6 wt.% through

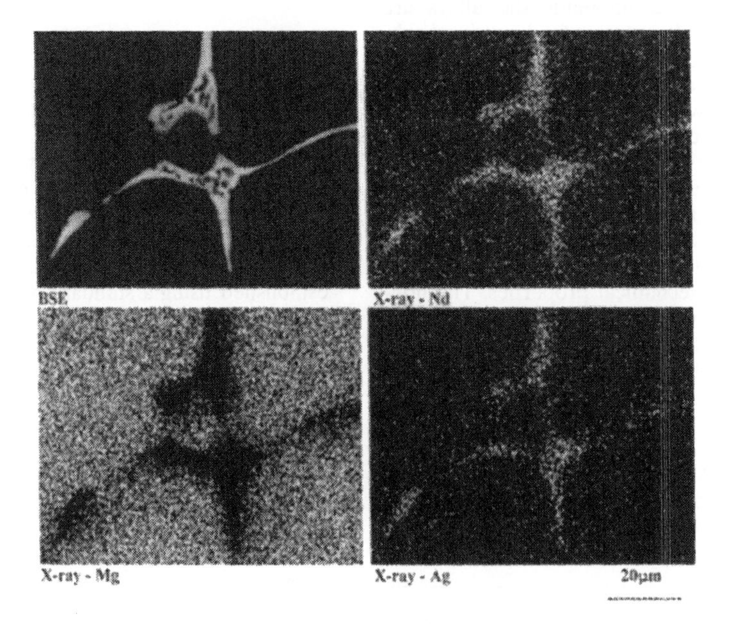

Figure 2 The backscattered electrons image (BEI) and the distribution of Nd, Mg Ag in a microarea of MgAg2.5Nd2.5 alloy by means of SEM + EDX analysis

addition of Zr, which refines the grain from 200 μm in the binary alloy to 15 μm [*Figure 1(c)*]. The alloy with the both additions reveals equiaxed grains of solid solution (10–50 μm in diameter) with a network of eutectics in their boundaries which covers approximately 19% of the area. The quantity of eutectics seems to be overestimated due to the same effect as in the binary alloy [*Figure 1(d)*].

The structural investigations of the binary alloy show that distribution of silver across a dendritic cell changes from 0.11 at% in the centre up to 13.86 at% of silver in the eutectics. The results of SEM + EDX quantitative analysis confirm that almost the whole amount of neodymium enters the ternary eutectics containing, apart from Mg, 3.31 at% Ag and 3.83% Nd. The mapping of Nd, Mg and Ag in the light area of eutectic precipitate surrounded by the α solid solution visible in the BSE image can be seen in *Figure 2*.

Figure 3 contains an example of a linear analysis across two cells of MgAg2.5Zr0.6 alloy with well visible Zr cores of 1.43 at% inside the dark solid solution and light areas enriched in silver in the boundaries. It can be seen that distribution of silver follows that of zirconium. *Figure 4* houses the EDX spectra of solid

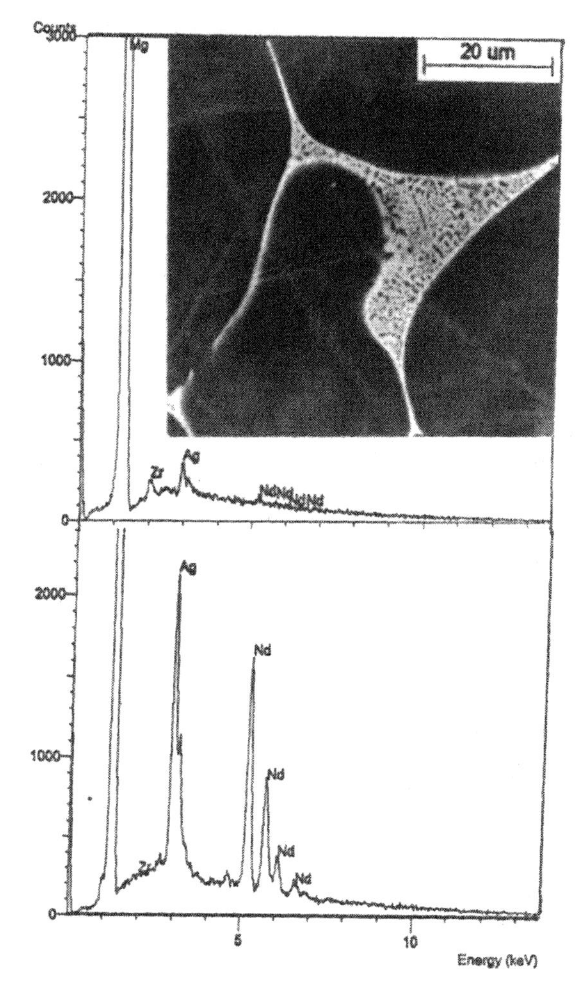

Figure 4 The BEI image together with the EDX spectra of the solid solution (upper one) and eutectics in the bottom spectrum of the cast MgAg2.5Nd2.5Zr0.6 alloy

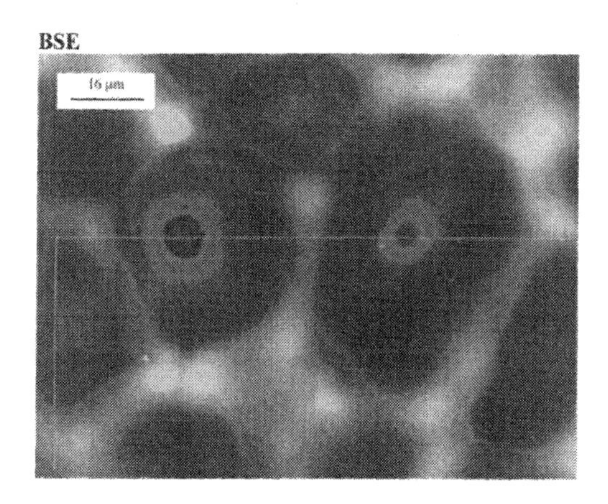

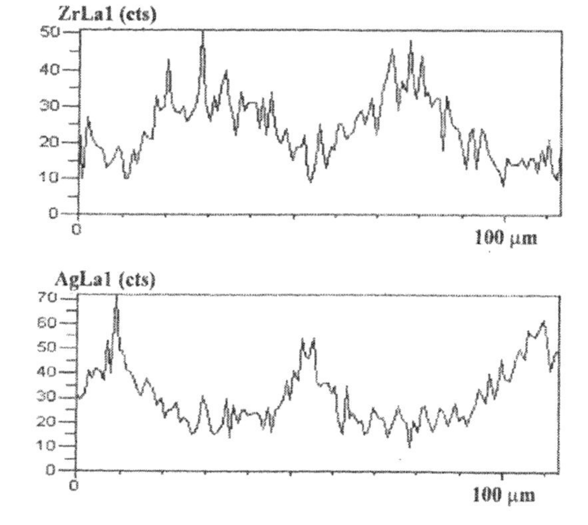

Figure 3 The linear analysis of Zr and Ag along the marked line in the backscattered electrons (BSE) image, across two cells of MgAg2.5Zr0.6 wt.% alloy

solution and eutectics in the MgAg – 2.5 Nd – 2.5 Zr-0.6 alloy. It can be seen that the eutectics consists of Mg, Ag, Nd while the solid solution is enriched in zirconium.

The TEM observation and X-ray diffraction analysis of the as cast binary alloy proved that the eutectic compounds consisted of an unequilibrium phase according to the results given by Podosek *et al.*[4], who observed the orthorhombic ϵ'' ($Ag_{17}Mg_{54}$) phase, not the hexagonal $AgMg_4$ or $AgMg_3$, suggested by the phase diagram to exist in the Mg–Ag binary alloys from the solid solubility range.

In the alloy with neodymium addition the compound detected should have been $Mg_{12}Nd$ with magnesium substituted by silver without destroying the crystal structure, due to reasonably small difference in the atomic radii of silver (144 pm) and magnesium (160 pm)· In *Figure 5(a,b)*, the diffraction patterns taken from the precipitates shown, suggested that there were the $Mg_{41}Nd_5$ phase and $Mg_{12}Nd$ one, respectively. Although the X-ray analysis allowed us to establish the appearance of both compounds, the EDX analysis de-

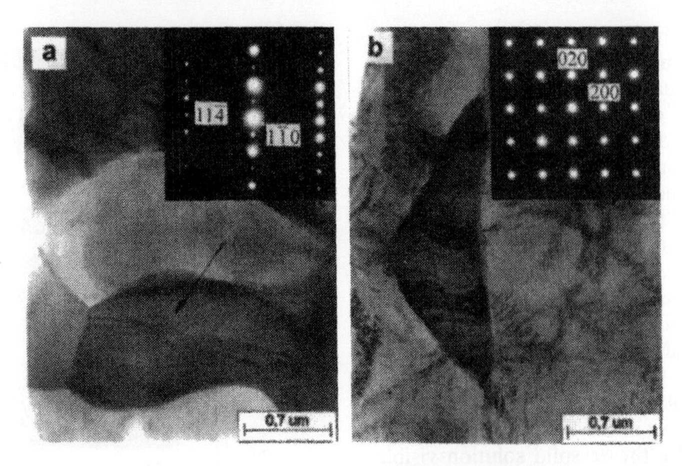

Figure 5 The TEM microstructures of two eutectic precipitates with relevant diffraction patterns in the cast MgAg2.5Nd2.5 alloy; (a) $Mg_{41}Nd_5$ phase; and (b) $Mg_{12}Nd$ phase

rived compositions (i.e.: 6.74 at.% Ag, 6.14% Nd and 87.07% Mg), which suggested rather the $Mg_{12}Nd$ phase. The results are in agreement with the data given by either Nayeb-Hashemi and Clark[8] who showed both phases or Okamoto[9] who observed only $Mg_{41}Nd_5$. Zirconium was found to be in the solid solution in amounts from 0.11 at.% (0.41 wt.%) up to 0.2 at.% (0.74 wt.%) in cores and it did not affect the phases. The investigations of the alloy with both additions confirmed the above by the SEM and TEM/EDX analysis.

The influence of heat treatment on hardness and structure of the Mg–Ag alloys

The hardness of cast samples differed by approximately 10 units and it fell systematically during heating at 450°C from 50 to 67 HV_5 down to 43–55 and after 16 h it reached a plateau. However, the samples of the quaternary alloy which were also heated at 530 and 550°C revealed plateaux after 2 h already. It followed from the microstructures of the alloys after the heat treatment, that only two alloys: the binary and that with zirconium are deprived of the second phase (which occurred in small amounts in these as cast alloys), while the remaining two were not. This explains such a small change of hardness during heat treatment, which was mainly caused by the decrease of solid solution differentiation in the grains, as long as silver addition is considered, based on the results obtained in the EDX analysis. It reached the average alloy composition in the binary and the zirconium bearing alloys, while its content was 0.41 at.% (1.8 wt.%) Ag in the alloy with neodymium and 0.51 at.% (2.14 wt.%) in the quaternary one.

The EDX analysis confirmed that neodymium dissolved in the solid solution of Ag in Mg after heat treatment to 0.10 at.% in the ternary alloy to its maximum solid solubility in magnesium which is 0.2 at.% according to Park and Wyman[10] in the alloy with both additions.

Figure 6(a,b) presents microstructures of two alloys with zirconium after solution heat treatment, in large magnifications. Zirconium rich coring is visible in both

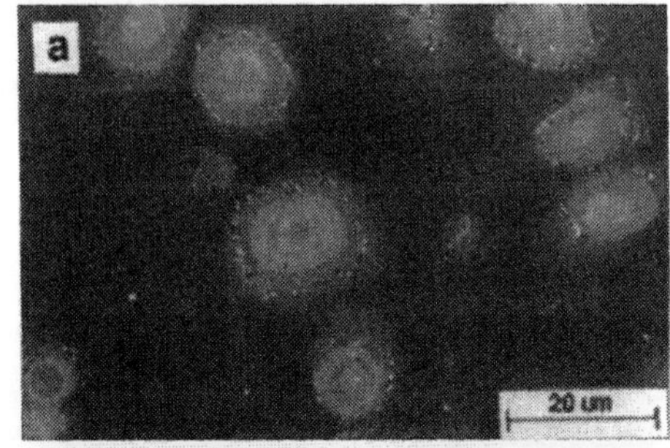

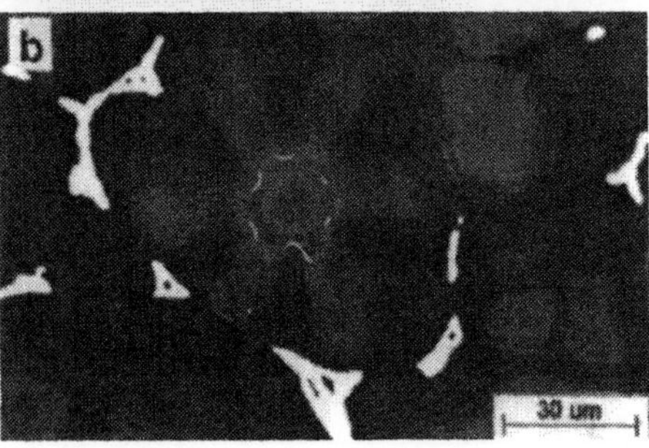

Figure 6 The BEI images of two zirconium bearing alloys after heat treatment (a) MgAg2.5Zr0.6 for 16 h at 450°C; and (b) MgAg2.5Nd2.5Zr0.6 for 4 h at 530°C

alloys and the Zr content can be even 1.22 at.% in the lightest areas of the ternary alloy while the outer grey parts are Zr depleted, below the limit of EDX detection. The fine bright particles encircling the cores seem to be Ag–Zr particles.

The TEM observations and X-ray diffraction showed uniform α solid solution in the binary alloy and the ternary zirconium bearing alloy and the $(Mg + Ag)_{12}Nd$ compound together with some amount of Mg_3Ag equilibrium phases in the alloy with neodymium and in the alloy with neodymium and zirconium and such a structure seems to provide the best mechanical properties of the alloys after ageing according to Payne and Bailey[1], Arakcheeva *et al.*[6] and Kamado *et al.*[7]

Conclusions

The structure of Mg–Ag alloys depends on the solidification rate in such a way that above 10 K/s, the grain size as well as the amounts of non-equilibrium precipitates decrease.

The applied solution heat treatment led to homogenisation of the MgAg2.5 alloy and to occurrence of a strong Zr coring in the alloys with zirconium. In the alloys with neodymium some amount of the ternary eutectics remained in accordance with the phase diagram.

The ϵ'' ($Ag_{17}Mg_{54}$) phase, $(Mg,Ag)_{41}Nd_5$ phase and $(MgAg)_{12}Nd$ were detected in the as cast MgAg2.5, MgAg2.5Nd2.5 and MgAgNd2.5Zr0.6 wt.% alloys, respectively. After the heat treatment they must have transformed into the equilibrium ones: Mg_3Ag and ($MgAg_{12}Nd$) phases, although from the X-ray investigations it does not seem as easy as that and obviously requires further investigations.

Acknowledgements

This work was supported by the Polish State Committee for Scientific Research under the Grant no. 7 T08 B018 09.

References

1 Payne, R. J. M. and Bailey, N., *Journal of Institute of Metals*, 1959–60, **80**, 417–427
2 Whitehead, D. J., *AFS Transactions*, 1961, **69**, 442
3 Polmear, I. J., *Materials Science and Technology*, 1994, **10(1)**, 410–427
4 Podosek, M., Rakowska, A., Ciach, R., In *Proceedings of the Third International Magnesium Conference Manchester* 10–12 April, 1996, ed. G. W. Lorimer. The Institute of Materials, 1997, pp. 545–555
5 Arakcheeva, A. V., Karpinskii, O. G. and Koleanichenko, V. E., *Sov. Phys.-Crystall., translated from Krystallografiya*, 1988, **33(6)**, pp. 907–908
6 Lagowski, B., *Journal Meier, AFS Transactions*, 1964, **72**, 310–320
7 Kamado, S., Tsukuda, M., Tokutomi, I., Hirose, K., *Journal of Japanese Institute of Light Metals*, personal communication
8 Nayeb-Hashemi, A. A. and Clark, J. B., *Bulletin of Alloy Phase Diagrams*, 1984, **5(4)**, 349–358
9 Okamoto, H., *Binary Alloy Phase Diagrams Updating Service*, JPE, 1991, **12(2)**
10 Park, J. J. and Wyman, L. L., *WACD Technical Report*, 1957, **33**, 57–59.

Materials & Design, Vol. 18, Nos. 4/6, pp. 285–291, 1997

0261-3069/98 $19.00 + 0.00

PII: S0261-3069(97)00065-4

Technical Report

Light weight design with light metal castings

D. Brungs

Honsel Aktiengesellschaft, Postfach 1364, D-59870 Meschede, Germany

Received 25 August 1997; accepted 29 August 1997

The article gives a number of examples of light weight design with aluminium and magnesium castings. The high perfection of the high-pressure die casting HPDC technology and new design opportunities for net shape or near net shape components will be presented. New applications of particle reinforced light metals offer additional potential for weight saving and better technical, economical and environmental performance.

Keywords: light weight design; aluminium; magnesium; castings

Introduction

In vehicles of all types, weight reduction is a question of crucial importance. Along with cost and environmental considerations, all newly developed components must have a weight advantage in relation to conventional solutions.

In order to achieve the best technical and economical performance, one has to consider the interrelationship between design, material and production process for the particular application.

The following article gives a number of examples of light weight design with aluminium and magnesium castings. The high perfection of the high-pressure die casting HPDC technology and new design opportunities for net shape or near net shape components will be presented.

New applications of particle reinforced light metals offer additional potential for weight saving and better technical, economical and environmental performance.

High pressure die casting HPDC technology

Conventionally manufactured high-pressure die castings cannot be applied for welded structures or age-hardening alloys. The reason is, under high pressure, the presence of entrapped gas, either in pores or in the metal matrix. Melting of the metal during welding leads to blisters, causing poor-quality, porous welding seams after solidification. Annealing prior to age-hardening isn't possible either, since the expanding entrapped gas would lead to surface blistering.

New HPDC technologies have been developed to produce weldable and age-hardenable die castings. Such new technologies are, for instance:

- Vacuum casting
- Squeeze casting
- Thixo casting

The principle of these new technologies are:

Vacuum casting

- Evacuated die cavity to less than 50 mbar pressure
- High filling velocity, turbulent filling mode
- High pressure during solidification
- Reduced gas inclusions due to vacuum

Squeeze casting

- Casting die not evacuated
- Reduced and controlled filling velocity to ensure laminar filling mode
- High pressure during solidification
- Reduced gas inclusions due to laminar filling mode

Thixo casting

- Casting die not evacuated
- Casting temperature between solidus and liquidus temperature (semi-solid-status)
- Reduced gas inclusions due to casting in the semi-solid metal status.

Design opportunities with aluminium HPDC

By varying the heat treatment of such castings, the mechanical properties can be tailored to a special application, such as

- High strength, medium elongation for high static loads
- Medium strength, high elongation for medium and dynamic loads (*Figure 1*)

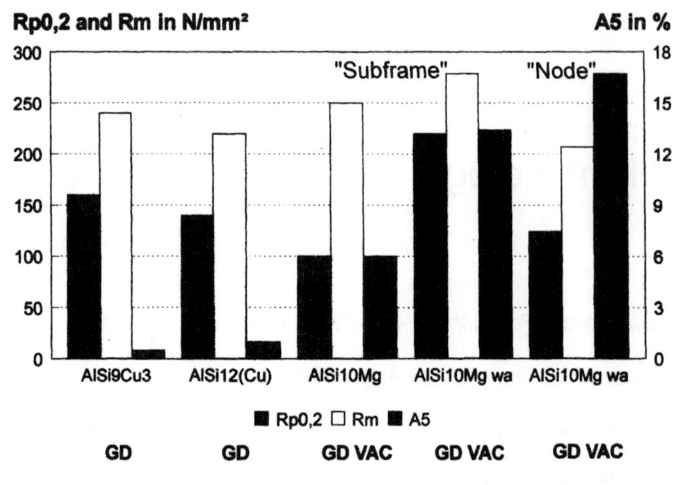

Figure 1 Mechanical properties of castings

Simultaneous development of the production parameters, in particular the composition of the alloy and heat treatment techniques, have allowed the use of aluminium die castings as joint elements (nodes) in the spaceframe of the AUDI A8 (*Figure 2*).

The thin-walled joint elements are reminiscent of deep-drawn sheet components and they also behave similarly when subjected to deformation in a crashtest.

With the use of welding to join the aluminium die-casting joints with aluminium extrusions, a completely new principle in designing structural bodywork components has been created.

The application of aluminium die castings in the chassis area has also become a reality. The desired mechanical characteristics—high yield strength and tensile strength combined with good elongation—are obtained through suitable alloy composition and heat treatment of the die castings which are manufactured using the Vacuum HPDC process (*Figure 3*).

Crash tests, tests of deformation characteristics, as well as drive and continuous-running tests, demonstrate conclusively that high safety demands made on the chassis components are met in full.

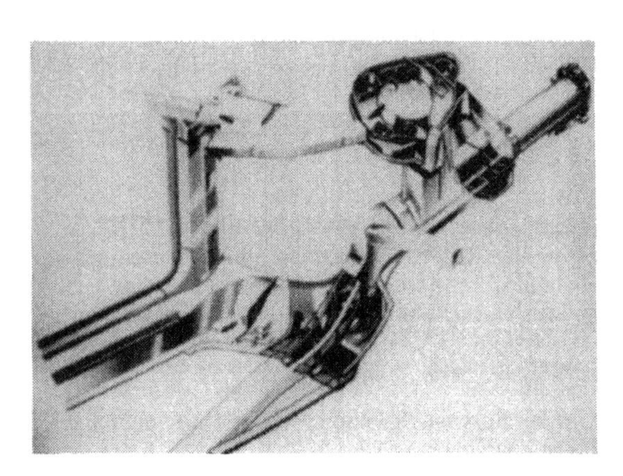

Figure 2 Structural details of the aluminium space frame

Cross Beam-Front

Side Parts-Rear

	Cross Beam	Side Part
Alloy:	GD-AlSi10Mg T6	GD-AlSi10Mg T6
Process:	Vacuum HPDC	Vacuum HPDC
Wall-Size:	2,5 - 5 mm	2,5 - 4 mm
Dimensions:	800 x 600 mm	350 x 300 mm
Weight:	4,0 kg	1,5 kg

Figure 3 Aluminium chassis components

Mechanical properties can be varied by different heat treatments depending on the design characteristics (*Figure 4*).

Design opportunities with magnesium HPDC

In contrast to aluminium, the magnesium die-casting process has the following features:

- Lower wall thicknesses possible, as magnesium has better die-filling behaviour
- Longer die lifespan, as magnesium has no tendency to alloy with the steel of the die
- Shorter cycle times due to lower heat content
- However: closed furnace system for melting and casting with inert gas protection is necessary.

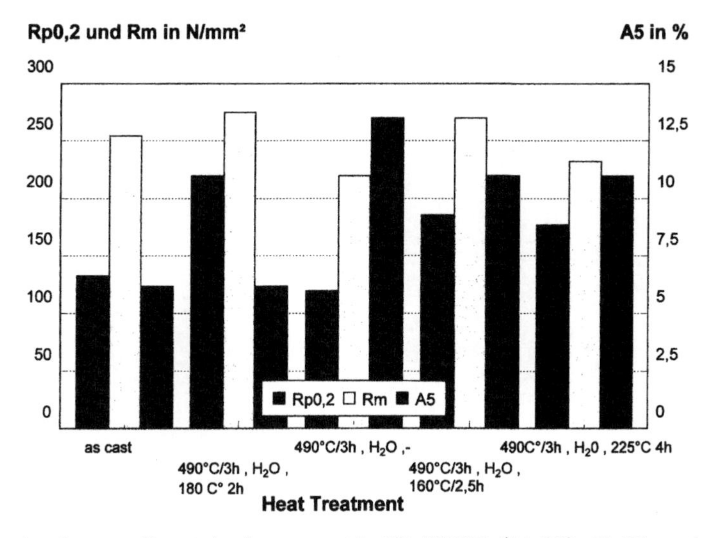

Figure 4 Mechanical properties of vacuum die cast chassis component in GD-AlSi10Mg (AA 360) with different heat treatment

The magnesium alloys which are most suitable for producing thin-walled, large-surface structures are MgAl9Zn1 (AZ91HP) and MgAl6Mn (AM60HP). The mechanical properties of magnesium die castings in the as-cast state are comparable to those of AlSi10Mg (360) after heat-treatment to the T6 state (*Figure 5*). In view of good crash properties of the alloy AM60HP, this is particularly well suited for bodywork applications. In general, these alloys based on high-purity (HP) magnesium have a good resistance to surface corrosion equal to that of the aluminium die-casting alloy AlSi9Cu3 (226).

Magnesium body parts, such as the front section of AUDI 200 and tank cover of Mercedes-Benz SLK demonstrate the design opportunities with the HPDC process for integrative castings with a number of other functional elements, for example, mounting flanges, bores, passages and reinforcement webs, all of which are integrated in the design of the casting (*Figures 6* and *7*).

An example for light metal design without increased cost is the seat in the new double-decker high-speed train TGV in France.

The previous steel-sheet structure meant a weight of 28 kg for each individual seat. As on alternative to this design, magnesium die castings were developed for connection with aluminium extrusions (*Figure 8*) The side supports were designed as thin-walled magnesium die castings, with the result that it was possible to include the desired mounting functions in the near-net-shape castings. The thread required for attachment to the extrusions was incorporated using self-cutting bushes without mechanical machining.

This allowed the total mass of the seat (including foam padding and textile fabric material) to be reduced to just 14 kg.

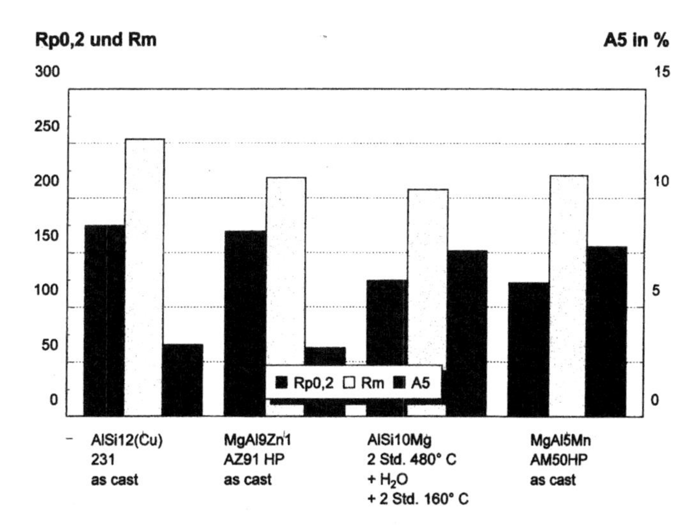

Figure 5 Mechanical properties of HPDC magnesium body parts

Front Section

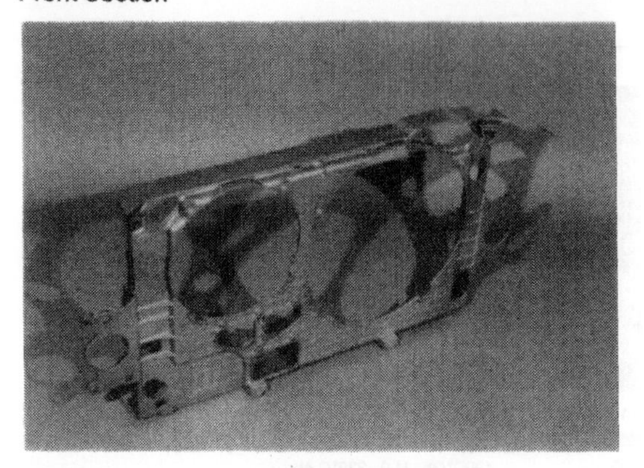

Alloy:	GD-MgAl9Zn1 (AZ91HP)
Process:	HPDC
Wall-Size:	3-6 mm
Dimensions:	1200 x 500 x 120 mm
Weight:	2,8 kg

Figure 6 Magnesium body parts

Tank Cover

Alloy:	GD-MgAl6Mn (AM 60)
Process:	Vacuum HPDC
Wall-Size:	2,5 - 5 mm
	Fins up to 35 mm Height for Increased Stiffness
Dimensions:	1280 x 450 mm
Weight:	3,2 kg

Figure 7 Magnesium body parts

Seats in the High-Speed French Train TGV

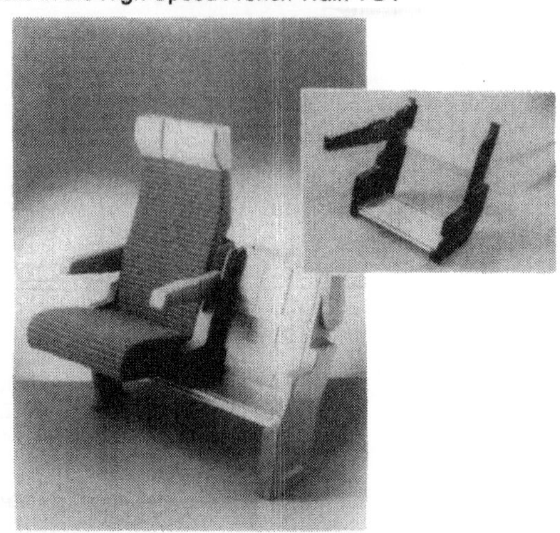

Side Parts and Arm-Rests:	GD-MgAl9Zn1 (AZ91HP)
Seat and Back :	AlMgSi0,5 (6063) - Extrusions
Sheet Cover:	AlMg - Type
Weight:	14 kg (inclusive Foam and textile fabric)
	28 kg (original Steel Sheet Design)

Figure 8 Light weight design

In spite of the fact that magnesium castings and aluminium extrusions have a higher price per kilogram than steel components, their ease of assembly and the amortization of the tooling costs for a limited number of 45,000 seats made it possible to achieve a balance in cost between the light-metal and a steel design.

Particle reinforced aluminium alloy applications

Si-particle reinforced aluminium for cylinder liners

Si-particle reinforced aluminium alloys with Si contents well above the eutectic level of AlSi12 and with fine particle structures and distributions are manufactured using the spray-compaction process (*Figure 9*). In this process, the aluminium melt which dissolves the high

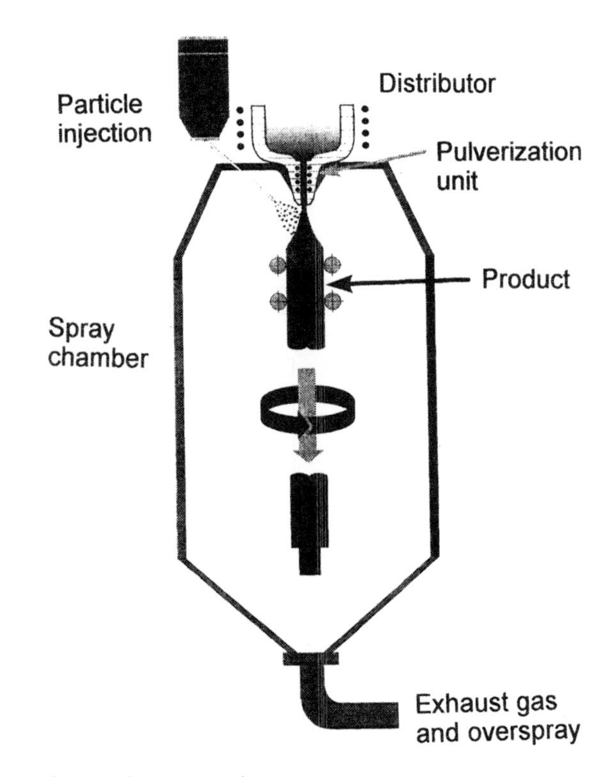

Figure 9 Spray compacting

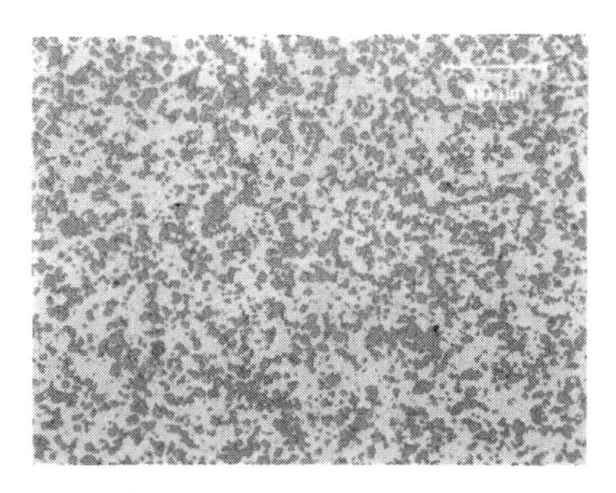

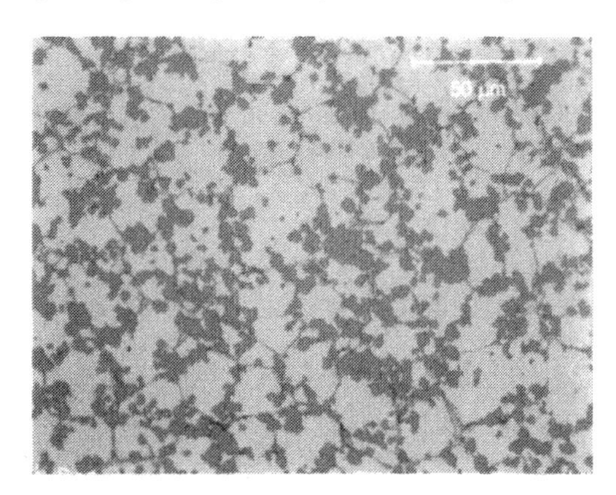

V= 200 : 1

V= 500 : 1

Si - Particle Size 2...15 µm

Figure 10 Aluminium with 25% silicon

Si-content, is pulverized. The spray jet is directed at a rotating plate and moved back and forward at the amplitude of the plate's diameter. The fine metal droplets solidify at a very high speed between the nozzle and the plate. This causes the silicon to precipitate in the form of extremely fine particles. The typical structure of an alloy with 25% silicon shows an extremely fine even precipitation form (*Figure 10*). The spray-compaction process is used for the production of round billets which are then extruded to tubes and further processed to achieve the final dimensions of the cylinder liner inserts.

Through the use of cylinder liners (either of grey iron or particle-reinforced aluminium), it became possible to produce aluminium engine blocks by die casting using the alloy AlSi9Cu3, the most commonly used die-casting material. The special advantages of the Si-particle reinforced aluminium cylinder liners are as follows:

- Low weight
- Same expansion behaviour as the basic engine-block material
- Improved exhaust quality
- Better conditions for recycling

The engine-block cylinder-liner system for the V6 engine block is a single-material utilization (*Figure 11*).

- An economical base standard alloy
- Latest material technology at the functional surfaces

6 - Cylinder V-Motor

Alloy:	GD-AlSi9Cu3
	with Si-Particle Reinforced Aluminium Liners
Process:	HPDC
Wall-Size:	6-30 mm
Dimensions:	500 x 500 x 450 mm
Weight:	23 kg

Figure 11 Aluminium HPDC motorblock

SiC-particle reinforced aluminium for brakes

SiC-particle reinforced aluminium materials are manufactured by metallurgical methods by introducing SiC particles into the base molten aluminium alloy. Aluminium materials with between 10 and 20% SiC particles are known by the name of DURALCAN (*Figure 12*).

This type of alloy is distinguished among other things by particularly high wear resistance.

In view of the high concentrations of SiC-particles in the melt (which tend to sink to the bottom of the melt because of their greater weight), special melt-treatment and processing methods had to be developed to make this material suitable for casting. In collaboration with a user, brake disks with weights of over 60 kg for rail vehicles were developed which are now being produced in series (*Figure 13*).

SiC-particle reinforced aluminium brake disks have an extremely high wear resistance which is superior even to that of iron materials. The good conductivity of the base material ensures that the thermal stressing of the brake disks never becomes critical. The reduction in weight is also of decisive importance. Through the application of aluminium brake disks, the weight reduc-

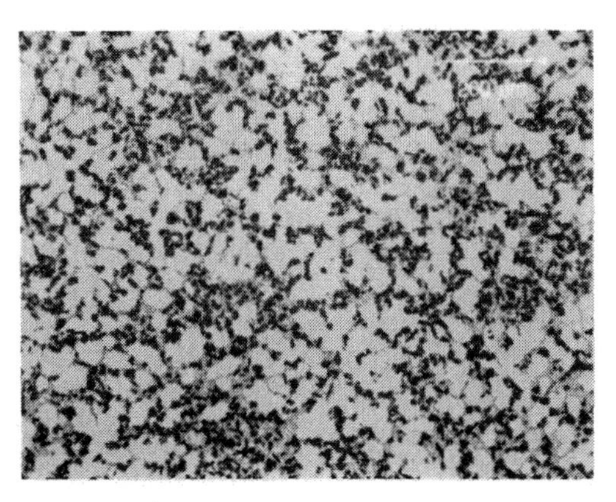

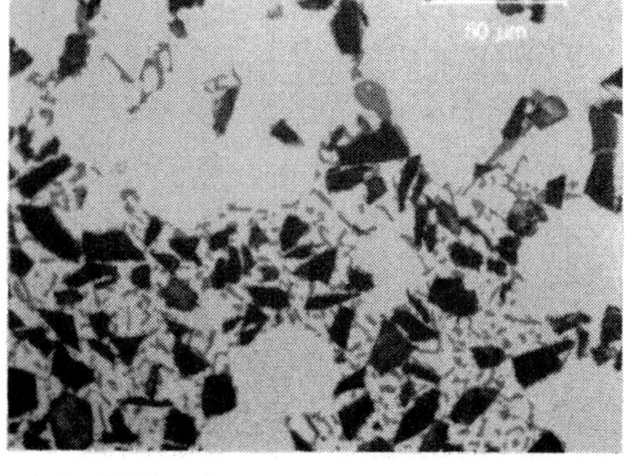

V= 100 : 1

V= 500 : 1

Alloy F3S.20S

Figure 12 Aluminium with 20% siliconcarbide

tion on one axle of the rail vehicle is around 200 kg (with four brake disks per axle) in comparison to brake disks of ferrous materials.

The aluminium brake disks are in use for example on the urban railway system of Copenhagen and on the ICE (Intercity Express) operating between Berlin and Munich.

In view of the above development, it may be expected that brake disks of particle reinforced aluminium will also find uses in car brake systems. The thermal stressing of car brake disks with current vehicle weights and speeds preclude the use of this material at the present time. However, in alternative designs, e.g. electric vehicles, it has a high degree of potential for future use.

Besides the sand-casting process, the permanent-mould process could also be used for the manufacture of larger series of internally ventilated brake disks.

However, brake disks and drums without undercuts can also be produced using the die-casting process. A particularly promising production method would also be the thixo process which takes place below the liquidus temperature, thus eliminating the risk of demixing. Naturally, the bar material supplied for the thixo casting process would also have to possess the desired Si-C-particle distribution. This could easily be ensured by the continuous-casting process used to produce this type of bar material.

Brake Disc for High-Speed German Train ICE

Alloy:	G-AlSi10Mg + 20% SiC
Process:	Sand Casting
Wall-Size:	20-36 mm
Dimensions:	650 mm diameter x 150 mm
Weight:	50 kg

Figure 13 Aluminium brake discs

Summary

New light-weight materials and advanced casting technologies are opening up new opportunities for weight reductions in vehicles. The key to intelligent light metal concepts is the design potential offered by

- new and improved alloys:
- Advanced production technologies
- Tailored heat treatment and processing technologies.

The high perfection of the high-pressure die casting technology opens new design opportunities for applications in body structures and chassis components. Welded assemblies, combining aluminium die castings with extruded and rolled products, respectively, are opening new design opportunities.

Mechanical properties can be tailored to specific requirements. Net shape or near net shape die castings reduce manufacturing cost.

For magnesium die castings, there are special opportunities in thin-wall body structure applications. Higher material cost for magnesium alloys can be compensated by intelligent casting design with multiple integrated structural functions.

Particle reinforced light metals lead to new alloy properties. Si-particle reinforced aluminium for cylinder liners and SiC-particle reinforced aluminium for brake disks offer additional potential for weight saving and better technical, economical and environmental performance.

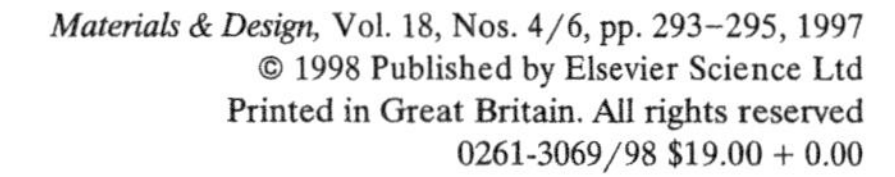

Materials & Design, Vol. 18, Nos. 4/6, pp. 293–295, 1997
© 1998 Published by Elsevier Science Ltd
Printed in Great Britain. All rights reserved
0261-3069/98 $19.00 + 0.00

PII: S0261-3069(97)00066-6

Abnormal grain growth in Al of different purity

B. B. Straumal[a,c,*], W. Gust[a], L. Dardinier[a], J. -L. Hoffmann[b], V. G. Sursaeva[c], L. S. Shvindlerman[c]

[a]*University of Stuttgart, Institut für Metallkunde, Seestr. 75, D-70714 Stuttgart, Germany*
[b]*CRV PECHINEY, ParcEconimique Centr'Alp, BP27, F-38340 Voreppe, France*
[c]*Institute of Solid State Physics, Chernogolovka, Moscow District, Moscow 142432, Russia*

Received 18 August 1997; accepted 25 August 1997

The transition from normal to abnormal grain growth has been studied in four Al alloys of various purity (2N, 3N, 4N and 5N). The temperature and time for the onset of abnormal grain growth depend strongly on the deformation and homogenization treatment. Generally, the formation of large grains before cold rolling makes easier the transition to abnormal grain growth during the subsequent annealing. The abnormal grain growth can take place only above a certain temperature which decreases with increasing alloy purity. The onset time of the abnormal grain growth decreases with increasing temperature. It can be qualitatively explained by the dissolution of submicron particles of a second phase. © 1998 Published by Elsevier Science Ltd. All rights reserved.

Keywords: grain growth; abnormal growth; grain boundary phase transitions

Introduction

The use of aluminium based alloys steadily increases. One of the important industrial problems is the *improvement of the surface quality* of several products which depends critically on the *uniformity of their grain structure*. The formation of very large grains or a large scatter of the grain size in a material can evoke a non-uniform deformation and recrystallization texture and, therefore, can cause intolerable shade fluctuations of the end product. The formation of a heterogeneous grain structure at any step of the aluminium transformation, from homogenization to intermediate annealing, makes the material unacceptable. Unfortunately, nowadays there exists no technology which would not allow a spontaneous formation of very large grains (so-called abnormal grain growth). This phenomenon has been well known since the nineteen-thirties[1–4]. However, the tendency to use pure aluminium for the production of industrial alloys, especially for deep drawing, has reviewed the old problem of control and/or suppressing the abnormal grain growth

Experimental

The materials used in our studies were 5N (99.999 wt.%), 4N (99.99 wt.%), 3N (99.92 wt.%) and 2N (99.00 wt.%) Al alloys. The main impurities in these materials are listed in *Table 1*. The alloys were produced by Pechiney CRV as hot rolled blocks $25 \times 25 \times 2$ cm. The blocks are cut out, homogenized in air (Table 2), machined and cold rolled to a thickness of 2.5 mm with a reduction of 63–90% and a high number of passes (over 20). During the rolling the plates are periodically cooled in liquid nitrogen in order to keep their temperature below about -10°C and to prevent the recrystallization reaction. The cold rolled bands of 4N, 3N and 2N Al were annealed in order to obtain a fully recrystallized structure without a deformed matrix (with a mean grain size of 200–500 μm). The 5N material was annealed at 350°C for 30 min in order to prevent the grain growth. The regimes of homogenization and recrystallization annealings are listed in *Table 2*.

The recrystallized bands were cut into pieces with the dimensions 4×6 cm. These pieces were then annealed in an air furnace at various temperatures from 350 to 650°C. Some specimens were annealed only once and others were annealed several times at the same temperature in order to avoid the influence of repeated etching. After annealing the samples were etched for 1–2 min in a solution of 10 ml HF, 15 ml HCl and 90 ml H_2O in order to reveal the grain structure. The microstructure was photographed, and the mean grain size d was determined on 400–500 grains by the intersection method, using image analysing optical microscopy and polarization contrast. The measurements were repeated after each new annealing in the same area in order to diminish the influence of the difference of the starting grain size.

Results and discussion

The most important feature of the grain growth in the Al alloys studied is the *transition from normal to abnormal grain growth*. At the beginning of annealing, the

*Correspondence to B. B. Straumal. Tel.: +49 711 1211276; fax: +49 711 1211280; e-mail: straumal@vaxww1.mpi-stuttgart.mpg. de and straumal@song.ru

Table 1 The main impurities in the Al alloys studied (concentrations are given in wt. ppm (10^{-4} wt.%))

Alloy	Fe	Cu	Si	Mg	Mn	Cr	Zn	Ti	Ga	Total
5N Al	2	1	7	1	< 1	1	< 4	< 1	2	10
4N Al	10	40	7	9						100
3N Al	170	40	300	10	10			240		770
2N Al	5000		1600		37					10 000

Table 2 The thermal treatment of the alloys studied

Material	Homogenization	Recrystallization
5N(1)	No	10 min, 350°C
5N(2)	70 h, 650°C	10 min, 350°C
4N(1)	No	10 min, 450°C
4N(2)	30 h, 570°C	10 min, 450°C
4N(3)	70 h, 650°C	10 min, 450°C
3N(1)	No	10 min, 450°C
3N(2)	30 h, 570°C	10 min, 450°C
2N	30 h, 570°C	10 min, 450°C

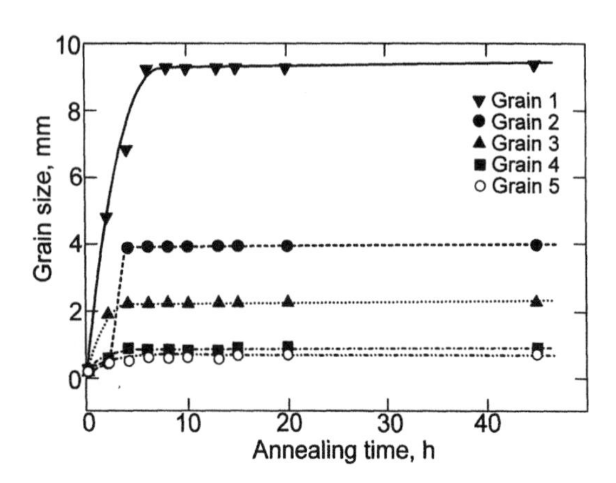

Figure 2 Time dependence of the mean grain size in 4N Al at 500°C (homogenization at 650°C for 70 h) measured in various 'old grains'

grain structure is uniform and the scatter of the grain size is low. Nevertheless, the cold rolled microstructure reflects the grain structure existing already before the deformation. The newly recrystallized grains are homogeneous inside an old grain of the former matrix, but the mean grain size varies greatly between the old grains. Later, the new recrystallized grains grow inside the old grains. For some of them, the growth is faster and some big new grains appear, having often the same shape as the former old grain (*Figure 1*). Thus, inhomogeneity develops, and the grain size distribution starts to be bimodal.

When changing the homogenization regimes (*Table 2*), we have observed that the size of the 'old grains' formed in the sample after the homogenization, prior to the cold rolling affects critically the onset of the abnormal growth. The larger the size of the 'old' grains, the easier the transition to the abnormal grain growth. Generally, the 'old' grain boundaries, i.e. the borders between the colonies of the new recrystallized grains, stop effectively the abnormal grain growth. The abnormal growth in different colonies ('old grains') starts after different time (*Figure 2*).

Even in the case of very large recrystallized grain colonies (big 'old grains'), there exists a temperature, below which a stagnation of the normal grain growth occurs, without transition to abnormal growth. This feature was already observed earlier[5,6]. The data plotted in *Figure 3* show that the temperature of the onset of abnormal grain growth decreases with increasing purity of the alloys studied. The data[7] for the 99.6 wt.% Al fit that curve well.

At a constant impurity content, the onset time of the abnormal growth decreases with increasing temperature (*Figure 4*) at least for 3N Al. This dependence can be explained by the dissolution of submicron intermetallic particles during the grain growth. During the homogenization at the temperature close to the melting point, all precipitates dissolve in the solid solution. New precipitates can build during the maintaining at room temperature and cold working (most possibly Al_3Fe because the Fe content in 3N Al is above the solubility of Fe at $T > 560°C$[8]). The grain boundaries fixed at the precipitates can become free during the grain growth, according to Zener's idea, due to the partial dissolution of ankering particles. According to[9], the radius r of the precipitate changes with the dissolu-

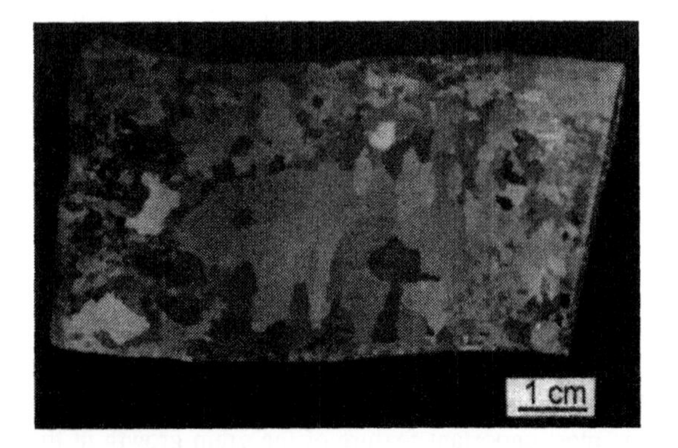

Figure 1 Microstructure of a sample after the onset of abnormal grain growth. 5N Al, homogenization at 650°C for 70 h; deformation 63%; thickness 2 mm; annealing at 450°C for 5 min

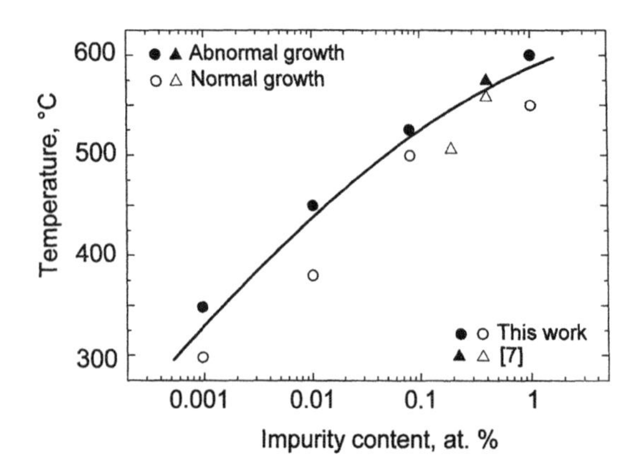

Figure 3 Dependence of the barrier temperature for the abnormal grain growth on the impurity content of the Al alloys studied. The literature data for the 99.6 wt.% Al are also presented[7]

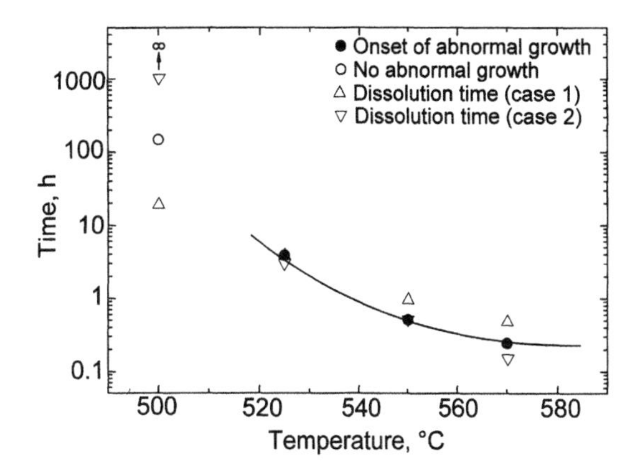

Figure 4 Temperature dependence of the onset time for abnormal grain growth in 3N Al and of the calculated dissolution times for the Al_3Fe precipitates

tion time t as follows:

$$r^2 = r_0^2 - 2\alpha D_m t,$$

where r_0 is the starting radius of a precipitate, D_m is the Fe diffusivity of Fe in Al[10] and

$$\alpha = (c_0 - c)/(c_{Al_3Fe} - c_0)$$

where c is the Fe concentration in the solid solution, c_0 is the Fe solubility at the annealing temperature, and c_{Al_3Fe} is the Fe content in Al_3Fe. One can estimate the time of the complete dissolution of a particle, having the radius r_0 before annealing: $t = r_0^2/(2\alpha D_m)$. In *Figure 4* the estimation data are shown for case 1 ($r_0 = 0.2$ μm, $c = 20$ wt. ppm) and case 2 ($r_0 = 0.1$ μm, $c = 57$ wt. ppm). If $t \rightarrow \infty$, no abnormal grain growth can happen. The estimations from this simple model fit well the experimental data on the onset time of abnormal grain growth (*Figure 4*).

Conclusions

The following conclusions can be drawn from the present studies.

- The time and the temperature of the onset of abnormal grain growth depend strongly on the deformation and homogenization treatment.
- In the alloys studied, the abnormal grain growth proceeds only above a certain temperature which decreases with increasing alloy purity.
- The onset time of the abnormal grain growth decreases with increasing temperature.
- This decrease can be qualitatively explained by the dissolution of submicron particles of a second phase.
- The abnormal grain growth in high-purity Al can be suppressed with the aid of a suitable combination of deformation, heat-treatment and micro-alloying.

Acknowledgements

The financial support of the TRANSFORM programme of the German Federal Ministry for Education, Science, Research and Technology (under contract BMBF 03N9004), the Volkswagen Foundation (under contract VW I/71 676), the NATO (under contract HTECH.LG.970342) and the Russian Foundation of Basic Research (under contract 950205487a) is gratefully acknowledged.

References

1 G. Gottstein, Rekristallisation metallischer Werkstoffe, DGM, Oberursel, 1984
2 V.Ju. Novikov, *Secondary Recrystallization*, Metallurgia Publishers, Moscow, 1990
3 S.S. Gorelik, *Recrystallization of Metals and Alloys*, Metallurgia Publishers, Moscow, 1978, (in Russian)
4 F. Haessner, *Recrystallization of Metallic Materials*, Dr. Riederer Verlag, Stuttgart, 1978
5 Straumal, B., Risser, S., Sursaeva, V., Chenal, B. and Gust, W. *J. Phys. C*, 1995, **5** (IV), 233–241
6 Mayer, C., Sursaeva, V., Straumal, B., Gust, W. and Shvindlerman, L. *phys. stat. sol. (a)*, 1995, **150**, 705–713
7 Dahl, O. and Pavlek, F. *Z. Metallkde.*, 1936, **28**, 266–277
8 Edgar, J.K. *Trans. AIME*, 1949, **180**, 225–229
9 Aaron, H.B., Fainstein, D. and Kotler, G.R. *J. Appl. Phys.*, 1970, **41**, 4404–4411
10 Beke, D.L., Gödény, I., Szabo, I.A., Erdélyi, G. and Kedves, F.J. *Philos. Mag. A*, 1987, **55**, 425–432

Materials & Design, Vol. 18, Nos. 4/6, pp. 297–301, 1997
© 1998 Published by Elsevier Science Ltd
Printed in Great Britain. All rights reserved
0261-3069/98 $19.00 + 0.00

PII: S0261-3069(97)00068-X

The dispersion mechanism of TiB_2 ceramic phase in molten aluminium and its alloys

A. Jha[a,*], C. Dometakis[b]

[a]*The School of Materials, Clarendon Road, University of Leeds, Leeds LS2 9JT, UK*
[b]*Department of Materials Engineering, Brunel University, Kingston Lane, Uxbridge UB8 3PH, UK*

Received 5 August 1997; accepted 18 August 1997

Titanium diboride (TiB_2) ceramic particulates are dispersed in molten aluminium and its alloys for grain refining and for making cast metal1matrix composites. For producing cast MMC, the dispersion of the ceramic phase via in-situ aluminothermic reduction of K_2TiF_6 and KBF_4 flux mixture with molten aluminium and, via the addition of exogenously formed TiB_2 with the fluoride flux has been studied at 900°C. In this article, the aspects of interfacial energy that govern the dispersion and agglomeration of TiB_2 particulates are examined. The Gibbs-adsorption interface equation is particularly employed to define and to quantify the change in the surface energy as a function of the alloying element concentration and, consequently the effect of interfacial energy on the nucleation rate of TiB_2 formed via metallothermic reduction reaction and the size of the ceramic phase is also explained. © 1998 Published by Elsevier Science Ltd. All rights reserved.

Keywords: aluminium alloys; Al-alloy MMC; TiB_2 and dispersed ceramics

Introduction

Aluminium alloy matrix reinforced with the particulates of TiB_2 as a ceramic phase has been identified as a possible alternative to the Al–SiC metal–matrix composites[1]. The original article of by Wood et al.[1,2] examined the microstructural and wear properties of Al–TiB_2 cast metal–matrix composites with less than 8 vol.% dispersion of 5–20 μm TiB_2 particulates. The method for ceramic phase dispersion adopted was based on the extension of the grain-refining master alloy manufacturing process which was simultaneously studied by Lee et al.[3]. In the grain-refining master alloy fabrication, as described by Dometakis et al.[4], the complex fluorides K_2TiF_6 and KBF_4 are reduced via aluminothermic reduction: $K_2TiF_6 + 2KBF_4 + 10/3\ Al = TiB_2\ (alloy) + 3.3(KF \cdot AlF_3) + 2AlF_3$ and yield TiB_2. The freshly formed TiB_2 at the metal–flux interface rapidly disperses in metallic aluminium. The microstructure of the dispersed TiB_2 is dependent on the interfacial energy conditions. In commercial alloys namely A356 (Al–7Si–0.3Mg) and Al–4.7Cu–0.8Mg-0.8Mn) segregate at the interdendritic boundaries by a particle pushing mechanism[1,2]. The dispersion characteristics of TiB_2 in molten aluminium and its alloys have not been studied in the context of surface energy, which is by far the most important factor governing the dispersion of TiB_2 in liquid aluminium.

In this article, we have studied the importance of surface tension or the interfacial energy on the dispersion of in-situ formed TiB_2 in two different types of alloy matrices: commercial purity aluminium and Al–8%Mg–1%Zr alloys. For this investigation, the Gibbs-adsorption equation has been examined for describing the role of alloying elements on the interfacial energy of aluminium alloys. We have also explained the importance of controlling other types of interfaces present during the dispersion process. Examples for the dispersion of exogenously formed TiB_2 particulates in aluminium alloys are also discussed briefly.

The control of size and size distribution of the ceramic phase dispersed in a light-alloy matrix, e.g. Al-alloy is of an engineering importance for fabricating materials with high specific modulus and strength via casting technique. The understanding of the mechanism of dispersion of particulates in Al-alloy will enable us to design better materials for structural applications.

Experimental

Commercial purity (CP) aluminium and Al–Mg (8%)-1%Zr alloys were selected as matrix materials. Each matrix composition was first melted and cooled in a partially reducing atmosphere of Ar–4%H_2 in order to minimize the oxygen contact with molten metal. A limited number of experiments were also carried out in an oxidizing atmosphere for making comparison between the cast microstructures. The melting was carried out inside a radio frequency coil as described by Dometakis et al.[4]. After the first melting and cooling cycle, a reasonably flat metal–gas interface was formed inside the alumina crucible, as shown in *Figure 1(a)*. The masses for K_2TiF_6 and KBF_4 fluxes were calculated for a selected dispersion volume, e.g. 15% TiB_2

* Correspondence to A. Jha

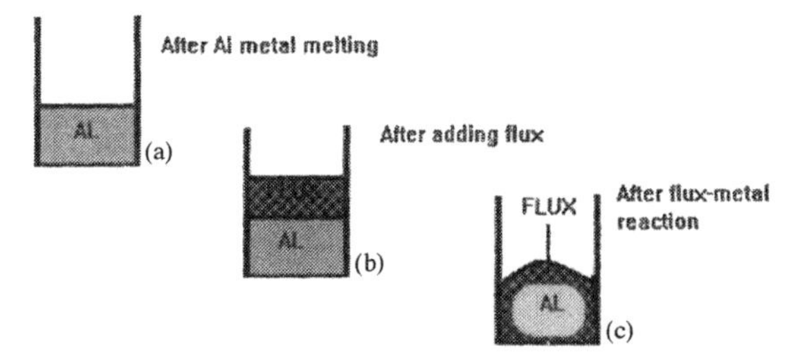

Figure 1 Schematic diagram of crucible containing aluminium and flux. (a) Flat surface of frozen aluminium. (b) Flat surface of Al-metal covered with fluoride flux. (c) Fluoride flux + Al ingot after reaction

in Al-alloy and added on the top of the frozen aluminium surface (see *Figure 1(b)*). Each flux mixture was then melted together with the Al-alloys at 900°C as shown in *Figure 1(c)* in a partially reducing atmosphere (Ar–4%H_2). In the case of exogenous TiB_2, the flux mixture was sodium cryolite ($3NaF \cdot AlF_3$) and MgF_2 in a proportion of 4:1. The cryolite/MgF_2 mixture was mixed with TiB_2 particulates of average particle size (25 μm). Each flux reaction and flux treatment experiment lasted for 15 min at 900°C. After this period, the ingot was allowed to solidify by switching-off the radio frequency coil power supply. The ingots weighing between 15 and 20 g were obtained and these were prepared for microscopic examination and the results are discussed below. A limited number of small-scale casting trials with uniform dispersion of TiB_2 were also melted using 1 kg of aluminium and the microstructures were reproduced in a Ar–4%H_2 atmosphere above 800°C.

Results

The microstructures of in-situ dispersed TiB_2

The microstructures of dispersed TiB_2 formed via in-situ reaction of fluoride fluxes with commercial purity aluminium are compared in *Figure 2*. The composition of fluoride flux was the only variant in the flux-assisted TiB_2 dispersion experiments. *Figure 2(a)* shows the back scattered image of the banded structure of TiB_2 in commercial purity aluminium. When the MgF_2 and Li_2TiF_6/$LiBF_4$ mixture partially replaced the K_2TiF_6/KBF_4 mixture, the resulting microstructures are compared in *Figure 2(b)* and *(c)*, respectively. The absence of the banded TiB_2 structure was confirmed in the latter two cases. The alloy matrix analysis confirmed the presence of Mg and Li between 0.25 wt.% and 0.35 wt.%. These two elements were therefore transferred during aluminothermic reduction reaction from the fluoride flux mixture to the aluminium matrix and, consequently the composition of CP aluminium changed to Mg-rich and Li-rich aluminium alloys, respectively. The particulates of TiB_2 were found to be fairly coarse (5–10 μm) in the CP aluminium with Mg as an alloying element (see *Figure 2(b)*), whereas the average particle size was below 0.5 μm when 0.25 wt.% Li was present in the CP alloy matrix. In both microstructures, there is an evidence for coalescence, however, the extent of coalescence is significantly higher

(a)

(b)
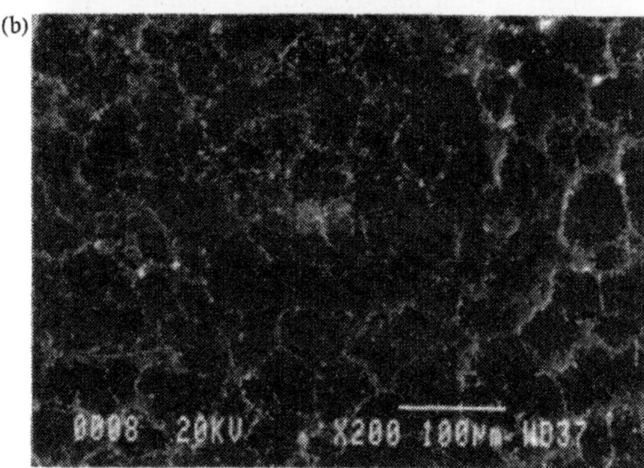

(c)
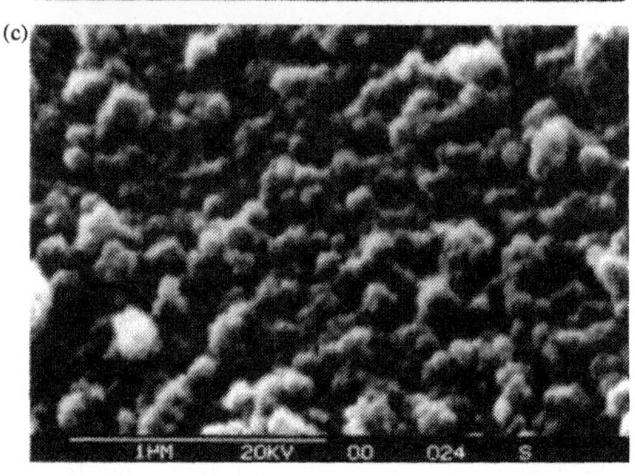

Figure 2 Microstructure of commercial purity aluminium containing TiB_2. (a) Back scattered image of banded layers of TiB_2 in molten aluminium after reaction with $K_2TiF_6 + KBF_4$ flux. (b) Microstructure of TiB_2 segregating at the grain boundaries in CP aluminium treated with 90% $K_2TiF_6 + KBF_4$ mixed with 10% MgF_2 flux. (c) Effect of the replacement of 10–15% $K_2TiF_6 + KBF_4$ flux mixture by 10–15% $Li_2TiF_6 + LiBF_4$ on the microstructure of TiB_2 in CP aluminium

for CP aluminium with Li. *Figure 3* shows a uniformly dispersed microstructure of 2–5 μm TiB_2 in the Al–Mg–Zr alloy matrix. There is neither the presence of banded structure nor is there any evidence for the presence of any particle–particle coalescence.

The microstructures of exogenously-formed TiB_2 in CP aluminium

The dispersion of TiB_2 particulates without any flux in the CP aluminium was found to be impossible under oxidizing conditions. No significant volume percent of TiB_2 was observed in the microstructure due to the presence of a tenacious and coherent film of alumina

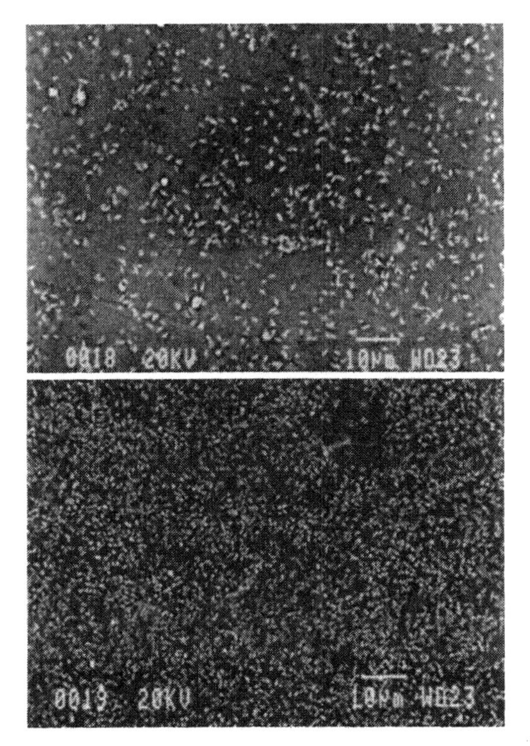

Figure 3 Microstructure of uniformly dispersed TiB_2 facetted particulates in Al-8%Mg-1%Zr alloy. Flux used was $K_2TiF_6 + KBF_4$

at the gas–metal interface. The alumina film was eliminated by using sodium cryolite mixed with MgF_2. The microstructure of coarse particulates in *Figure 4* can be compared with the morphological features of TiB_2 in *Figures 2* and *3* which varies sharply from the in-situ microstructure. In *Figure 4*, the matrix–particulate interfacial regions were also analyzed for Al and Ti by energy-dispersive X-ray technique and it confirmed that the matrix consisted of a significant amount of elemental Ti which may be present as the Al_3Ti intermetallic phase. The cavity formed around each particle suggests that TiB_2 particulates may have reacted with the matrix, as it is well known from the grain-refining reaction: $4Al + TiB_2 = AlB_2 + Al_3Ti$. Therefore Ti should be present in the matrix which is shown in *Figure 4*.

Discussion

The dispersion of TiB_2 particulates in liquid aluminium and its alloys is greatly influenced by the surface tension of the alloy. In other words, the interfacial energy between the liquid alloy and the ceramic particulates is one of most important factors whether the particulates of ceramic phase would coalesce or disperse. Besides the matrix–particulate interfacial energy, there are three additional interfaces and each one can have an effect on the dispersion of the ceramic phase in the molten aluminium metal. They are the flux–atmosphere, flux–crucible and metal–flux interfaces. We have not considered the crucible–flux boundary for the in-situ reaction because the TiB_2 particulates form via aluminothermic reduction at the interface and they are rapidly transferred into the molten matrix. Each of these factors is discussed in the context of the microstructural evidences shown above. The flux-crucible boundary could become significant in the case of exogenously formed TiB_2 in view of the low interfacial energy of the graphite–TiB_2–Al interface[5].

The flux–atmosphere interface is important from the point of view of reducing the oxygen contamination of the metal–flux interface. The presence of oxygen and moisture promotes the oxidation of aluminium alloys and prevents the dispersion of TiB_2. In a commercial grain-refining master alloy production process, the fluoride flux is melted in air together with aluminium metal and cast in air containing moisture. The presence of moisture in the flux and in air therefore adversely affects the dispersion of the ceramic phase by forming an inhibitive layer of alumina; and the extent dispersion then becomes inversely dependent on the concentration of moisture in the melting atmosphere and flux during the melting process. Alumina is thermodynamically more stable than TiB_2 and fluoride flux and its removal is only possible by the presence of a cryolite flux for example. An upper limit for oxygen as alumina in the flux must be determined from the Al_2O_3-cryolite phase diagram at a given temperature[6] for avoiding the saturation condition of Al_2O_3 at which alumina will start forming an inhibitive layer.

The metal–flux interface is by far the most important factors for understanding the origin of dispersion of TiB_2 in molten metal. In the in-situ process, the exclusion of TiB_2 particulates by the layer of alumina formed at the surface of liquid aluminium is minimized by forming cryolite as a by-product of the dispersion reaction shown above. Whereas when the exogenous TiB_2 particles are present, the dispersion condition necessitates the presence of cryolite which helps in reducing the formation of alumina at the metal–flux interface. This is because alumina has a finite solubility in cryolite, melts at a given temperature and an example of the solubility diagram is shown by Dewing et al.[6].

Besides the molten flux acting as a sink for alumina formed, it also envelopes molten aluminium and physically cuts-off the contact between the metal and the crucible. This can only be explained by considering the interfacial conditions that satisfies the energetics: $\sigma_{Al/Flux}$ is smaller than $\sigma_{Al/crucible}$ and $\sigma_{flux/crucible}$, where σ is the interfacial energy in J m^{-2}.

$$\frac{d\sigma}{d\ln X_i} = \frac{d\sigma}{d\ln a_i} = -\Gamma_i.R.T \qquad (1)$$

The molten flux acts as a reservoir for the surface-active elements which greatly modify the flux–metal and TiB_2-metal interfacial energies. It is well known from Korol'kov's review[7] that the interfacial tension of molten aluminium drops significantly with the addition of solute elements Mg and Li. At a concentration of 0.5 wt.% of the solute elements, the interfacial energy of pure aluminium at its melting point drops from 0.92 to 0.72 Jm^{-2} and 0.58 Jm^{-2}, for Mg and Li, respectively[7]. It is also important to note that when CP aluminium was used as a reducing agent for fluxes containing MgF_2 and LiF in the forms of Li_2TiF_6 and LBF_4, these two fluorides are also reduced by Al to their elemental forms and they rapidly dissolve in molten aluminium by

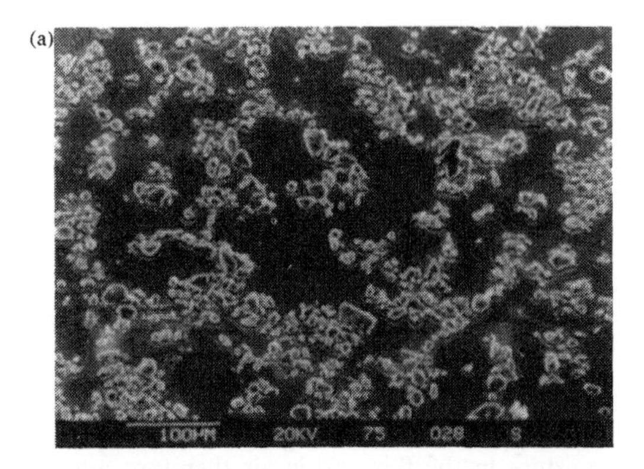

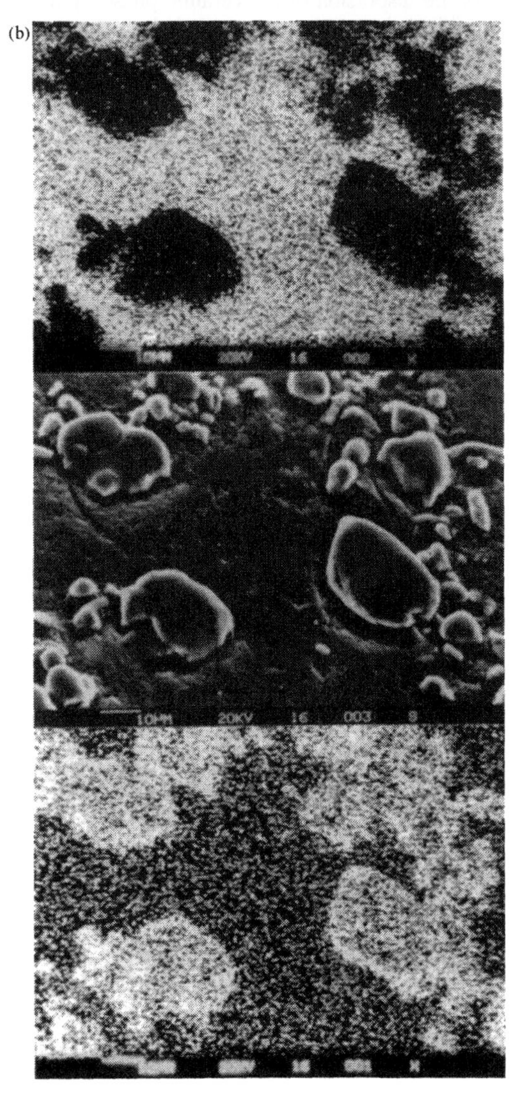

Figure 4 Microstructure of exogenously formed TiB_2 added in CP aluminium with the help of sodium cryolite + MgF_2 flux. (a) Low magnification micrograph showing segregation at grain boundaries. (b) Aluminium and Ti-metal energy dispersive X-ray micrograph

lowering its Gibbs free energy and the interfacial energy between the metal and ceramic phase. The latter is mathematically described by the Gibbs-adsorption Eq. (1)[7]. Γ_i is called the surface concentration (number of atoms adsorbed per m^2 per g-atom) of the solute atom i, R is the universal gas constant (8.314 J mol^{-1} K^{-1}), T is the absolute temperature and a_i is the thermodynamic activity of the solute atom i in the aluminium alloy which is expressed by a product of a constant h_i and X_i, i.e. $a_i = h_i X_i$, where h_i is the Henrian activity coefficient. In this relationship, X_i is the atomic fraction of the solute in the solvent. Clearly for a solute 'i' to be surface active, it is essential that the slope of σ against $\ln a_i$ must be negative. Otherwise the adsorption of the solute must not occur. Both Mg and Li satisfy the Gibbs adsorption condition and lower the value of σ for molten aluminium and also promote a high nucleation rate during the process of flux reaction[4]. This is the reason that the average size of TiB_2 nucleated is significantly smaller with the presence of Li than when Mg and (Mg and Zr) are present as solute elements in the solvent aluminium.

The role of Zr as a minor solute element in the Al–Mg–Zr alloy during the dispersion process is rather complex. Zirconium atoms serve two important purposes while present in the metal with Mg. It is one of the strongest boride formers amongst all the solute and solvent elements present and, therefore, it promotes the formation of $(Ti,Zr)B_2$ in preference to other complex borides. However, the coexistence of Mg with Zr reduces the surface adsorption of Mg on to $(Ti,Zr)B_2$ surface compared to $(Ti,Al)B_2$ surface. Magnesium as solvent with less 0.2 at.% concentration of Zr promotes a tendency to demix with β-Zr in the Mg–Zr phase diagram[8]. The behaviour of Ti and Zr as solute atoms in Al-alloy is similar to those observed in the Mg–Zr alloy. As a result, the magnesium atoms prefer to be associated with the aluminium as the solvent atoms and favour the conditions for surface desorption of Mg on Zr-rich TiB_2 surface and associative interaction of Mg with aluminium in the matrix to form Mg_3Al_2 intermetallic phases (cf. Eq. (1)). This is the main origin for achieving virtually ideal dispersion condition of $(Ti,Zr)B_2$ in the Al–Mg–Zr alloys.

On the basis of Eq. (1), the ceramic particle coalescence arises as a result of the adsorption of the surface-active species on to the surface of TiB_2-metal interface by lowering the interfacial energy of pure aluminium. Because the particles of ceramic phase form an interface with the liquid aluminium, the interfacial energy is further reduced by multiple particle–particle coalescence.

Conclusions

The mechanism of dispersion and coalescence of TiB_2 can be explained by using the Gibbs adsorption (Eq. (1)) which connects the interfacial energy with the bulk thermodynamic properties of the solute atoms in the solvent aluminium phase. CP aluminium with Mg and Li as solute atoms promote the formation of TiB_2 by high nucleation and low growth which is favoured by the lower surface energy conditions. The ceramic particles, however, cluster together in order to reduce the interfacial energy. The presence of zirconium promotes the dispersion of TiB_2 by allowing Mg atoms to be desorbed from the boride surface which is explained from the miscibility behaviour of solutes in the Mg–Zr,

Al–Zr and Al–Mg binary phase diagrams and from Eq. (1).

Acknowledgements

The authors acknowledge the support from MERCK, and the SERC for studentship support.

References

1 Wood, J. V., Davies, P. and Kellie, J. L. F., *Materials Science and Technology*, 1993, **9**, 830–840
2 Wood, J. V. et al., *Cast Metals*, 1995, **8**, 57–64
3 Lee, M. S. and Terry, B. S., *Materials Science and Technology*, 1991, **7**, 608–612
4 Dometakis, C., Jha, A., Riddle, R. and Smith R., Proceedings of the 9th International Symposium on Rapid Quenched and Metastable Materials (RQ9), 25–30 August 1996. eds. P. Duhaj, P. Mrafko and P. Svec. Elsevier, Amsterdam, 1997, pp. 60–67
5 Watson, K. D. and Toguri, J. M., In *Proceedings of the International Symposium on Extraction, Refining and Fabrication of Light Metals*, Ottawa, Ontario, August 18–21, 1991. eds. M. Sahoo and P. Pinfold. Pergamon Press. N.Y., 1991
6 Dewing, E. W., *Metallurgical Transactions*, 1972, **3**, 495
7 Korol'kov, A. M., In *Casting Properties of Metals and Alloys*. Consultants Bureau, New York, 1960, p. 37
8 Massalski, T. B., In *Binary Alloy Phase Diagrams, American Society for Metals*, 1986, **1567**, pp. 130, and as in[3]

Materials & Design, Vol. 18, Nos. 4/6, pp. 303–307, 1997

0261-3069/98 $19.00 + 0.00

PII: S0261-3069(97)00069-1

Influence of the heat treatment on the structure and mechanical properties of the AlZnMgLi alloy

Andrzej Klyszewski*

Institute of Non-ferrous Metals, Light Metals Department in Skawina, ul. Pilsudskiego 19, 32-050 Skawina, Poland

Received 18 August 1997; accepted 29 August 1997

The influence of lithium on the structure and properties of AlZn5Mg1 alloy after heat treatment has been determined. Lithium added in the amount up to 2/ increases strength properties of the alloy.

Keywords: AlZn5Mg1 alloy; lithium; structure; properties; heat treatment

Introduction

Research of the alloy of AlLiCuMgZr type has been carried out for a long time and such alloys as Lital A (8090) and Lital B (8091), from which bands, sheets and extruded sections are made, are used in aero- and space technology.

Influence of Li addition on the construction alloys of the AlZnMg type (series 7000) has also been the subject of research.

In the above alloys, the phase $MgZn_2$ (η), which is formed as a result of the disintegration of a saturated solid solution is the final phase. After an addition of lithium the second hardening phase of Al_3Li (δ) is formed. The $MgZn_2$ phase appears at temperatures < 150°C, i.e. considerably lower than the temperature of the formation of the Al_3Li precipitation in the alloys with lithium (170°C).

The aim of the work was to estimate the influence of lithium on the properties of the, PA 47 alloy produced in Poland, and to determine the optimum parameters of the heat treatment, the result of which are the hardening phases, and therefore the high mechanical properties.

Materials and experimental procedures

The subject of this research was AlZnMg (PA 47) alloy, with three different contents of lithium.

The chemical constitution is presented in *Table 1*.

The alloy ingots (130 mm diameter) were extruded in WMD Czechowice–Dziedzice with the use of a 6-hole matrix. Rods of 10 mm diameter were obtained and subjected to further experiments. The following heat treatment was applied: hyperquenching from the temperature of 530°C (they were kept in this temperature for 45 min) in cold water and naturally and artificially aged at temperatures 90, 120, 150, 180 and 200°C. The time of ageing ranged from 6 min to 400 h. On the basis of the obtained ageing curves, different variants of single- and two-stage heat treatments were determined. The above heat treatments were performed on strength samples.

Tables 2 and *3* show parameters of the applied procedure and the obtained mechanical properties of the alloys.

The diagrams in *Figures 1–6*, on the other hand, show the influence of the ageing time at temperatures 150°C and 180°C on the mechanical properties of the alloys. The structure of the alloys was observed with the TESLA electron microscope. The requirements of PN-79/H-88026 standard for the PA 47 alloy had to be considered at the beginning of the experiment.

$$R_m = 360 \text{ MPa}, R_{0,2} = 280 \text{ MPa}, A_5 = 10\%, \text{HB} = 100$$

It must be pointed out, that in the examined range the values of hardness HV and HB are to a large extent comparable.

Results and discussion.

Mechanical properties of alloys after a single-stage ageing are listed in *Table 2*, and the diagrams in *Figures 1–6*, show their changes according to ageing time.

After the analysis of the diagrams, it was found that the most favourable mechanical properties were obtained in the alloy containing 1.82% of lithium. The

*Tel.: +12 76 4088/776672; fax: +12 76 4776

Table 1 Chemical composition of the examined alloys wt.%

Alloy symbol	Li	Zn	Mg	Mn	Fe	Si	Cu	Zr	Ti
1.1	1.15	4.60	1.15	0.20	0.15	0.18	0.12	0.14	0.03
1.8	1.82	4.50	1.20	0.26	0.22	0.10	0.07	0.15	0.04
2.5	2.50	4.55	1.15	0.22	0.20	0.15	0.08	0.14	0.04

highest properties in this alloy, i.e. $R_m = 530$ MPa, $R_{0,2} = 460$ MPa, $A_5 = 3\%$ were acquired after ageing at a temperature of 150°C for 48 h. High properties for 1.8 were also obtained after ageing at a temperature of 180°C for 8 h ($R_m = 517$ MPa, $R_{0,2} = 440$ MPa, $A_5 =$ 3,0%). The prolongation of the ageing time at the temperature of 180°C caused a decrease in the mechanical properties of the alloys.

Almost all variants of the applied heat treatments produced much higher mechanical properties than the required ones in the standard for the PA 47 alloy. On the other hand the lower elongation was observed, when compared with the values for the PA 47 alloy.

The course of the ageing curves, and the obtained results of mechanical properties after single-stage ageing and the observation of the structure in the electron microscope showed that the increase of hardness and mechanical properties of alloy at temperatures lower than 150°C was caused by precipitation liberation of GP zones and then of the transient phases η'. However, the δ' (Al_3Li) phase and not the η' phase contributed mainly to the increase of the strength of alloys at temperatures higher than 150°C, because the growth of the η' phase, connected with the loss of coherence takes place at these temperatures. Probably, the highest strength of alloys acquired at 150°C was the result of the formation of hardening precipitates (η' and δ' phases). But this requires further detailed observations.

Some tests of two-stage heat treatment (ageing) were undertaken to create favourable conditions for the phases η' and δ' formation and to achieve in this way high and favourable properties of the alloys. Parameters of the heat treatment and the results of these tests are presented in *Table 3*.

Similarly as before, the 1.8 alloy has the highest strength (450–520 MPa) at the high yield point (360–430 MPa) and elongation (4.5–6.5%). Therefore as a result very favourable properties were achieved.

The highest yield point was achieved in the 2.5 alloy, which at the lower tensile strength in comparison with the 1.8 alloy and as a consequence of a small reserve of plasticity, led to low values of elongation (1.5–2.5%).

It can be generally said that the change of temperature of the first stage of ageing and changes of ageing times did not influence the change of mechanical properties in a decisive way.

In the structures of alloys, which were two-stage aged, both hardening phases were observed, the η' ($MgZn_2$) phase was coagulated and distributed irregularly, quite opposite to the δ' (Al_3Li) phase, which, though small and homologously distributed, determined the strength of the alloys (*Figures 7–9*).

Table 2 Mechanical properties of alloys after single-stage ageing (Hyperquenching 530°C/45' in cold water)

Parameters of heat treatment	Alloy symbol	R_m [MPa]	$R_{0,2}$ [MPa]	A_5 [%]
150°C/8 h	1.1	434	330	6.5
	1.8	487	385	4.6
	2.5	425	379	2.9
150°C/12 h	1.1	414	300	6.3
	1.8	490	409	4.0
	2.5	430	380	2.5
150°C/24 h	1.1	443	350	6.5
	1.8	487	374	5.5
	2.5	456	371	5.0
150°C/48 h	1.1	458	362	7.4
	1.8	530	467	3.2
	2.5	471	435	2.0
180°C/8 h	1.1	428	348	6.5
	1.8	517	440	3.0
	2.5	448	400	1.5
180°C/12 h	1.1	357	254	8.5
	1.8	478	382	5.0
	2.5	489	444	2.5
180°C/24 h	1.1	362	240	8.5
	1.8	467	367	5.5
	2.5	435	415	2.0
180°C/48 h	1.1	355	255	8.2
	1.8	463	352	6.2
	2.5	446	425	2.5

Table 3 Mechanical properties of alloys after two-stage ageing (hyperquenching 530°C/45′ in cold water)

Parameters of heat treatment	Alloy symbol	R_m [MPa]	$R_{0,2}$ [MPa]	A_5 [%]
90°C/8 h + 180°C/12 h	1.1	365	256	8.0
	1.8	509	412	5.0
	2.5	447	418	1.5
90°C/12 h + 180°C/18 h	1.1	365	262	9.3
	1.8	515	433	5.0
	2.5	470	425	2.0
120°C/8 h + 180°C/12 h	1.1	390	278	8.0
	1.8	448	358	6.5
	2.5	461	435	2.0
120°C/12 h + 180°C/18 h	1.1	397	272	7.7
	1.8	490	376	5.5
	2.5	476	407	2.5
150°C/8 h + 180°C/12 h	1.1	364	247	8.4
	1.8	522	413	4.2
	2.5	482	412	2.6
150°C/12 h + 180°C/18 h	1.1	380	252	8.5
	1.8	512	430	4.5
	2.5	486	432	2.5

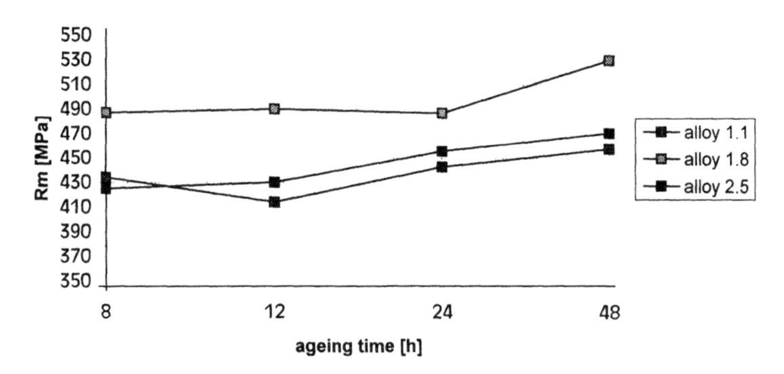

Figure 1 Influence of ageing time in the temperature of 150°C on tensile strenght of the tested alloys

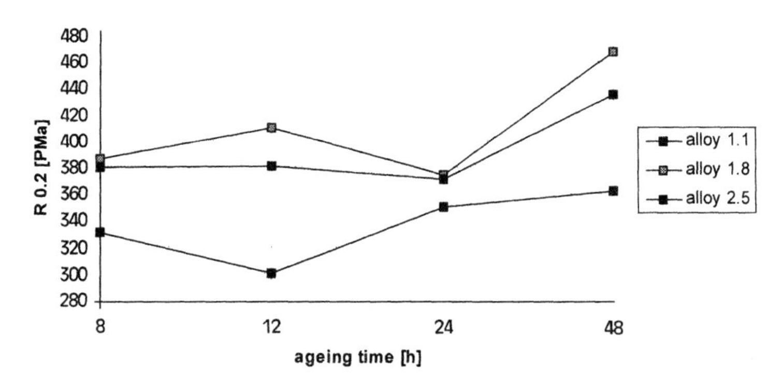

Figure 2 Influence of ageing time in the temperature of 150°C on the yield point of the tested alloys

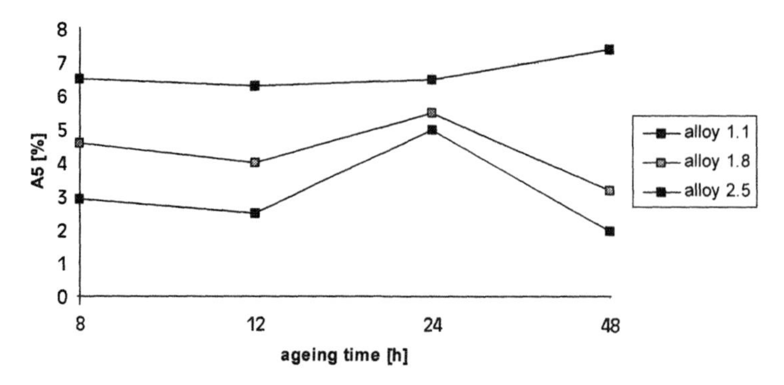

Figure 3 Influence of ageing time in the temperature of 150°C on the elongation of the tested alloys

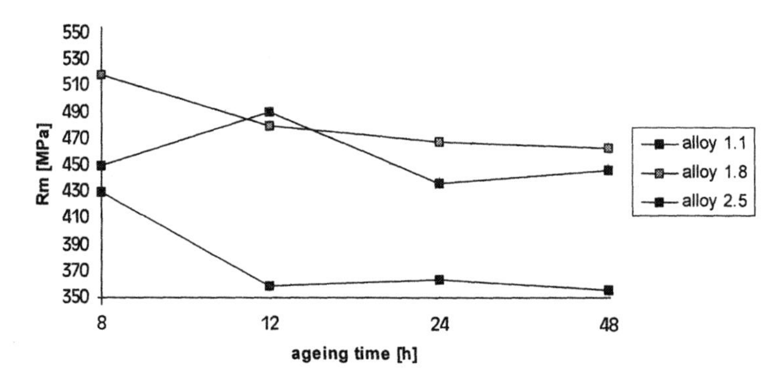

Figure 4 Influence of ageing time in the temperature of 180°C on tensile strength of the tested alloys

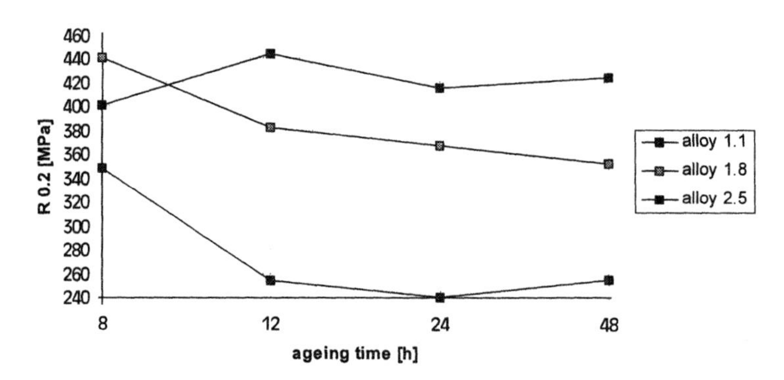

Figure 5 Influence of ageing time in the temperature of 180°C on the yield point of the tested alloys

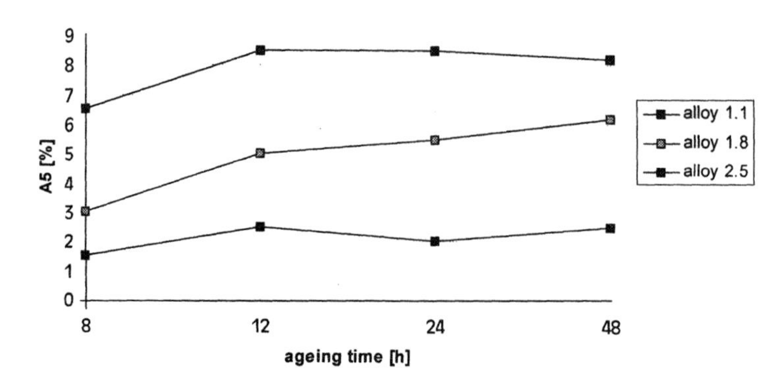

Figure 6 Influence of ageing time in the temperature of 180°C on the elongation of the tested alloys

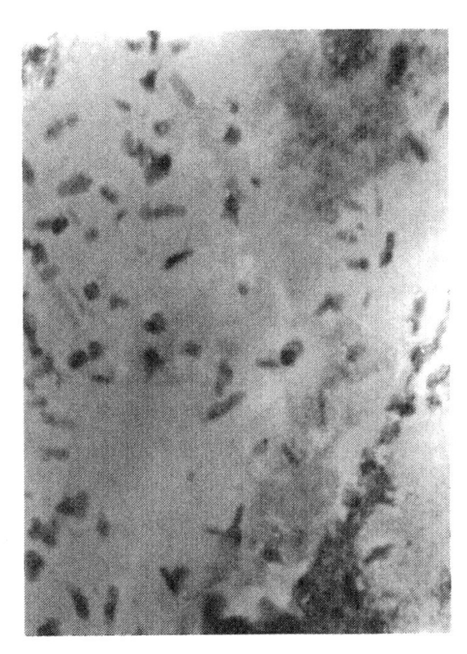

Figure 7 Microstructure of the 1.8 alloy, two-stage aged at 150°C/12 h. Magnification: 24,000 ×

Figure 8 Electron diffraction obtained from the area visible in *Figure 7*

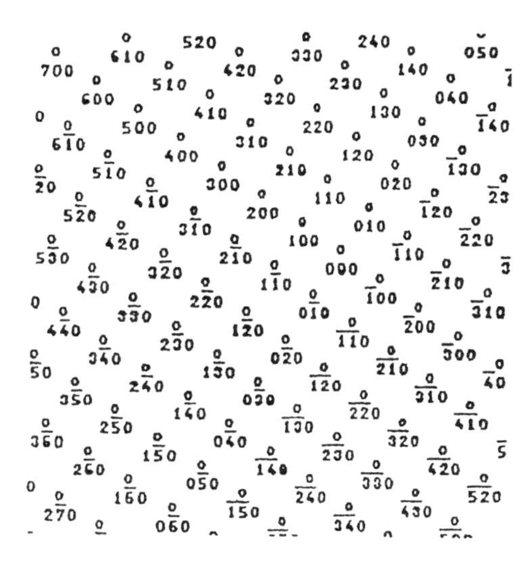

Figure 9 Solution of the electron diffraction shown in Fig. 8, obtained from the phase Al_3Li

Conclusions

(1) After heat treatment the most favourable mechanical properties were obtained for the alloy which contained 1.82% of lithium: tensile strength $R_m = 510–530$ MPa, yield point $R_{0,2} = 440–460$ MPa, elongation $A_5 = 4\%$.

(2) Increase of lithium content above 2% decreases considerably the elongation of the alloy, while the tensile strength at the level obtained at lower lithium content is maintained.

(3) It results from the observation of the structure, that two phases occur in AlZnMgLi alloys, during the ageing process. At ageing temperatures below 150°C, the η' ($MgZn_2$) phase dominates, but at higher temperatures above 150°C, the increase in the alloy strength properties is caused by the δ' (Al_3Li) phase.

Materials & Design, Vol. 18, Nos. 4/6, pp. 309–313, 1997

0261-3069/98 $19.00 + 0.00

PII: S0261-3069(97)00070-8

A new corrosion protection coating system for pressure-cast aluminium automotive parts

H. Schmidt*, S. Langenfeld, R. Naß

Institut für Neue Materialien gem. GmbH, Im Stadtwald, Geb 43 A, D-66 123 Saarbrücken, Germany

Received 22 July 1997; accepted 26 August 1997

A simple to employ corrosion protecting coating based on sol-gel derived new nanocomposites for Al alloys has been developed. This composite coating reacts to the Al surface by formation of a thermodynamically stabilized interface. Moreover, for pressure cast Al, the coating process can be used for sealing the pores at the same time.

Keywords: corrosion protecting coating; sol-gel derived nanocomposites; al alloys

Introduction

Corrosion, in general, is an undesired reaction and affects various materials, such as metals, glasses or ceramics. Whereas in metals, corrosion is mainly combined with an oxidation reaction, on glasses corrosion takes the form of a dissolution of the glassy network. The initial reaction in this case, in general, is an ion exchange where alkaline ions are exchanged against protons[1]. In ceramics, the corrosion mainly takes place in the grain boundaries as a first step followed by a degradation of the microstructure dissolution of the ceramic materials, especially in hotmelts. A simple way for preventing any corrosion would be a hermetic sealing of the surface by an appropriate coating with materials completely stable against diffusion of corrosive molecules, such as water, oxygen, acids or bases or a combination of these. Only these properties can be provided by purely inorganic materials like metals, glasses or ceramics or a combination of them. However, two difficulties appear. First, these coatings have to be brought onto the surface in a pinhole-free way, for example, by liquid metal coating (such as zinc plating) or dip coating into glass melts. Coatings grown upon the surface by electrochemical processes or anodizing, in general, do not have completely pinhole-free surfaces. On the other hand, the cited coating techniques cannot be employed in many cases due to the temperature sensitivity of the substrate or the properties of the coating, for example, the brittleness in the case of ceramic or glass coatings.

So the idea came up to combine inorganic and organic properties, as shown by Dow Corning in 1992[2], the principle of which is shown in *Figure 1*.

The results show that aluminum does not show any corrosion even after 1000 h of salt spray test but as clearly can be seen from *Figure 1*, this process is costly

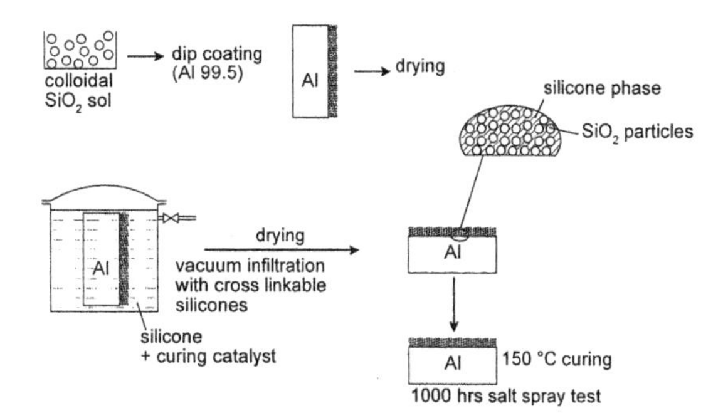

Figure 1 Schematics of a nanocomposite coating on aluminum

since it provides a series of different steps. So it never had a breakthrough on the market, but the basic idea to use inorganic–organic composite materials to combine both diffusion barrier and, as a result from the inorganic network, the required flexibility, for example, on coatings on metals, is worthy of investigation. In this paper, the development on an inorganic–organic composite materials for corrosion protection of aluminium alloy surfaces is shown. Inorganic–organic composite materials, meanwhile, have a interesting history and are employed in many other areas[3–5].

Basic principle and thermodynamical consideration

The synthesis of materials by the so-called sol-gel process[6] opens up the possibility of the fabrication of inorganic–organic polymers, which are multifunctional materials. This allows the incorporation of different functions, which is of interest for making tailored materials for protective coatings. The basic principles of such structures are shown in *Figure 2*.

*Correspondence to Dr H. Schmidt, Tel.: +49 681 3025013/14; fax: +49 681 3025223; e-mail: schmidt@imn-gmbh.de

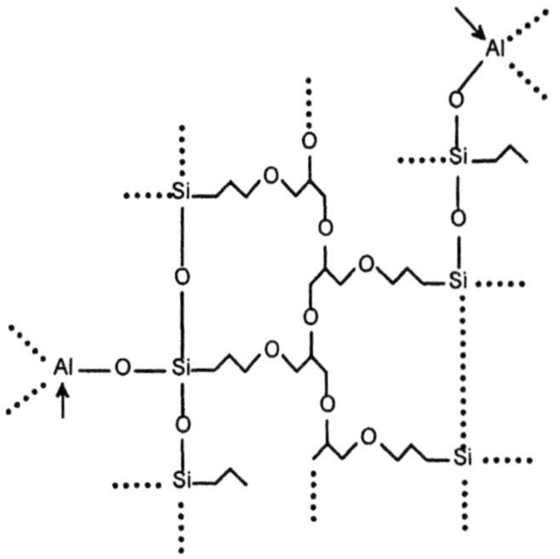

Figure 2 Structural principles of inorganic-organic composite materials

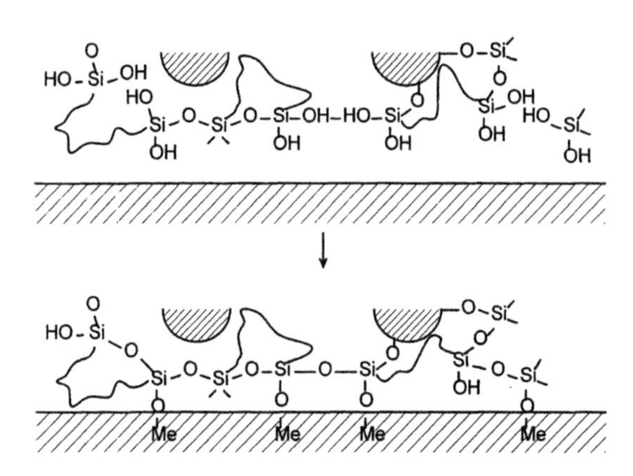

Figure 3 Concept of the interface design for corrosion-protective coatings on aluminium

Due to the fact that the described network in *Figure 2* can be modified by additional components, for example, components reactive to aluminum surfaces in the form of SiOH or SiOR groups, which then form stable bonds to the surface, these composite systems can be used as an approach for corrosion protection. It is postulated that the formation of Si–O–Al links should be thermodynamically favored. In *Table 1*, data for the free energy for several Si-O–Al containing compounds are given.

The interesting fact is that the alumina silica mixed compounds altogether show much lower free energy values than the boehmite, which is one first step of the oxidation of a aluminum surface in the presence of moisture. This leads to the concept that if it is possible to form alumina/SiO_2 interfaces, they should be stable against corrosion.

The resulting concept of this approach is shown in *Figure 3*.

The dashed areas represent nanoparticles. These nanoparticles seem to be suitable to be incorporated into the inorganic–organic backbone in order to provide mechanical stability, as already shown elsewhere[7]. In combination with the thermodynamical stabilization of the surface, another thermodynamical effect also can be taken into consideration, which is closely linked to the structure of these inorganic–organic composite materials. As depicted in *Figure 4*, if organic polymers only are used as a coating, in the case that water is penetrating into the interface and oxidizes the aluminium, due to the flexibility of the organic coating, an interfacial corrosion layer can be formed in the voids created by the organic network flexibility. In opposition to this (right side of the drawing), if an inorganic backbone is formed, which creates much more inflexibility, even in the case that one of the Si-O–Al bonds is hydrolysed, the 'corrosion product', in this case OH groups, cannot be transported away and according to the law of mass actions an equilibrium is produced between a hydrolyzed and an unhydrolysed bond. So the interface is 'immobilized' and the corrosion cannot propagate.

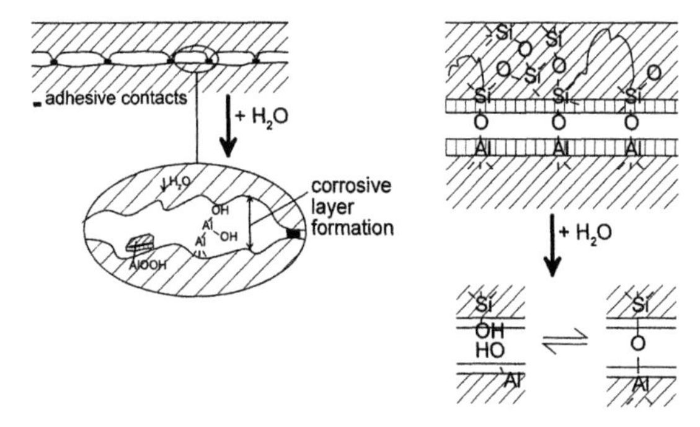

Figure 4 Differences in the interfacial behavior of a 'flexible' metal-to-polymer interface and an 'immobilized' composite interface

Table 1 Free energy values of several alumina and silica compounds

Thermodynamics
Gibbs free energy of alumina and silica modifications and alumo-silicates

Type		$213\Delta_{Gf}$(kJ mol^{-1})
Mullite	$3Al_2O_3 \cdot 2SiO_2$	−6901
Sillimanite	$Al_2O_3 \cdot SiO_2$	−2616
Cormierite	αAl_2O_3	−1675
Boemite	AlOOH	−985
Quartz	SiO_2	−910
AlOH + HOAl → Al-O-Al + H_2O	$\Delta H = -48$ kJ mol^{-1}	
AlOH + HOSi → Al-O-Si + H_2O	$\Delta H = -462$ kJ mol^{-1}	

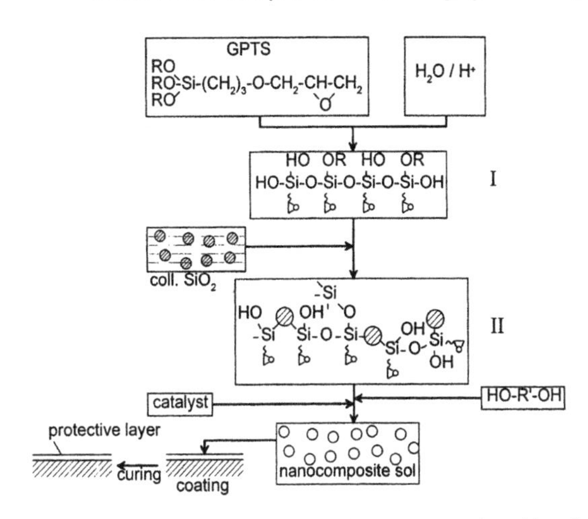

Figure 5 Schematics of the synthesis of a nanocomposite sol based upon GPTS, nano-scaled SiO_2 particles and sols, application process of these liquid sols onto metal surfaces and curing to protective layers

Experimental

The schematics of the experimental is shown in *Figure 5*. At first, the pre-hydrolyzation of 3-glycidyloxypropyltrimethoxysilane (GPTS) was started by adding acidic water (1–3 mol of water for every mol of silane). This mixture was stirred for 2–24 h at room temperature. A colloidal solution of SiO_2 particles (5–15 wt.% SiO_2 related to the amount of silane) either in 2-propanole or in water was added to the solution. After 10–60 min of continued stirring, a solution of a diolic compound (e.g. bisphenol-A, bisphenol-S) either in ethanol or in ethyl-isopropyl-ether was added. The molar ratios of diol: GPTS were varied between 20–40 mol%. Finally the synthesis was finished by addition of an alkaline catalyst (e.g. *N*-methylimidazole; 2–5 wt.% related to the amount of GPTS) as initiator for the organic crosslinking reaction. Pre-cleaning, i.e. degreasing, of the aluminum samples was carried out at 60°C for 2 min in an ultrasonic bath using a commercially available alkaline solution (0.8 wt.% of Metax R-1570 in deionized water). Immediately afterwards the samples were rinsed with deionized water for 1 min and dried at 80°C for 10 min in an electric oven. The process is carried out in a way that an organoalkoxysilane (GPTS) is reacted to an oligomeric compound (I) and filled with SiO_2 nanoparticles to obtain a ligand nanocomposite (II). A catalyst to crosslink the epoxy grouping is added, and, after coating, the crosslinking process takes place at $T = 100–150°C$.

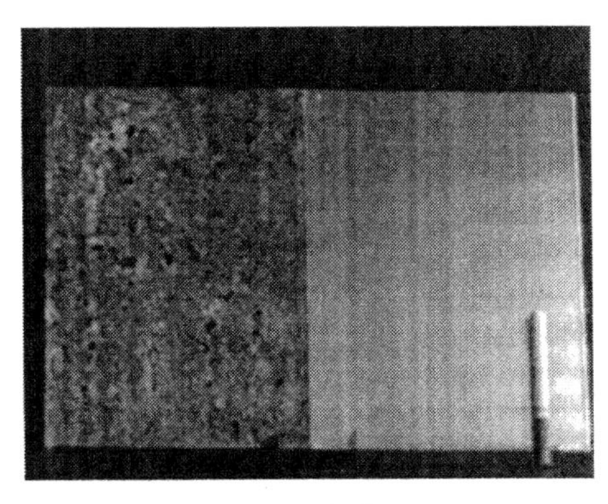

Figure 7 Comparison of uncoated (left) and nanomer-coated (right) Al 99.5 after 120 h CASS test (salt spray, copper chloride, acetic acid, pH 3); film thickness: 12 μm; to demonstrate the proportion, a cigarette was placed on the right

Results

Systems prepared according to the experimental procedure can be as so-called nanocomposite sols and used for coating techniques on aluminum. The coating thicknesses used in these cases are in the range of several μm (5–10 μm). This is a very thin coating compared to organic polymer coating, but due to the high scratch-resistance, the mechanical stability of these coatings is in the range of the best numbers known from anodizing processes, and it is expected that, due to the thermodynamical stabilization of the interface, even this thin type of coatings provides sufficient corrosion protection. After coating and curing, the effect of the coatings has been tested by CASS tests. In *Figure 6*, two

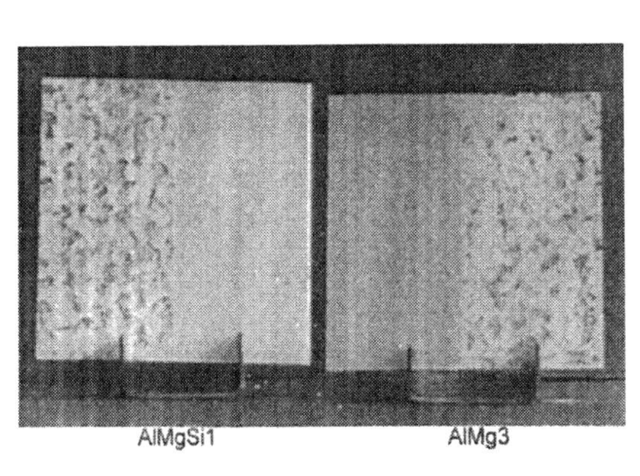

Figure 6 Comparison of uncoated and nanomer-coated aluminum alloys after 120 h CASS test (salt spray, copper chloride, acetic acid, pH 3); AlMgSi1 (left) is half-coated on the right; AlMg3 (right) is half-coated on the left; coating thickness: 15 μm; corrosion can only be noticed in the uncoated area

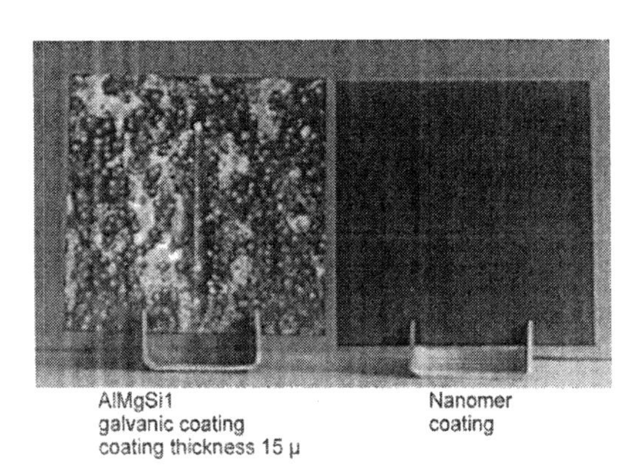

Figure 8 Comparison of commercial protective coatings (left: galvanic coating; coating thickness: 15 μm) and nanomer-coatings (right: film thickness: 10 μm) on AlMgSi1 after 240 h CASS test (salt spray, copper chloride, acetic acid, pH 3)

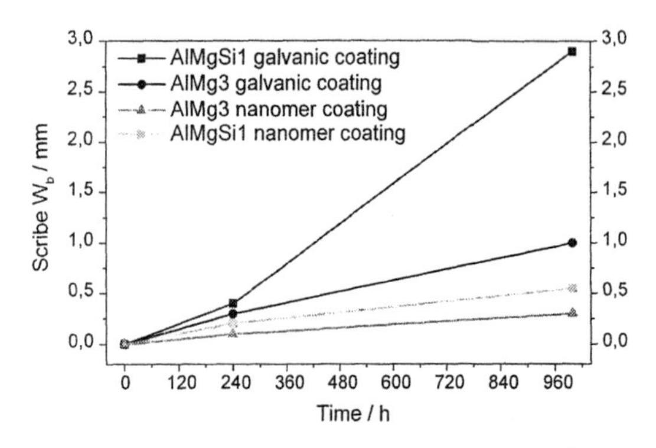

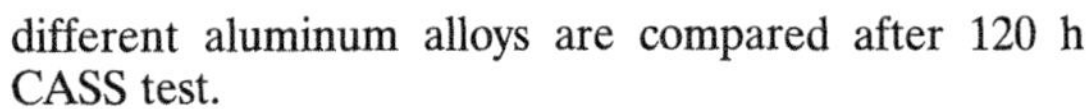

Figure 9 Comparison of commercial protective coatings (galvanic coatings; thickness: 15 μm) and nanomer-coatings (thickness: 10 μm) on different aluminum alloys (AlMgSi1, AlMg3) concerning scribe expansion versus time after salt spray test according to German standard DIN 50021

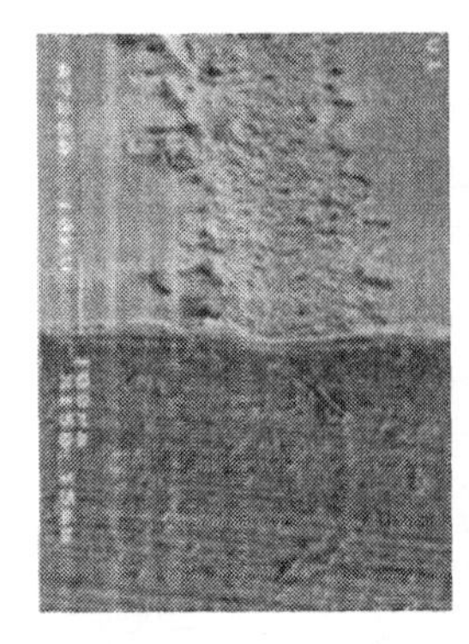

Figure 10 Electron microscopy of a nanomer-coated aluminum sheet (Al 99.5) at an artificially generated scratch (width: 200 μm) after 1000 h salt spray test according to German standard DIN 50021; film thickness: 10 μm; no scribe at the coating-metal-interface can be noticed

different aluminum alloys are compared after 120 h CASS test.

The experiments clearly show the high protective power in this type of coating. In *Figure 7*, 99.5 aluminum is shown also in the 120-h CASS test, and no corrosion is observed. In *Figure 8*, the 240-h CASS test is shown, and it also clearly depicts the good corrosion resistance.

In *Figure 9*, different coatings are compared by the scribe expansion vs. time. The coatings are compared to galvanic coatings (anodizing processes), and it clearly shows the much better corrosion protection. In *Figure 10*, it is shown that even after a 1000-h salt spray test (DIN 50 021), no scribe expansion can be observed. This is consistent with the hypothesis that due to the thermodynamic stabilization, no damage of the interface, even if the scribe goes through to the substrate, occurs. Based on these investigations and results, a process has been developed to coat pressure-cast aluminum alloys from recycling aluminum with traces of manganese, copper, iron, magnesium, nickel, lead, tin, titanium and zinc with a silicon content between 10.5 and 13.5%. These type of alloys, in general, show an undesired porosity after casting. The developed process is shown in *Figure 11*. It shows that it is possible to seal and coat the aluminum parts in one and the same step by vacuum infiltration. The obtained corrosion-protective properties are the same as described above. So, summarizing, one can say that this new type of corrosion-protective system, which easily can be employed, shows overall properties equal or superior to anodizing processes. The advantage is that, as also shown in the last figure, no chromatizing or phosphatizing processes are required. In addition to this, the coatings show excellent abrasion-resistant properties. For example, the Taber abrader test (1000 cycles, 500 g, Al_2O_3 rubber wheel) leads to hardly visible damage.

Conclusion

The investigations show that it is possible by using a thermodynamical concept of stabilized interfaces to employ highly effective corrosion-protective coatings

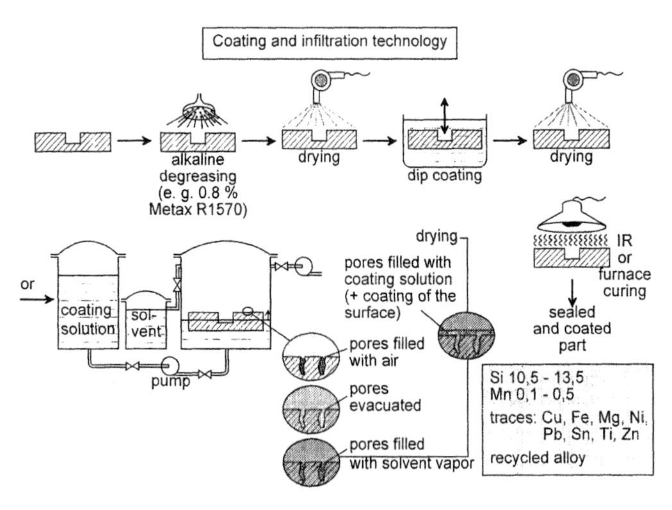

Figure 11 Schematics of cleaning, coating, sealing and curing of porous pressure-cast aluminum parts; cleaning was carried out easily only by alkaline degreasing; coating and sealing can be achieved by an one-step vacuum infiltration process

on aluminum alloys by simple wet-coating techniques. Due to the fact that the required coating thicknesses are rather low, it is possible to end up with low-cost highly effective coating techniques. The concept is not only restricted to aluminum, but has the potential to be employed to various other metals.

References

1 Scholze, H., *Glas: Natur, Struktur und Eigenschaften.* Springer, Berlin, 1988

2 Gentle, T.E. and Baney, R.H., *Materials Research Society in Symposium Proceedings*, 1992, p. 274

3 Kasemann, R., Schmidt, H. and Wintrich, E., *Materials Research Society in Symposium Proceedings*, 1994, **346**, 915–921 (1994)

4 Schmidt, H., Relevance of sol-gel methods for synthesis of fine particles. *KONA Powder and Particle*, 1996, **14**, 92–103

5 Schmidt, H., Entwicklung, Mikrostrukturierung und Anwendung von Keramik-polymer-Nanokompositen. In *Proceedings Werkstoffwoche 1996, Symposium 9.* DGM-Informationsgesellschaft mbH (in print)

6 Brinker, C. J. and Scherer, G. W., *Sol-Gel Science: the Physics and Chemistry of Sol-Gel Processing.* Academic Press, Boston, 1990

7 Geiter, E., PhD Thesis, University of Saarland, Saarbrücken, 1997

ELSEVIER

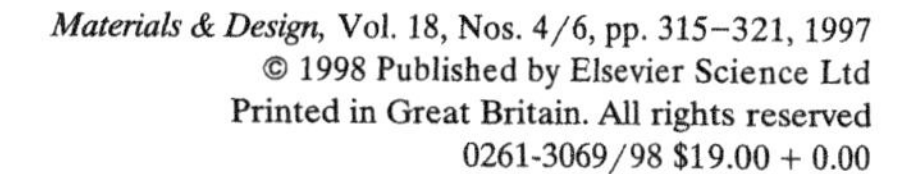

Materials & Design, Vol. 18, Nos. 4/6, pp. 315–321, 1997

0261-3069/98 \$19.00 + 0.00

PII: S0261-3069(97)00071-X

Lightweight near net shape components produced by thixoforming

G. Hirt[a,*], R. Cremer[a], T. Witulski[a], H.-C. Tinius[b]

[a]*EFU GmbH, Jagerhausstrasse 22, 52152 Simmerath, Germany*
[b]*Krupp-Gerlach GmbH, Homburg/Saar, Germany*

Received 18 June 1997; accepted 18 July 1997

Recent thixoforming development work at EFU has been concentrated on: (1) improved machine concepts for serial production, (2) die design and prototype manufacturing from various light metals including magnesium and metal matrix composites, (3) numerical simulation for process optimization, (4) accessing process benefits through the appropriate use of design options (specially designed steering knuckle demonstrator). A pilot thixoforming system consisting of four EFU heating modules with patented sensor systems, an ABB robot for billet handling and a Frech 5,8 MN real time controlled high pressure die casting machine has been used to produce various prototype parts. The process parameters have been examined with incomplete die filling experiments and numerical simulation. To exploit the full potential of the thixoforming process, a steering knuckle for a compact passenger car has been redesigned, thixoformed and tested. The weight of the new part is approx. 50% below that of a conventional forged steel design, despite identical functional capabilities.

Keywords: thixoforming; developments; concepts; production

Introduction and state of the art

Thixoforming has come to play an increasing role as an alternative to classic manufacturing techniques such as casting and forging. One reason for this is the growing demand for high-strength aluminium components for lightweight automotive designs. A necessary condition for the manufacture of low-cost lightweight aluminium parts is the ability to process this expensive material into near net shape components of complex geometry and high strength. Thixoforming provides the engineer with a vastly broader range of design options than forging—possibly even broader than conventional casting processes, for instance, when it comes to wall thicknesses and wall thickness variations. With manufacturing costs now on a level with high-grade casting applications, high mechanical component qualities can be achieved, and applications in the field of highly-integrated safety critical components have become possible.

The basic principles of the process, which consists of the feedstock production, billet reheating and semi-solid casting operations, are described in detail in refs.[1–3] and elsewhere in the literature. In the production of the feedstock (usually by continuous casting with electromagnetic stirring), the objective is to achieve a globular structure of maximum grain fineness. In this context EFU has gathered positive results with a variety of aluminium casting alloys, wrought alloys and MMCs in the 76–200 mm diameter range[4,5].

*Correspondence to Dr G. Hirt, Tel.: +24 73 601548; fax: +24 73 601545, e-mail: efugmbh@aol.com

For reheating the billet (previously cut to a precise length) to a temperature in the solidus-liquidus interval, inductive heating has evolved into the method of choice since it is best suited to achieve quick, uniform and accurate heating results. In addition to traditional revolving type furnaces, this is achieved increasingly with modular-type units in which the billet remains in one coil during the entire heating cycle[6].

Most specimen parts and authentic application components produced in the world to date are still made from AlSi7Mg alloy. This material can be thixoformed into components weighing from under 10 g to more than 10 kg (e.g.[7,8]); products in the medium size range have already been manufactured in high-volume applications exceeding 2 million units per year, especially in the US. Apart from a reliable process technology, which should provide real-time control of the injection curve, it will become necessary in future applications to include simulation tools for optimizing the injection parameters and die design.

Plant technology

Billet reheating system

In the modular furnace type preferred by EFU, the billet is not transferred from one coil to the next but remains in the same individually controlled coil for the entire heating cycle. Depending on the required cycle time, it is possible to install between 1 and 16 coils. Since this design requires each heating module to have its own frequency converter, the recent availability of low-cost converters with 'intelligent' control functions and large control ranges was a prerequisite for an economically feasible implementation of this concept[9], which provides maximum precision and flexibility. As an additional advantage, EFU's patented sensor technology[6] to control the softness of the billet can be included into the system. This closed loop control system increases the efficiency of the total plant significantly, because it is now possible to maintain each billet in its current state in the case of an exterior interference (e.g. one affecting the die) and to continue the reheating cycle after the problem has been remedied. A pilot plant operating on this principle was installed at EFU in the spring of 1996 (*Figure 1*). This is a four-coil system capable of reheating billets measuring up to 100 mm in diameter and 250 mm in length.

A master control system coordinates the operation of the individual heating modules, the robot, and the die-casting machine. The reheating furnaces are operated on a menu-controlled basis using COROS (Siemens). The various menu levels can be used to select different operating modes (set-up, manual, production), to load and store the parameter sets for various billet dimensions and alloys, and to visualize the plant status and heating curves for each billet.

Casting plant technology

As with other forming methods, plant technology plays a crucial role in the thixoforming process. When using standard machines a reproducible injection curve adapted to individual component needs can best be achieved with a pressure die-caster with real time injection control. The DAK 500 DC RC machine used by EFU in cooperation with Oskar Frech Co. has a closing force of 580 metric tons (*Figure 1*); its piston rod movement is controlled by a quick-acting control valve arranged on the secondary flow side of the casting piston. At a typical V_{max} of 2.5 m s^{-1}, the high dynamic response of this valve allows an acceleration from 20–80% V_{max} (and conversely, a deceleration from 80 to 20% V_{max}) at a rate of 4 m s^{-1} [10]. This is sufficient in any case for meeting the requirements of a form-specific injection curve for thixocasting. The high pressure die casting machine, together with the heating system described in Sec. 2.1 and a handling robot provide a versatile system with modular features which allow a flexible equipment layout, depending on the required number of coils and floorspace availability (*Figure 2*).

Forming process

Process potential and demonstration components

In thixoforming, a semi-solid metal is injected into a closed die. In this condition, the reheated billet is stable similar to a solid as long as it is not stressed, but flows similar to a liquid when subjected to shear stresses. Compared to plastic deformation in solid state forging the semi solid metal has almost no flow restrictions. Thus very complex shapes with thin ribs, hollow cross sections, undercuts and difficult mass distribution can be formed in <u>one</u> forming step. In comparison to high quality casting methods, like squeeze-casting or vacuum die casting, the unique advantage of thixoforming is, that the solidification shrinkage is significantly reduced which gives the designer more freedom towards thicker cross sections and wall thickness changes. Other process advantages are reduced gas entrapment due to laminar flow and significantly reduced thermal die loading. Compared to castings thixoformed components exhibit none (or minimum) of porosity and inclusions, which results in high mechanical properties, pressure tightness and the possibility to apply thermal treatment. During the last few years at least the following four categories of components have

Figure 1 Four-coil pilot system for billet reheating (background) with ABB robot and Frech DAK 500 DCRC real time controlled high pressure die casting machine (left)

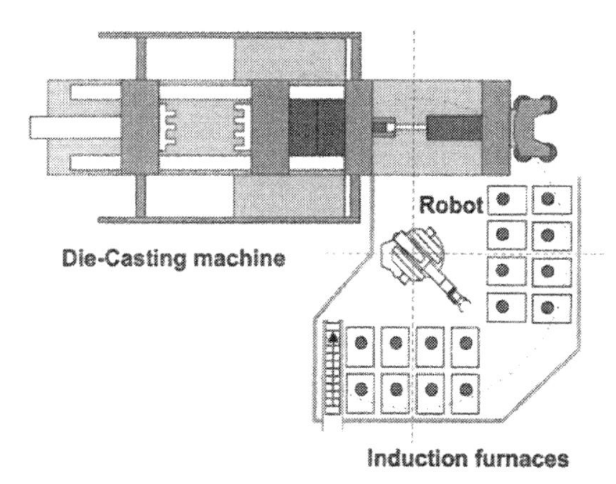

Figure 2 Typical plant layout

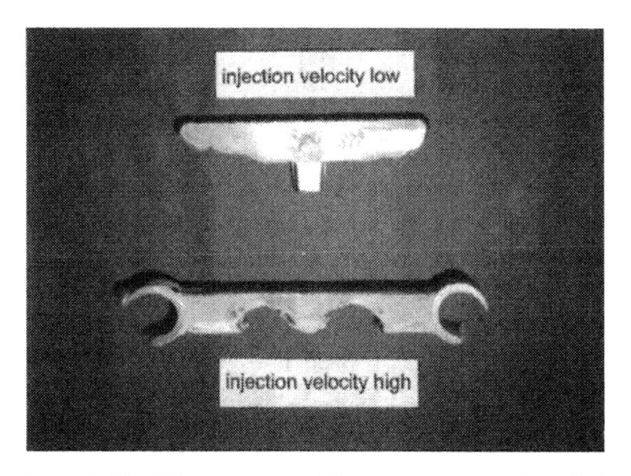

Figure 4 Die filling pattern at different casting injection velocitiesInjection velocity: low (top), high (bottom)

been identified, for which the thixoforming process may be of particular interest:

- Components subject to high pressures (e.g. brake cylinders)
- Thick-walled components subject to high loads (e.g. suspension parts)
- Thin-walled structural components (e.g. space frame nodes)
- Components made of special materials such as metal matrix composites, which are expensive, difficult to machine and known for their poor casting properties

In all of these categories EFU has manufactured a number of initial demonstration components (*Figure 3*) which were kept deliberately straightforward in some cases:

- The ribbed 'hat section' shown in *Figure 3*(1) has a wall thickness between 2.5 and 3 mm. This is a typical example of a thin-walled structural component which should ideally achieve high tensile and yield strengths even without heat treatment to avoid distortion

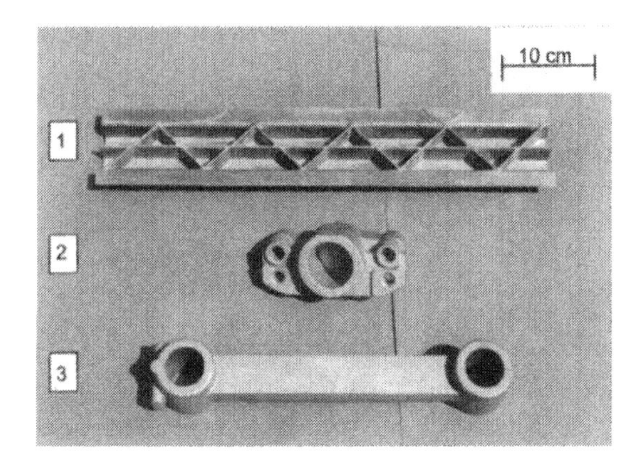

Figure 3 Demonstrator parts produced by thixoforming: (1) rib component, (2) wheel cylinder housing, (3) suspension arm

- The brake cylinder housing [*Figure 3*(2)] produced in cooperation with the Frech Co. is a typical pressure-loaded component. In this application thixoforming avoids the porosity of thick-walled component sections which is almost impossible to avoid with other casting techniques
- In the case of the suspension link [*Figure 3*(3)] it is possible to achieve wall thicknesses between 6 and 25 mm, depending on the mould insert selected. Typical design elements like wall thickness changes and meeting flow fronts behind sliding cores can be examined with this part

Simulation

The die design (gating system, overflows, cooling, venting) is still very much an empirical matter at this point, and compared to conventional processes the available experience is naturally limited. The same applies to the selection of optimum casting parameters (piston speed curve, point of changeover from speed to pressure control, die temperature, etc.). The importance of these parameters is clearly evident from the case of the simple suspension link shown in *Figure 3*(3). At a low piston speed, the die fills evenly from the inside out (*Figure 4*). If the piston speed becomes too high, an advance jet of molten metal is formed and may cause air inclusions when deflected by back pressure. The determination of optimum casting parameters may involve costly and time-consuming trial runs, especially with sophisticated component geometries.

In the meantime EFU has gathered some initial experience with the numeric simulation of the die filling process, which is expected to accelerate the development process and provide advance evidence of the feasibility of the die design. EFU performs these simulations using the Magmasoft FDM package. In the first simulations conducted at Magma Co., the material behaviour was described using the Oswald-deWaele model in which the viscosity is a function of temperature and shear-rate:

$$\nu = \nu(T; \dot{\gamma}) = m \cdot \dot{\gamma}^{n-1} \quad (1)$$

where ν = kinematic viscosity, $\dot{\gamma}$ = shear rate, T = temperature, m and n are temperature-related parameters

The calculations performed with material characteristics taken from the literature[11] showed a surprisingly good qualitative coincidence with the actual form filling process (cf. *Figure 5*). However, other comparison parameters such as the injection pressure and slurry temperature could not be correctly determined with these material characteristics. Therefore EFU has started to determine adequate viscosity values using a capillary-type viscosimeter[12].

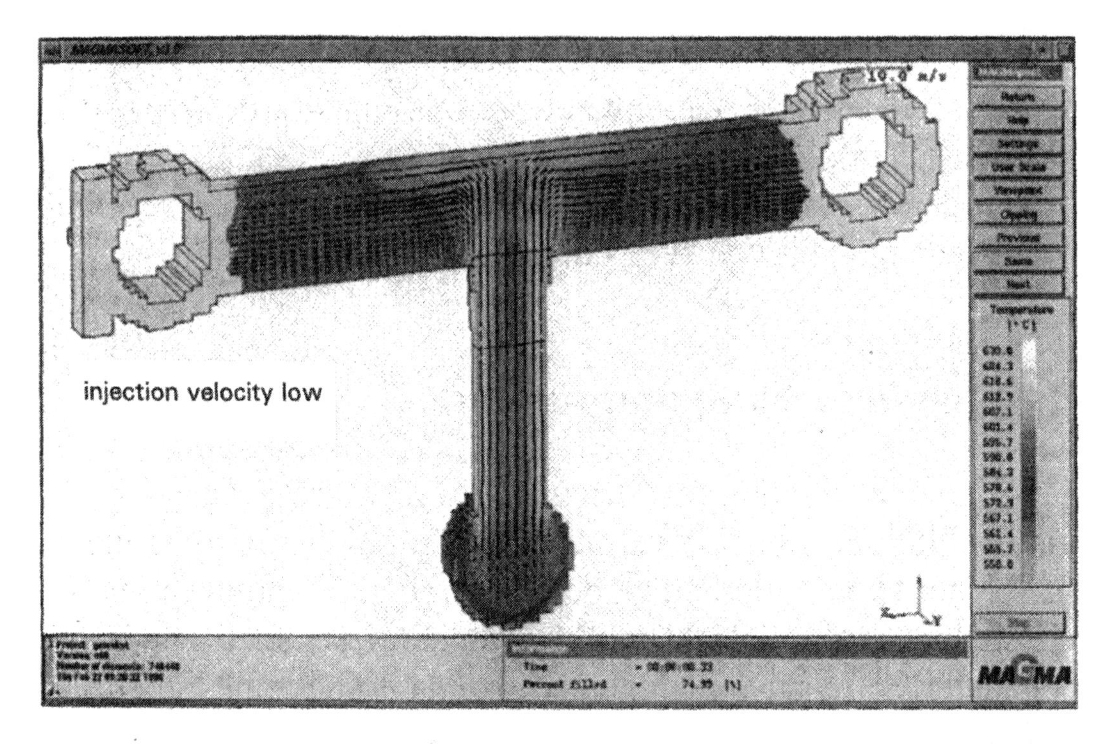

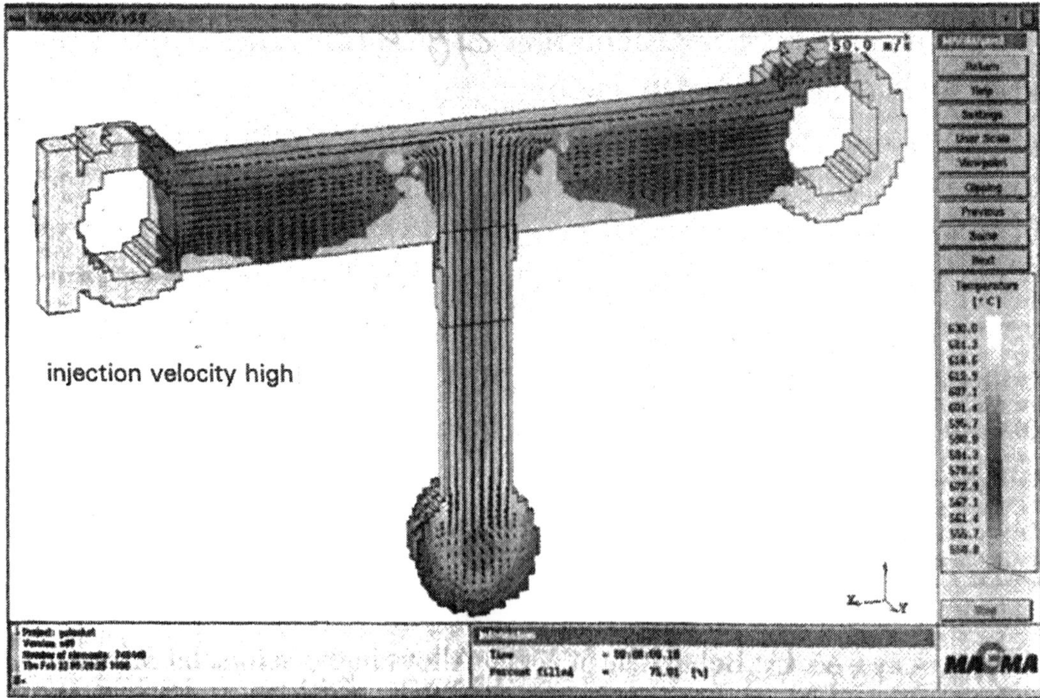

Figure 5 Die filling simulation results for the parameters used in *Figure 4*. Top: $\upsilon = 0.3\ \mathrm{m\ s^{-1}}$, bottom $\upsilon = 1\ \mathrm{m\ s^{-1}}$

Accessing process benefits through the appropriate use of the broader range of design options

In order to exploit the full potential of the thixoforming process, it is necessary to adopt a component design geared to the specific process and application requirements, while making use of the expanded range of design options offered by this technology. Thixoformed aluminium components are often intended to replace steel forgings or nodular cast iron components. Apart from a high yield strength and fracture toughness, the stiffness of the component will be an important boundary condition in this case—especially allowing for the fact that the modulus of elasticity of aluminium is only one-third that of steel. If there is a simultaneous need for weight reduction, the product must be redesigned by exploiting the typical thixoforming design options of combining thick-walled and thin-walled areas with stiffening ribs and undercuts.

A decisive step in this direction is the steering knuckle for a compact passenger car which was thixoformed by EFU under a contract with Krupp-Gerlach GmbH. The steering knuckle was chosen as an example to demonstrate the thixoforming potential. The conventional part is forged from a special C35 steel quality (Youngs' modulus $= 2.1 \cdot 10^5$ N mm^{-2}, yield stress $>$ 400 N mm^{-2}, tensile strength = 700–800 N mm^{-2}). On the basis of prescribed loading conditions, geometrical restrictions and given functionality the part was redesigned by Porsche Engineering (under a contract with Krupp-Gerlach GmbH) also considering the process-specific boundary conditions defined by EFU. For this redesign the following lower limit data were selected for thixoformed and heat-treated A 356 aluminium alloy:

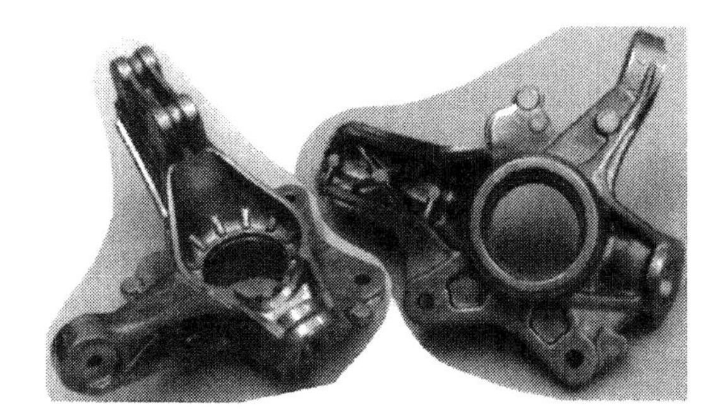

Figure 6 Thixoformed steering knuckle (A 356)

Youngs' modulus	$= 0.7 \cdot 10^5$ N mm^{-2}
Yield stress	> 200 N mm^{-2}
Tensile strength	> 250 N mm^{-2}

However, additional data were required for the life cycle analysis under the special dynamic load characteristics related to the application. For this purpose, specimens were prepared from available thixoformed components and cyclic stress strain curves were recorded to determine the Wöhler stress-cycle diagram[13].

The result of the redesign was the rather complex structure shown in *Figure 6*, which involves hollow profiles with stiffening cross ribs as well as thick-walled sections according to functional requirements. As an

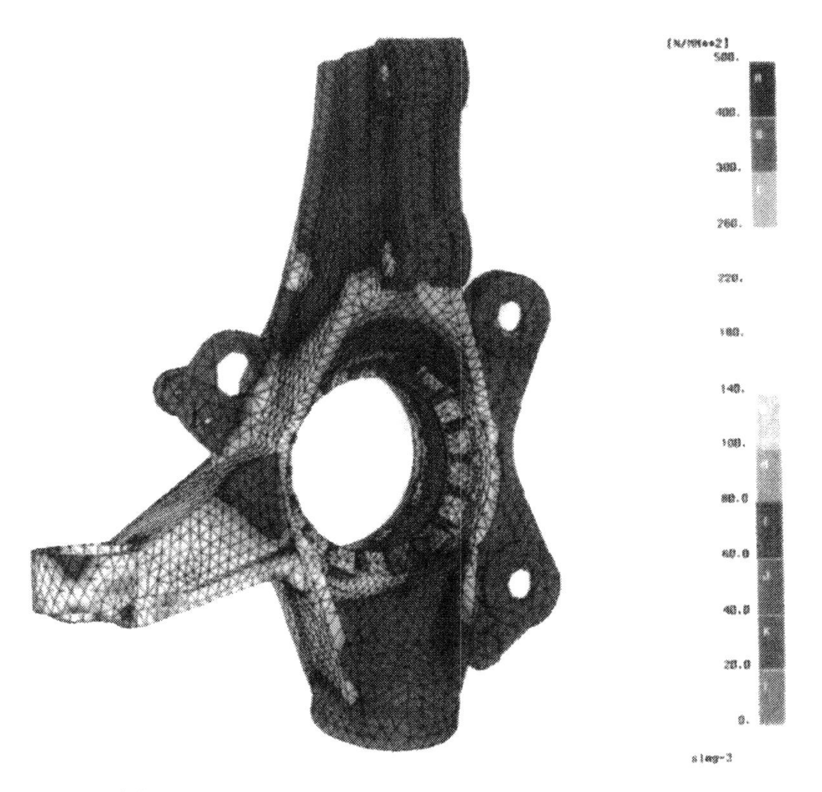

Figure 7 Finite element stress analysis

example *Figure 7* shows a Finite Element Stress analysis for a critical load case indicating that in this case large portions of the steering arm will be loaded up to the design limit.

After optimizing the die design and the forming parameters a larger number of test parts have been thixoformed, T6 heat-treated and machined to final shape. In some places small steel inserts were press-fitted according to surface hardness requirements at specific junction points. Finally the parts underwent extensive multiaxial testing according to the specified life cycle loading and proved to fulfil the specification. The weight of this part is approx. 50% below that of the forged steel part.

As mentioned above, in some cases steel may be required in certain areas to achieve sufficient surface hardness when substituting steel parts with Aluminium. Press-fitting of steel inserts however involves additional labour and costs. Therefore EFU successfully used a slightly modified die to demonstrate that thixoforming has the possibility to include steel inserts directly during casting. This additionally increases the design freedom for weight optimization.

Suitability of the thixoforming process for other aluminium alloys and magnesium

Apart from the standard thixoforming alloy AlSi7Mg, EFU has conducted trials with various other alloys. Several of the demonstration components mentioned earlier were manufactured from these alloys listed below and the experience gathered in these trials can be summarized as follows:

AlMgSi1. This wrought alloy can be thixoformed into products of very good mechanical properties[14]. However, its susceptibility to hot cracking and poor fusion of confluent melt fronts greatly restrict the application range.

AlSi17CuMg, AlSi25CuMg. Hypereutectic alloys are fairly well suited for the thixoforming process and may be of interest for components subject to wear (*Figure 8*).

Metal matrix composites. The production of feedstock from modified Duralcan material with subsequent thixoforming has been extensively investigated at EFU[15]. These materials can be processed very well by thixoforming, without the problems encountered with conventional casting methods. The suspension link, the rib type part, the brake cylinder (all shown in *Figure 3*) and some other components similar to brake disks were thixoformed from this alloy type. A low-cost future alternative might consist of materials with an increased content of TiB particles which are obtained by means of an in-situ process[16].

Magnesium. AZ 91 alloy feedstock can be produced by a suitable combination of chemical grain refinement and control of the solidification rate[17]. Billets of this type were sourced from North Hydro and the complex knuckle shape shown in *Figure 9* could be thixoformed from these billets.

Summary and outlook

In recent years, we have seen a continuous evolution in thixoforming equipment technology (billet heaters and casting machines) and accumulated experience through the manufacture of numerous specimen components. From the volume production applications realized to date it is evident that large production runs can be realistically implemented. In order to exploit the full potential of the process it will be necessary to demonstrate its reliability under volume production conditions (also with regard to safety-related components) and to consistently adopt component design principles accommodating the specific process and loading characteristics. Numeric simulations of the die filling process, although still flawed, will soon help to reduce development times and minimize die adapting costs.

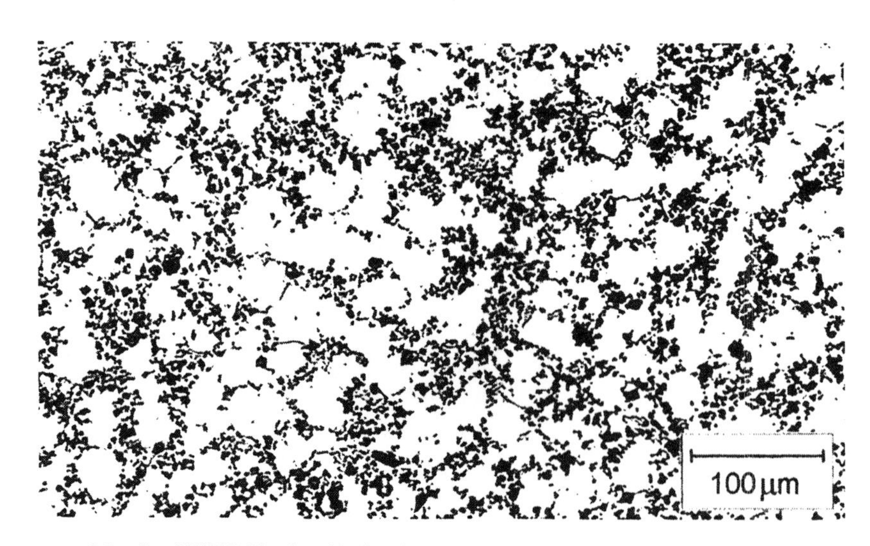

Figure 8 Microstructure of the alloy AlSi25CuMg after thixoforming

Figure 9 Magnesium demonstrator part (AZ 91)

References

1 Flemings, M.C., Behaviour of metal alloys in the semisolid state. *Metallurgical Transactions A*, 1991–957, **22A**

2 Kirkwood, D.H., Semisolid metal processing. *International Materials Reviews*, 1994, **39(5)**, 173

3 Hirt, G., Cremer, R. and Winkelmann, A., *Thixoforming—ein Verfahren mit Zukunft*[Thixoforming—a Process with a Future]; Symposium Neuere Entwicklungen der Massivumformung [Recent Developments in Solid Metal Forming], Fellbach near Stuttgart. DGM Conference Transactions: ISBN 3-88355-213-5, 1995, pp. 275–295

4 Zillgen, M., Hirt, G. and Engler, S., Vertikaler rheostrangguß von aluminium-legierungen [Vertical continuous rheocasting of aluminium alloys. *Metallurgical*, 1995, **49**, 808–813

5 Zillgen, M. and Hirt, G., Microstructural effects of electromagnetic stirring in continuous casting of various aluminium alloys. In *4th Int. Conf. on Semi-Solid Processing of Alloys and Composites*. Sheffield, England, Conference Transactions, 1996, pp. 180–186

6 Cremer, R., Winkelmann, A. and Hirt, G., Sensor controlled induction heating of aluminium alloys for semi-solid forming. In *4th Int. Conf. on Semi-Solid Processing of Alloys and Composites*, Sheffield, England, Conference Transactions, 1996, pp. 159–164

7 Chiarmetta, G., Thixoforming of automobile components. In *4th Int. Conf. on Semi-Solid Processing of Alloys and Composites*. Sheffield, England, Conference Transactions, 1996, pp. 204–207

8 Midson, S.P., The commercial status of semi solid casting in the USA. In *Int. Conf. on Semi-Solid Processing of Alloys and Composites*. Sheffield, England, Conference Transactions, 1996, pp. 159–164

9 Cremer, R., Hirt, G. and Schiele, J., *Verbesserte Anlagentechnik für die Bolzenerwärmung in das Solidus / Liquidus-Interval* [Improved equipment technology for heating billets to a temperature in the solidus/liquidus interval]. Elektrowärme International, 1996

10 Fink, R. and Blaesse, R., *Digitale Echtzeitregelung des Einpreßvorganges von Druckgießmaschinen*. Frech informiert Nr. 23, Schorndorf

11 Loué, W.R., Suéry, M. and Querbes, J.L., Microstructure and rheology of partially remelted AlSi alloys. In *Proc. 2nd Int. Conf. on the Semi Solid Processing of Alloys and Composites*. MIT, Cambridge, MA, USA, 1992, pp. 266–275

12 Hirt, G., Witulski, T. and Cremer, R., *Neuere Entwicklungen beim Thixogießen*. Fortschritte mit Gußkonstruktionen, 13.-14.03. Veitshöchheim, 1997

13 LBF-Report No 8295, *Dehnungsgeregelte Versuche mit bauteilentnommenen Proben aus AlSi7Mg0,3wa*. Darmstadt, January 1997

14 Hirt, G., Witulski, T., Cremer, R. and Winkelmann, A., *Thixoforming: Neue Chancen für Leichtbau in Transport und Verkehr* [Thixoforming: new opportunities for lightweight design in vehicle and transport engineering]. 10. Aachener Stahlcolloquium [10th Aachen Steel Symposium], Aachen, Conference Transactions, 1995

15 Hirt, G., Witulski, T. Zillgen, M., Semi solid forming of particulate reinforced alloys. In *4th Eur. Conf. on Advanced Materials and Processes*, Padua/Venice, Italy, Proceedings: ISBN 88-85298-22-2, 1995, pp. 125–130

16 Wood, J.V., McCartney, D.G., Dinsdale, K., Kellie, J.F.L. and Davies, P., Casting and mechanical properties of a reactively cast Al-TB2 alloy. *Cast Metals*, **8(1)**, 57–64

17 Haavard, G. and Hakon, W., *Method for production of thixotropic magnesium alloys*. European Patent Application No. 93109014.6, Publication Number 0 575 796 A1

Materials & Design, Vol. 18, Nos. 4/6, pp. 323–326, 1997
© 1998 Published by Elsevier Science Ltd
Printed in Great Britain. All rights reserved
0261-3069/98 $19.00 + 0.00

PII: S0261-3069(97)00072-1

Hypereutectic Al–Si based alloys with a thixotropic microstructure produced by ultrasonic treatment

V. O. Abramov[a], O. V. Abramov[b], B. B. Straumal[a,c,*], W. Gust[a]

[a]*University of Stuttgart, Institut für Metallkunde, Seestr. 75, D-70714 Stuttgart, Germany*
[b]*Institute of General and Inorganic Chemistry, Russian Academy of Sciences, Leninsky Prospect 31, Moscow 117907, Russia*
[c]*Institute of Solid State Physics, Chernogolovka, Moscow District 142432, Russia*

Received 17 June 1997; accepted 20 July 1997

The present investigation attempts to evaluate the effect of an ultrasonic treatment on the microstructure of hypereutectic Al–Si based alloys. In conventional casting solidified at a moderate cooling rate, the primary silicon crystallizes in the form of hexagonal plates joined together at the centre into star-shaped particles, as they appear in cross-section. During the ultrasonic treatment most of the silicon plates were disconnected and broken, forming spheroidized crystals. The ultrasonic treatment results in an increase of the plasticity and strength of the alloys. To study the thixotropic behaviour, ultrasonically treated and non-treated specimens were upset in an electrohydraulic press under semi-solid conditions. The investigation confirms great advantages of ultrasonically treated ingots of Al–Si based alloys upon deformation in the semi-solid state. © 1998 Published by Elsevier Science Ltd. All rights reserved.

Keywords: ultrasonic treatment; hypereutectic aluminium alloys; thixocasting; mechanical properties

Introduction

Spencer *et al.*[1] pioneered a study on the behaviour of metals and alloys in the semi-solid state at the Massachusetts Institute of Technology in 1971. Owing to the shear stress which acts on the semi-solid material during filling of the die, the viscosity of the material is reduced and the force required to fill even complex geometries is relatively low as compared to that needed for conventional processing. Depending on the material involved, the temperature level required to work the material is often significantly lower than in a conventional casting process, and the thermal stresses on dies and moulds are reduced[2,3]. The lower working temperature has also a favourable effect on the accuracy-to-size of the geometries and on the production of pore-free parts, since the solidification shrinkage is reduced due to a much less liquid fraction and the pressure maintained during solidification. This technology opens up completely new routes for processing innovative materials, e.g. particle- and fibre-reinforced composites, and for manufacturing composite or hollow parts. It is, however, questionable whether all the above mentioned potentialities of thixoforming can be exploited with aluminium alloys.

*Correspondence to B. B. Straumal. Tel.: +49 711 1211276; fax: +49 711 1211280; e-mail: straumal@vaxww1.mpi-stuttgart.mpg. de, straumal@song.ru

The future implementation of thixocasting in the industrial series production of high-grade components depends especially on the quality of the pre-material and efficiency of its production. In these processes, the alloys with a non-dendritic semi-solid structure can be used in order to obtain the desired fluidity and mechanical properties. The non-dendritic semi-solid structure can be produced by two routes which are: (i) a mechanical or magneto-hydrodynamic stirring, and (ii) an intensive ultrasonic treatment during solidification.

Systematic investigations to use ultrasound during the thixocasting have not been made up to now. The main aim of the present study is to investigate the possibility of use ultrasound as an effective alternative to electromagnetically produced thixotropic material. The possibility of obtaining a thixotropic structure by an ultrasonic treatment in Al–Si based alloys was investigated. Silicon is one of the few alloying elements that does not increase the density of the alloy. Due to the very high hardness of Si crystals dispersed in the matrix the wear resistance of Al–Si based alloys is very high. Aluminium and silicon constitute a simple binary eutectic system with the eutectic point at 12.2 at.% and 577°C. Most Al–Si based alloys being used are hypoeutectic, but hypereutectic alloys are attractive to the automotive industry and desirable for wear resistant applications, where high strength and low weight are also required.

The present investigation attempts to evaluate the

effect of an ultrasonic treatment on the microstructure of hypereutectic Al–Si based alloys.

Experimental

In our experiments, we used alloys with Si concentrations of 13.8 and 17.1 wt.%. The chemical composition of the alloys is given in *Table 1*.

At the higher silicon content, the alloy melting point increases due to the steep slope of the liquidus on the silicon-rich side. Primary silicon crystals coarsen substantially, particularly due to the wide solidification range of this alloy, and impair the fracture toughness of the casting. The fluidity of the alloys was also found to decrease beyond 18 wt.% Si. Silicon coarsening and fluidity reduction become more significant when the silicon content exceeds 18 wt.%, and it is almost impossible to obtain a good quality casting at a conventional foundry cooling rate.

The transmission of the ultrasonic vibration to the solidifying melt is not an easy problem because a resonator can rapidly fail under the effect of temperature and cyclic stresses[4]. Dobatkin and Eskin[5] used ultrasound intensities in the range of 7–20 W cm^{-2} and found that carbon steel resonators were dissolved quickly in liquid aluminium, whereas 18 wt.% Cr–9 wt.% Ti steel resonators have a life time of only 1–2 min. Niobium alloys proved to be more resistant[6]. Laboratory tests were based on the use of special benches where ultrasonic vibrations were fed to the solidifying melt by a top transmission method. An ultrasonic generator of 10 kW and a magnetostrictive transducer were used in those tests. The ultrasonic treatment was carried out in a cylindrical ceramic crucible of approximate dimensions 80 mm diameter and 160 mm height. The thermocouple was positioned to record the temperature of the melt between the centre of the charge and the crucible wall, i.e. 30 mm from the crucible wall.

To study the thixotropic behaviour, ultrasonically treated and non-treated specimens were upset in an electro-hydraulic press under semi-solid conditions. The specimens measuring $20 \times 20 \times 10$ mm were deformed at 580°C after heating and holding for 15 min. During the deformation the height of the specimens was decreased down from 10 to 5 mm. The variation of the deformation load during the upsetting was measured. Tensile tests were done with an Instron machine at a loading rate of 0.5 mm min^{-1}. The tensile specimens had a working length-to-diameter ratio equal to 10.

Table 1 Chemical composition of the alloys (in wt.%)

Alloy	Si	Cu	Mn	Al
AlSi13	13.8	3.1	0.3	Balance
AlSi17	17.1	4.5	0.55	Balance

Results and discussion

It was revealed that an ultrasonic treatment additionally superheats the melt in the core compared with the casting process without ultrasonic treatment (*Figure 1*). The microstructure changes in the solidifying metal are generally due to the processes in the melt and the two-phase liquid-solid zone, i.e. crystal nucleation, dispersion and mixing. These processes depend on the cavitation and streaming as well as processing factors and material properties. The shock waves appearing during the collapse of the cavitation bubbles caused some crystals at the solidification front to break down and move towards the liquid bulk. In addition to breaking down the growing crystals, the ultrasound also affects the nucleation rate. Simultaneously, an ultrasonic treatment of the melt reduced the undercooling from 2.1 to 0.3°C. The changes of cooling curves show that the nucleation takes place more easily. There are two possible mechanisms of the cavitation effect on the nucleation rate. Cavitation activates insoluble particles (e.g. oxides or ultrafine particles of some intermetallics) existing in the melt, and turns them into solidification sites. The fragments of destroyed dendrites also act as solidification sites. Another mechanism is the following[7]. During the expansion half-period, the bubble rapidly increases in size, and the liquid evaporates inside the bubble. The evaporation and expansion tend to reduce the bubble temperature. A decrease of the bubble temperature below the equilibrium temperature results in an undercooling of the melt at the bubble surface, and hence in the probability that a nucleus will be formed on a bubble. The ultrasound also prevents

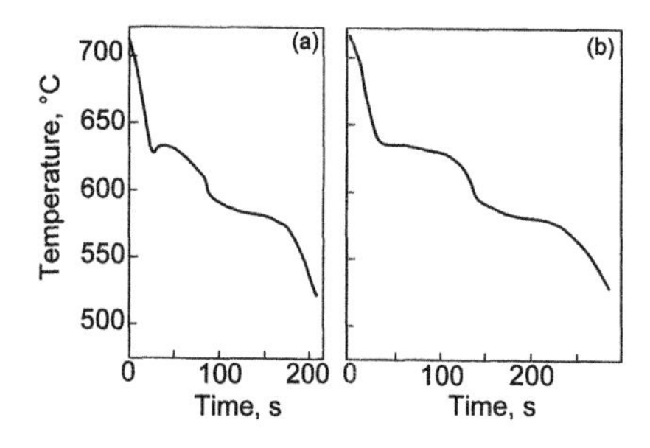

Figure 1 Cooling curves for AlSi17 alloys solidified conventionally (a) and during ultrasonic treatment (b)

Table 2 Effect of the casting technique on the mechanical properties of hypereutectic Al–Si based alloys

Alloy	Casting technique	Tensile strength [MPa]	Elongation [%]	Hardness [HB]
AlSi13	Conventional casting	185	0.5	98
AlSi13	Conventional casting	210	1.5	96
AlSi17	Conventional casting	160	0.5	102
AlSi17	Conventional casting	180	1.2	98

Table 3 Effect of an ultrasonic treatment (UST) on the true stress σ_s during deformation at 580°C

Alloy	σ_s, MPa		$\sigma_{sUST}/\sigma_{sCC}$
	Conventional casting (CC)	Casting with UST	
AlSi13	29	20	0.69
AlSi17	51	42	0.82

the agglomeration of individual nuclei into a polycrystal during their growth in the semi-solid alloy. At high temperatures, above the grain boundary wetting phase transition temperature, the liquid phase wets all grain boundaries and can stop the agglomeration of single crystals[8]. However, the liquidus temperature of the Al–Si based alloys studied is rather low, and a mechanical stirring or an ultrasonic treatment is required for a refinement of the microstructure.

In hypereutetic AlSi13 specimens prepared with the aid of conventional casting, the primary silicon crystallizes as hexagonal plates joined together at the centre into star-shaped particles, as they appear in cross-section (*Figure 2*). By contrast, the ultrasonic treatment refined the silicon crystals and distributed them uniformly over the cross-section. Most of the silicon plates were disconnected and broken during the ultrasonic treatment, forming spheroidized crystals. The microstructure of hypereutetic AlSi17 specimens prior and after the ultrasonic treatment is shown in *Figure 3*. It is clear from this figure that the primary Si crystals have faceted morphologies prior to the ultrasonic treatment. However, the ultrasonic treatment resulted in morphological changes of the primary Si crystals from faceted to spherical. A fragmentation of large primary crystals followed by aggregation of the fragmented Si is considered to be responsible for the spheroidization of the primary Si crystals. The ultrasonically induced structure changes improve the mechanical properties at room temperature. As shown in *Table 2*, as-cast hypereutectic Al–Si based alloys showed a modest increase of the strength and a decrease of the hardness. Typically, the ultrasonic treatment results also in an increase of the plasticity by a factor of 1.2–1.5. The increase in the ductility makes hot cracking during casting less probable. In *Fig. 4* is shown the variation of deformation load during the upsetting of the specimens by 50% at the temperature of 580°C. The quantitative assessment of the true stress σ_s upon the upsetting in the semi-solid state is given in *Table 3*. It is clear from *Figure 4* and *Table 3* that the maximum stress of the upsetting is considerably lower for ultrasonically treated alloys. Probably, in this case similar to the sliding of ultrafine grains during superplastic deformation in the solid state one can observe a thixotropic sliding of non-dendritic grains under semi-solid conditions.

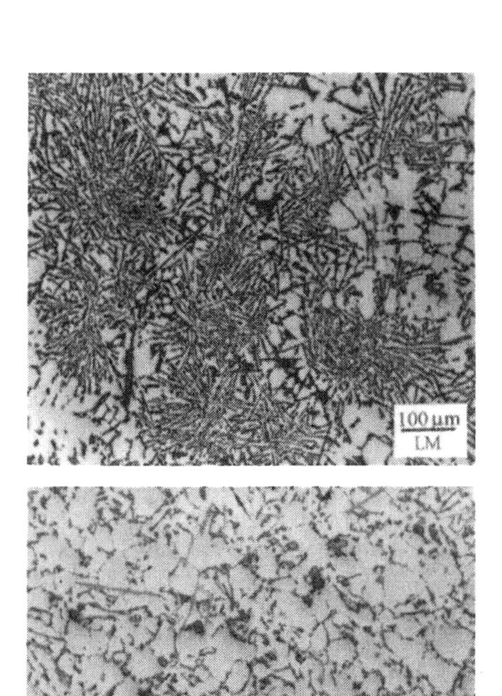

Figure 2 Microstructure of AlSi13 alloys solidified conventionally (a) and during ultrasonic treatment (b)

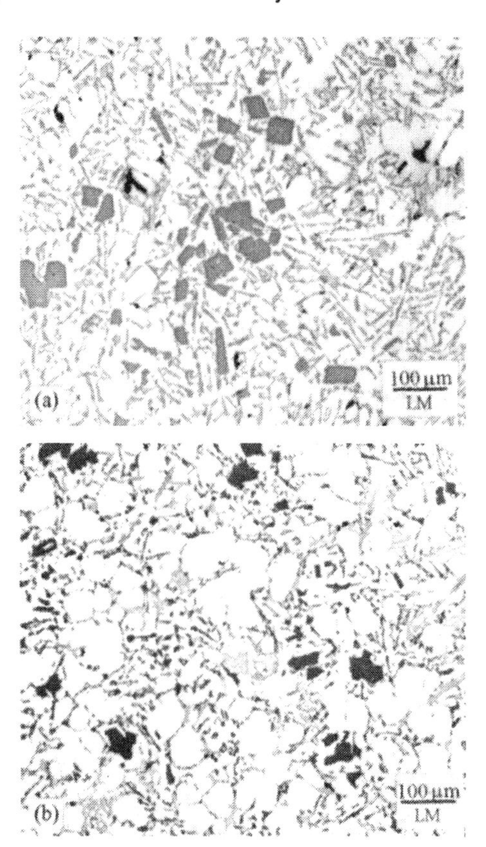

Figure 3 Microstructure of AlSi17 alloys solidified conventionally (a) and during ultrasonic treatment (b)

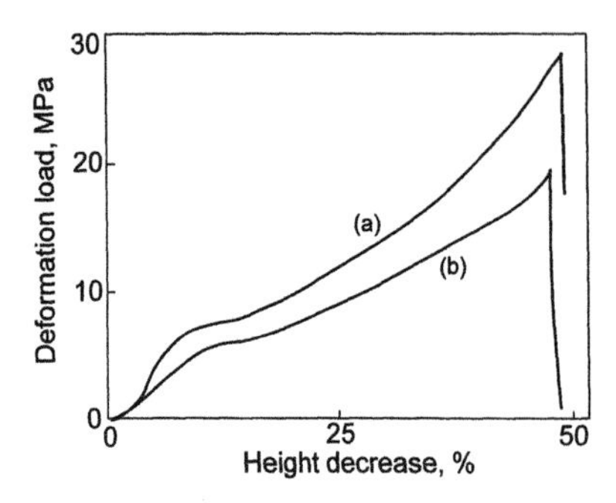

Figure 4 Variation of the deformation load upon upsetting of AlSi13 specimens at 580°C: (a) Conventional casting, and (b) casting during ultrasonic treatment

Conclusions

During the casting of Al–Si based alloys the ultrasonic field provides nucleation sites and destroys the growing crystals. This is an effective method for reducing the primary silicon particle size. The structure changes led to an increase of the strength and ductility. Our investigation confirms great advantages of ultrasonically treated ingots of Al–Si based alloys upon deformation in the semi-solid state.

Acknowledgements

The authors wish to thank the INCO-COPERNICUS programme (contract IC 15-CT96-0740), the NATO (contract HTECH.LG.970342) and the Volkswagen Foundation (contract VW I/71 676) for financial support.

References

1 Spencer, D.B., Merhabian, R. and Flemings, M.C., *Metall. Trans. 3*, 1972, 1925–1932
2 Gabathuler, J.P. and Ditzler, C., *Semi-Solid Processing of Alloys and Composites*. SRP, Sheffield, 1996, pp. 331–352
3 Kahl, W., *Metall 4*, 1994, pp. 295–306
4 Abramov, O.V., *Ultrasound in Liquid and Solid Metals*. CRC Press, Boca Raton, FL, 1994, pp. 251–256
5 Dobatkin, V.I. and Eskin, G.I., *The Effect of High Intensity Ultrasound on the Phase Interface in Metals*. Nauka, Moscow, 1986, pp. 87–93 (in Russian)
6 Sterritt, A., Bacon, M., Bell, F. and Mason, T.J., *Proc. Ultrasonic World Congress*, Part 2, Berlin, 1995, pp. 725–732
7 Kapustina, O.A., *The Physical Principles of Ultrasonic Manufacturing*. Nauka, Moscow, 1970, pp. 45–61 (in Russian)
8 Straumal, B., Gust, W. and Molodov, D., *Interface Science 3*, 1995, 127–132

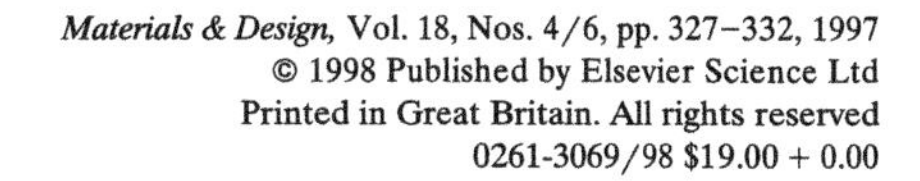
Materials & Design, Vol. 18, Nos. 4/6, pp. 327–332, 1997
© 1998 Published by Elsevier Science Ltd
Printed in Great Britain. All rights reserved
0261-3069/98 $19.00 + 0.00

PII: S0261-3069(97)00086-1

The effect of silver addition on 7055 Al alloy

C. W. Lee*, Y. H. Chung, K. K. Cho, M. C. Shin

Div. of Metals, Korea Institute of Science and Technology (KIST), P.O.Box 131, Cheongryang, Seoul, 130-650, South Korea

Received 15 September 1997; accepted 16 September 1997

The effect of silver addition on the microstructure and mechanical properties of the 7055 Al alloy was investigated. Mechanical properties of various alloys used in this study are correlated with the formation of constituent particles and microstructural changes of fine precipitates. According to the experiment, tensile strength of the 7055 Al alloy decreased by the addition of silver, although silver additions were observed to refine η'. The low tensile strength is attributed to the relatively low number density of η' in silver-bearing alloys in which a large amount of hardening elements, such as Zn and Mg, are consumed to form silver-bearing constituent particles. © 1998 Published by Elsevier Science Ltd. All rights reserved.

Keywords: microstructure; mechanical properties; 7055 aluminium alloy

Introduction

The 7055 Al alloy possesses a superior combination of high strength and low susceptibility to stress corrosion cracking (SCC)[1–4]. This alloy is gaining a commercial importance in aerospace industries where high specific strength is required. The 7055 Al alloy has a higher content of zinc and a higher density of η' when compared with the 7050 Al alloy.

It was well known that proper control of size and volume fraction of η' is essential for improving mechanical properties of 7000 series Al alloys. During the past 30 years, silver has been reported to promote the formation of fine precipitates in these alloys. According to the experimental results carried out by Polmear, the addition of 0.3 wt.% Ag to Al-5.8Zn-2.4Mg-1.8Cu alloy was found to be effective in facilitating the fine and uniform η' precipitates[5]. On the other hand, Kusui reported that silver had both positive and negative effects on mechanical properties of 7000 Al alloys, which were fabricated via powder metallurgy using the atomized powders[6]. He showed that silver promoted fine η' precipitates when silver content is lower than 0.1 wt.%. However, with increasing silver contents more than 0.1 wt.%, the tensile strength of the alloys decreased linearly.

The purpose of this study was to investigate the effect of silver in 7055 Al alloy. In this study, the effects of silver on the structure and properties of 7055 Al alloy which is produced by ingot metallurgy are studied. It is found that silver is concerned not only with fine η' precipitates but also with large constituent particles. The strength decrease in silver-bearing alloys was attributed to the increased volume of silver-bearing constituent particles.

*Correspondence to Dr C. W. Lee, Tel.: +82 2 9585439; fax: +82 2 9585449; e-mail: leewoo@kistmail.kist.re.kr

Experimental method

The alloys were made by using an induction furnace under Ar atmosphere. Size of the ingots was 50 mm × 50 mm in cross section and 120 mm in height. The chemical composition of the alloys was analyzed by inductively coupled plasma-mass spectrometry (ICP-MS). Nominal compositions of the experimental alloys are provided in *Table 1*. Alloy numbers 1 and 2 are high-Zn alloys and numbers 3 and 4 are low-Zn alloys. Number 1 and 3 are 0.37 wt.% silver-bearing alloys.

The alloys were homogenized at 460°C for 24 h and hot rolled into a sheet of 4 mm in thickness. Strips of each alloy were solution-treated at 477°C for 1 h and two kinds of aging treatments were applied. One was T77 heat treatment and the other was an aging treatment at 121°C for various lengths of times up to 90 h. T77 heat treatment was carried out by aging at 121°C for 40 h, at 182°C for 0.5 h and again at 121°C for 24 h.

Constituent particles of each alloy were observed by scanning electron microscopy (SEM) and analyzed by energy dispersive spectrometry (EDS), secondary ion mass spectrometry (SIMS). Over 30 results of EDS were sorted and averaged separately according to the particle types. In order to measure the volume percent of constituent particles in SEM images, an image analyzer was used.

Some microstructures of η' were observed by TEM. Size distributions of η' in the silver-bearing alloys were compared to that in silver-free ones. Mechanical properties of the alloys in T77 condition were measured by tensile test.

Results

The distribution of constituent particles is shown in *Figure 1*. Particle shapes are shown in *Figure 1(b)* and *(d)*. The constituent particles are classified into three

Table 1 Chemical composition of alloys used in the present work (wt.%)

Alloy no.	Elements					
	Zn	Mg	Cu	Ag	Zr	Al
1	8.76	1.94	2.44	0.37	0.06	Bal.
2	8.76	2.32	2.61	—	0.10	Bal.
6	7.58	1.96	2.15	0.37	0.12	Bal.
7	7.73	2.29	2.60	—	0.12	Bal.

Table 2 The average composition of constituent particles determined using EDS (wt.%)

Particle types	Elements					
	Zn	Mg	Cu	Ag	Fe	Al
Silver-bearing particles	21.9	11.3	27.5	11.3	—	28.0
Silver-free particles	14.9	7.8	39.3	—	—	38.0
Fe-containing paticles	2.4	1.2	30.3	—	13.4	52.7

types by their contrast observed in SEM images (arrow A and B in the *Figure 1*) and major elements found in EDS results (*Table 2*). The silver-bearing alloy shows a larger number of constituent particles than the silver-free one. Particles found in the silver-bearing alloy are Al-Zn-Mg-Cu-Ag type compounds. The other kind of particles found in the silver-free alloy was Al-Zn-Mg-Cu type compound. Another kind of particles found in both alloys was Al-Fe-Cu-Zn-Mg type compound.

The compositions of the particles of each type were averaged as shown in Table 2. The compositions of the particles were averaged by assuming that all kinds of particles in the same type were evenly distributed. The volume percent of constituent particles was measured by an image analyzer. Hardening elements exhausted by these constituent particles were calculated roughly with averaged composition and volume percent of the particles. The volume percent of constituent particles

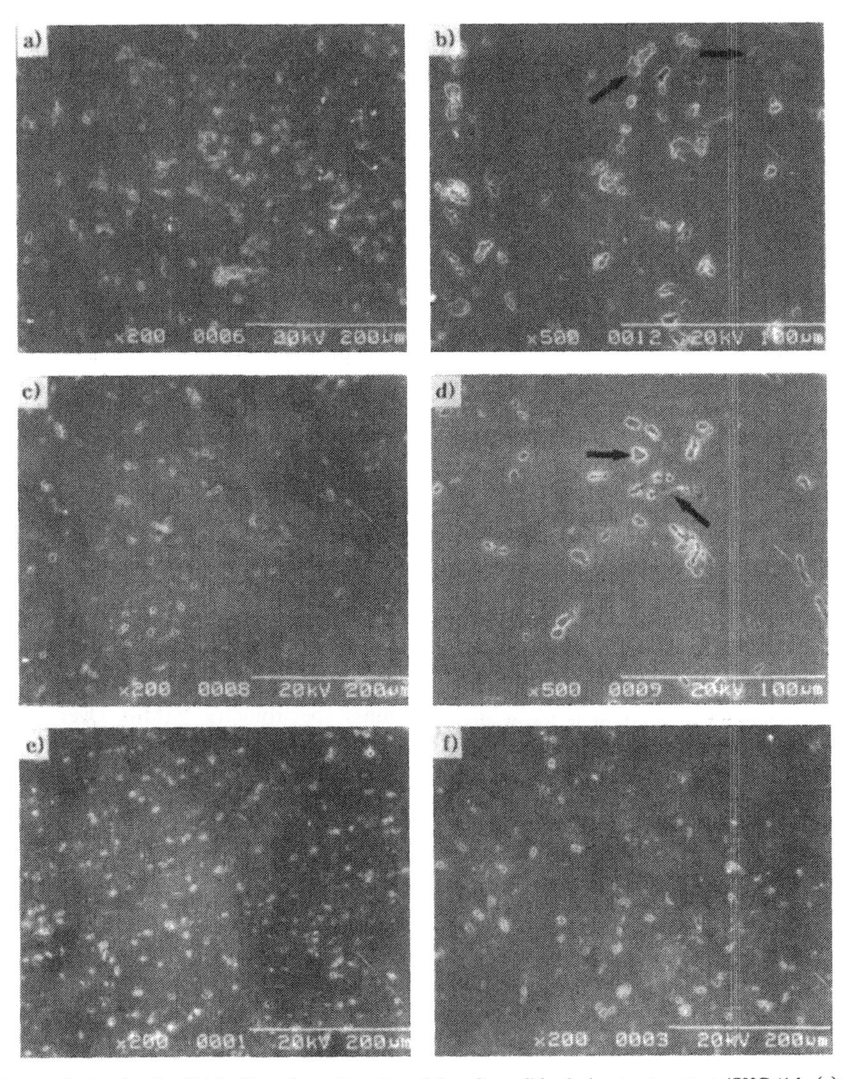

Figure 1 SEM micrographs showing the distribution of constituent particles after solid solution treatment at 470°C/1 h. (a) Silver-bearing alloy with high zinc content (×200); (b) silver-bearing alloy with high zinc content (×500); (c) silver-free alloy with high zinc content (×200); (d) silver-free alloy with high zinc content (×500); (e) silver-bearing alloy with high zinc content (×200); and (f) silver-free alloy with low zinc content (×200)

Table 3 Volume percent of the constituent particles measured by an image analyzer and the calculated contents of the various elements remained within matrix

Particle types	Alloy no.			
	1	2	3	4
(a) Volume percentage (%)				
Silver-bearing particles	5.1	—	2.48	—
Silver-free particles	—	3.2	—	1.75
Fe-containing particles	1.37	1.37	1.37	1.37

(b) Percentage of exhausted elements and elements remainng within matrix (%) Alloy no.		Elements		
		Zn	Mg	Cu
Exhausting elements	1	1.62	0.81	2.51
	2	0.65	0.31	2.23
	3	0.8	0.37	1.44
	4	0.35	0.90	1.45
Remaining within matrix	1	7.4	1.5	0.1
	2	8.4	2.0	0.3
	3	6.9	1.7	1.1
	4	7.4	1.9	1.1

measured and exhausted elements calculated were summarized in *Table 3*.

The element mapping of the particles by SIMS is shown in *Figure 2*. Each element in the particles seems evenly distributed all over the particle's inside in low resolutions. Detailed image of a constituent particle was observed by SEM as shown in *Figure 3*. Small particles were imbedded in a large constituent particle (as notified by arrow in the *Figure 3*).

Figure 4 contains dark field images of η' observed in [112](110) zone axis of matrix phase. Alloy 4 aged at 121°C for 30 h showed the finest η' precipitates among the observed specimens. The relative frequency of η' at various aging conditions is compared in *Figure 5*.

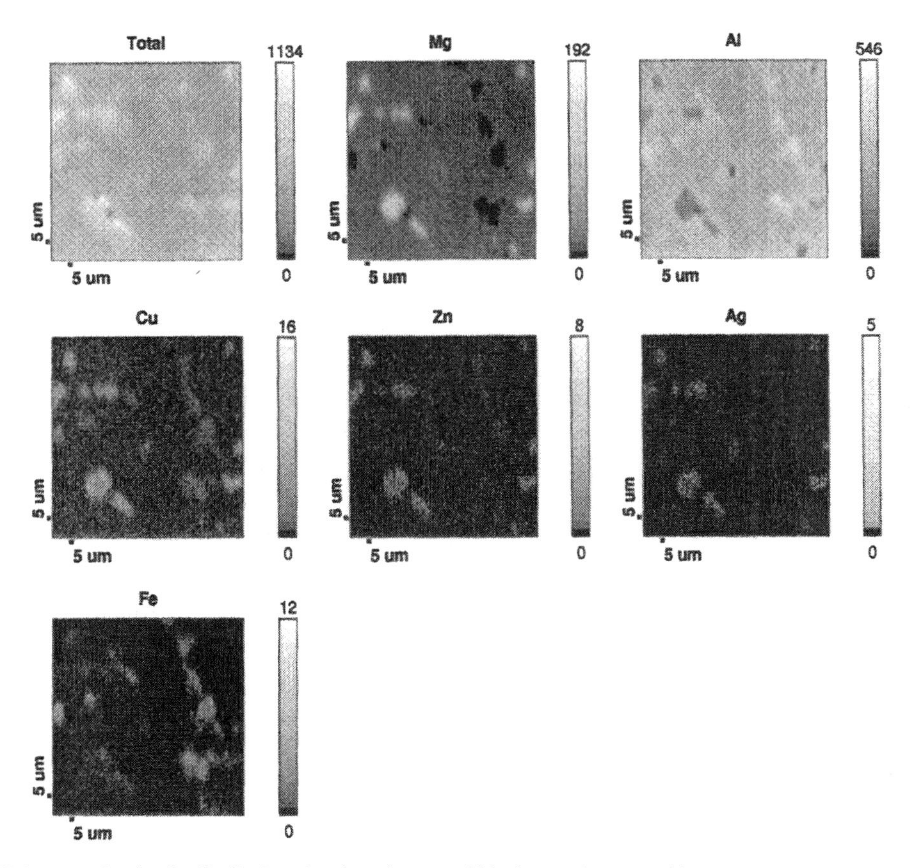

Figure 2 SIMS dot maps showing the distribution of various elements within the constituent particles

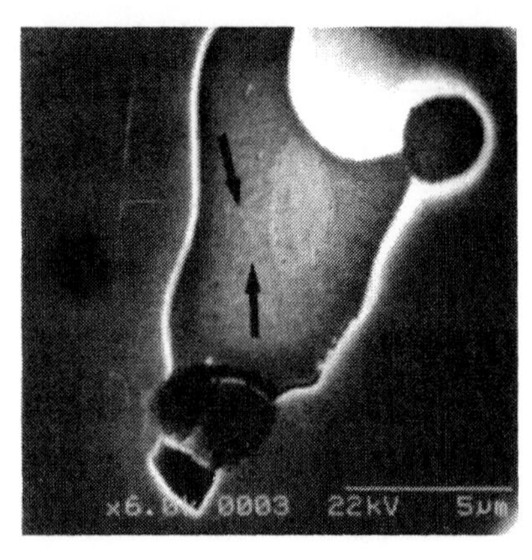

Figure 3 SEM micrograph showing detailed shape of a constituent particle

Difference in size distribution and its changes by prolonged aging are compared.

Table 4 shows tensile test results. Yield strength of the silver-bearing alloys was approx. 10 MPa lower than that of the silver-free alloys in the T77 condition. Hardness changes according to the aging time at 121°C are shown in *Figure 6*. The peak hardness of the silver-bearing alloy and that of the silver-free alloy were found at 30 and 45 h in aging time, respectively. In the low Zn alloys, silver seemed to accelerate the aging response and the peak hardness of the silver-bearing alloy appeared earlier.

Discussion

EDS results show three types of the particles. The first type is Al-Zn-Mg-Cu compound, the second type is Al-Zn-Mg-Cu-Ag compound and the third type is Al-Fe-Cu-Zn-Mg. The number of constituent particles found in the silver-bearing alloys are larger than that found in the silver-free ones. Silver is directly concerned with specific type of constituent particles (Al-Zn-Mg-Cu-Ag compound) and it means that silver promotes the constituent particle formation.

On the other hand, the volume and number of particles containing Fe (Al-Fe-Cu-Zn-Mg type compounds) are not changed by silver addition. The particles show regular compositions with no relation to silver addition. Therefore silver has no influence on the Fe-containing particles.

Tables 2 and *3* show the volume percent and the averaged composition of the particles. It also contains available hardening elements estimated by theoretical calculation. Hardening elements are far decreased in the silver-bearing alloys. For this reason, the silver-bearing alloys show lower strength than the silver-free ones.

The SEM image reveals the silver-bearing constituent particle in detail. Small particles are imbedded in a constituent particle (as shown by the arrow in the *Figure 3*). These small particles are not found in the silver-free alloys. However, these imbedded particles are too small to be identified. Considering the number increase of constituent particles in the silver-bearing alloys, these small particles are supposed to be connected with silver and to act as a seed of the constituent particles.

The element mapping results of the constituent particles by SIMS confirms that three kinds of particles are present. Silver in a particle is evenly distributed all over the particle's inside in low magnification. It means that silver is continuously diffused into the particles during the particle's growth. Therefore silver may also affect the growth rate of the particles.

The TEM image in *Figure 4* shows the change of η' when the specimens are aged for various length of time at 121°C. *Figure 5* shows the relative frequency of η'

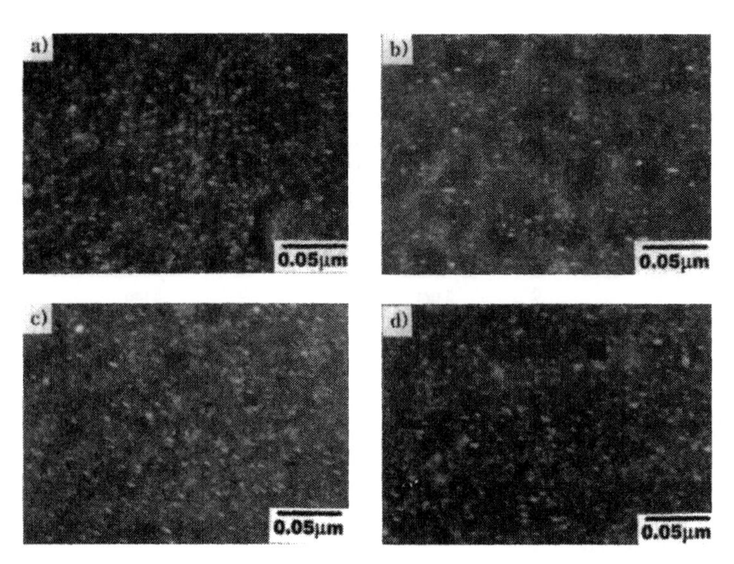

Figure 4 TEM dark-field images showing the variation in η' size as a function of the aging time in low zinc alloys. (a) Silver-bearing alloy aged for 30 h at 121°C; (b) silver-bearing alloy aged for 45 h at 121°C; (c) silver-free alloy aged for 30 h at 121°C; and (d) silver-free alloy aged for 40 h at 121°C

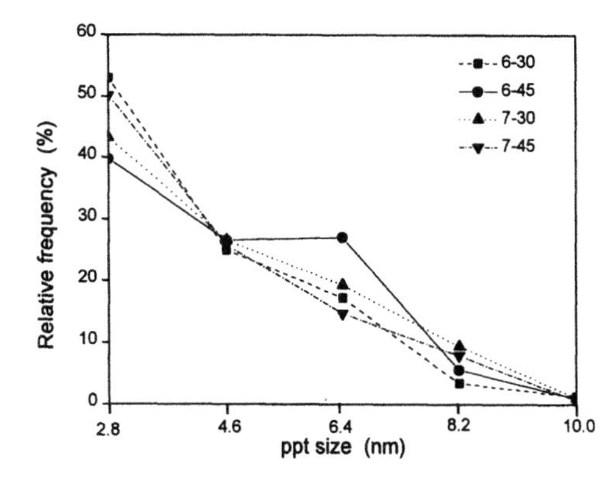

Figure 5 The size distribution of precipitates within the silver-bearing and silver-free alloys, which were aged at 121°C for 30 and 45 h, respectively

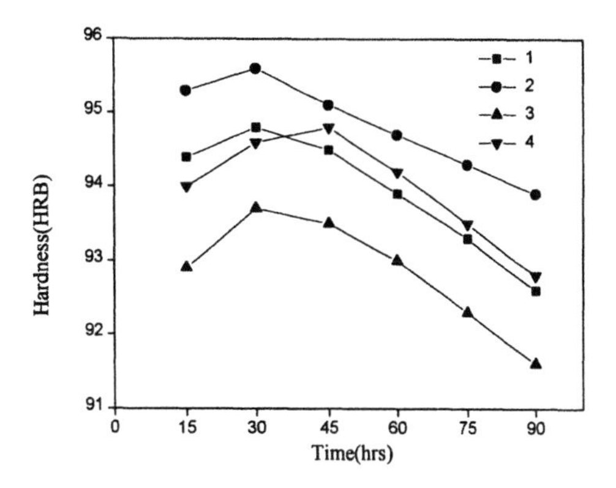

Figure 6 Variations in hardness as a function of the aging time at 121°C

Table 4 Tensile properties of silver-bearing and silver-free alloys in their T77 treatment condition (MPa)

Property	Alloy no.							
	1		2		6		7	
	TD	LD	TD	LD	TD	LD	TD	LD
Yield stress (MPa)	610	614	619	628	585	598	576	604
Tensile stress (MPa)	621	628	636	648	616	613	613	628
Elongation (%)	4.1	4.5	3.5	5.2	6.3	7.0	8.2	7.4

TD, transverse direction; LD, longitudinal direction.

precipitates measured in *Figure 4*. Relative portion of fine η' precipitates (2.0–3.0 nm in length) in the silver-bearing alloy aged for 30 h is larger than that in the same alloy aged for 45 h, whereas this trend is reversed in the silver-free alloy. The relative portion of mid-size precipitates (between 3.0 and 5.0 nm) in the silver-bearing alloy is increased when the aging time is prolonged from 30 h to 45 h. It means that silver-bearing alloy is beyond the peak aging condition when it is aged for 45 h. However, the silver-free alloy shows that relative portion of fine η' precipitates is still increasing even it is aged for 45 h. It means that silver-free alloy aged for 30 h has not yet reached the peak aging condition. It is consistent with the results of hardness test in *Figure 6*. The peak hardness of the silver-bearing alloy is found by aging for 30 h and that of the silver-free alloy is at 45 h. These results imply that the formation of η' precipitates is accelerated by silver.

Yield strength of the alloys decreases by approx. 10 Mpa by silver addition. It seems that the low strength in the silver-bearing alloys comes partly from the loss of hardening elements and partly from the coarse constituent particles themselves. Another reason for this strength decrease is the T77 heat treatment. In the course such heat treatment, large numbers of η' precipitates are transformed to η phase. In the silver-bearing alloys, silver accelerates the transformation rate of η and the yield strength can decrease more rapidly. These negative effects of silver on the yield strength seem to overwhelm the favorable effects of precipitate refining.

Conclusions

1. Silver accelerates η' precipitates in 7055 Al alloy.
2. Silver promotes the formation of large amounts of constituent particles. These constituent particles consume hardening elements, such as Zn, Mg and Cu, and decrease the remaining hardening elements in the matrix.
3. Silver has no influence on the Fe-containing constituent particles.
4. Yield stress of the silver-bearing alloy decreases by approx. 10 Mpa compared with that of the silver-free alloy in T77 heat treatment. The negative effect of silver on the yield strength seems to overwhelm the favorable effects of precipitate refining.

Acknowledgements

The authors wish to thank the Aluminum Korea Co. Ltd. for providing funds under the research project of 'Research on the 7000 series high strength Al alloys', Dr. S. Lee, project manager.

References

1 Chang, Y. C. and Howe, J. M. *Metallurgical Transactions A*, 1993, **24A**, 1461
2 Vietz, J. T., Sargant, K. R. and Polmear, I. J. *Journal of Institute of Metals*, 1963, **92**, 327
3 Hono, K., Sakurai, T. and Polmear, I. J., *Scripta Metals and Materials*, 1994, **30**, 695

4 Ringer, S. P., Yeung, W. U., Muddle, B. C. and Polmear, I. J., *Acta Materiale*, 1994, **45**, 1715
5 Polmear, I. J., *Journal of Metals*, 1968, **June**, 44
6 Kusui, J., Fujii, K., Yokoe, K., Yokote, T., Osamura, K., Kubota, O. and Okuda, H., *Materials Science Forum*, 1996, **217/222**, 1823

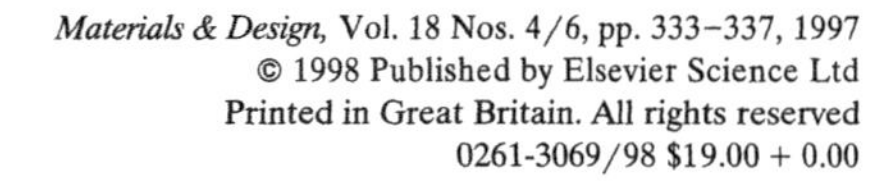
Materials & Design, Vol. 18 Nos. 4/6, pp. 333–337, 1997

0261-3069/98 $19.00 + 0.00

PII: S0261-3069(97)00073-3

Formation and relative stability of interstitial solid solutions at interfaces in metal matrix composites

Simon Dorfman[a,*], David Fuks[b]

[a]*Department of Physics, Technion-Israel Institute of Technology, 32000 Haifa, Israel*
[b]*Materials Engineering Department, Ben-Gurion University of the Negev, P.O. Box 653, 84105 Beer Sheva, Israel*

Received 5 August 1997; accepted 18 September 1997

On the example of Cu-C composite material the way to calculate the occupation of the interstitial positions and its temperature dependence is shown. The results obtained on the basis of non-empirical calculations indicate the preferable occupation of octahedral positions up to T ~ 1200 K. This confirms the structure of the interstitial solid solution. Within this model the influence of the alloying on the height of the diffusion barrier and on the temperature dependence of carbon diffusion in copper is calculated. © 1998 Published by Elsevier Science Ltd. All rights reserved.

Keywords: interstitial positions; temperature dependence; interstitial solid solution; metal matrix composites

Introduction

The stability of the copper matrix MMC with carbon (Cu/C) was investigated in refs.[1–3]. These works were carried out to check the possibility of replacing silver in electronic contacts by copper. By adding graphite fibers it is possible to improve properties of sliding due to the lubricating properties of carbon. The system Cu/C shows good thermal, electrical and stress properties in comparison with other Cu-based MMC. Carbon is insoluble in copper up to very high temperatures, its solubility does not exceed 0.02 at%. Thus a problem of wettability of carbon fibers by copper arises. This wettability is extraordinary small[4] and does not allow to fabricate the composite material. Consequently the interface bonding in copper-carbon composites is extremely weak[5]. In[5] the increase of interfacial strength by adding Fe and Ni to copper was studied. The influence of oxidation of copper coated carbon fibers on the thermal stability of the coating was discussed in[6]. Formation of very dilute Cu-C interstitial alloys at the interface may be diffusion-controlled. To predict the diffusion behaviour of carbon we have to study the structure and interatomic interactions in such alloys.

In order to write the expression for the energy of the interstitial atom in different positions it is possible to use the perturbation series on potentials in reciprocal space (PSP RS method). It is evident that this approximation will only roughly describe the basic particularities of the electronic spectrum of the solutions. This is especially true for solutions, where transition metals are chosen as alloying elements. But, as thermodynamic values are always obtained as a result of averaging over the spectrum, they are less sensitive to its particularities than, for example, optical characteristics.

Making use of the PSP RS method, we calculate the potentials of Cu-Cu, Cu-C, and C-C interactions in Cu-C alloy. The use of these potentials is twofold. The study of occupations of tetrahedral and octahedral interatomic positions by carbon atoms on the basis of statistical thermodynamics with the obtained potentials was carried out. Exploring this result and the same interatomic potentials we calculate the height of the diffusion barrier for a carbon atom which passes between the nearest interstitial positions and predict the diffusivity of carbon in copper. As a last step the changes of the diffusion of carbon atoms in copper matrix with different alloying elements were studied. The idea of 'averaged' or 'effective' interaction is used for the investigation of the influence of the third element upon the diffusion process in interstitial alloys.

Occupation of interstitial positions

Let us assume that interstitial atoms of carbon, C, occupy both octahedral and tetrahedral positions in the lattice. The number of octahedral positions is M_o and the number of the tetrahedral positions is M_t. It means that the total number of interstitial positions is $M = M_o + M_t$. The n atoms C are placed in the interstitial positions in the following manner: n_o atoms in octahedral positions and n_t atoms ($n_t = n - n_o$) in tetrahedral positions. The part of the total energy of the crystal depending on the number of interstitial atoms may be

* Correspondence to Dr S. Dorfman

written as

$$E = n_o u_o + n_t u_t \quad (1)$$

on assumption that C atoms do not interact with each other. This assumption is reasonable because the concentration of carbon is very small. Here u_o and u_t are the energies of C atoms in the octahedral and tetrahedral positions, respectively. The number of different permutations of C atoms on the interstitial positions is

$$L = \frac{M_o!}{n_o!(M_o - n_o)!} \cdot \frac{M_t!}{n_t!(M_t - n_t)!} \quad (2)$$

while the entropy of the system, S, is

$$S = k \ln L \quad (3)$$

where k is the Boltzmann constant. Substituting $n_t = n - n_o$ and making use of the Stirling formula, we obtain the configurational part of the free energy $F = E - TS$ in the form

$$\begin{aligned} F = n_o u_o + (n - n_o)u_t - kT\{&\ln M_o! - n_o \\ &(\ln n_o - 1) - (M_o - n_o)[\ln(M_o - n_o) - 1] \\ &+ \ln M_t! - (n - n_o)[\ln(n - n_o) - 1] \\ &-(M_t - n + n_o)[\ln(M_t - n + n_o) - 1]\} \end{aligned} \quad (4)$$

Applying the equilibrium conditions $(\partial F/\partial n_o)_{T,n} = 0$, the following equation for the equilibrium numbers n_o and n_t may be obtained

$$\frac{(M_t - n_t)(n - n_t)}{(M_t - n + n_o)n_o} = \frac{(M_o - n_o)n_t}{(M_t - n_t)n_o} = e^{\frac{u_o - u_t}{kT}}. \quad (5)$$

It is easy to see that when the temperature increases, the system of C-atoms approaches the uniform distribution of C-atoms on the interstitial positions. In this case (at $T \to \infty$) we obtain

$$\frac{n_t}{n_o} = \frac{M_t}{M_o}$$

To clarify the temperature dependence of n_o and n_t, Eq. (5) may be solved giving

$$n_o = \frac{M_o + n - M_t\mu - n\mu \pm \sqrt{(M_o + n + M_t\mu - n\mu)^2 - 4M_o n(1-\mu)}}{2(1-\mu)}. \quad (6)$$

Here we use the equality $n_t = n - n_o$ and $\mu = \exp[(u_o - u_t)/kT]$. Using the definition of partial concentrations of atoms in octahedral and tetrahedral interstitial positions $c_o = n_o/n$ and $c_t = n_t/n$, the following result may be obtained

$$c_o = \frac{\alpha + \beta \pm \sqrt{(\alpha+\beta)^2 - 4\beta}}{2\beta},$$

$$c_t = 1 - c_o \quad (7)$$

Here $\alpha = 1 + (M_t/M_o)\mu$; $\beta = \gamma(1-\mu)$; and $\gamma = n/M_o$. For the relatively small n, this result has to be transformed to the Boltzmann distribution. This condition leads to the consideration that only the minus sign before the square root in Eq. (7) has to be left and in this case from Eq. (5) it follows

$$c_o = \frac{1}{1 + \frac{M_t}{M_o}\mu}; \quad c_t = \frac{1}{1 + \left(\frac{M_t}{M_o}\mu\right)^{-1}}. \quad (8)$$

When the value γ is small the value of β is also small and the square root may be expanded in Taylor's series. Then from Eq. (7) in the limiting case $\gamma \to 0$ we get $c_o = 1/\alpha$ which coincides with Eq. (8).

In the case of the fcc lattice with N lattice sites $M_o = N$ and $M_t = 2N$. Thus the final expressions for the concentrations c_o and c_t have the form:

$$c_o = \frac{1}{1 + 2e^{\frac{u_o - u_t}{kT}}};$$

$$c_t = \frac{1}{1 + \frac{1}{2}e^{\frac{u_t - u_o}{kT}}}. \quad (9)$$

Diffusion of interstitial atoms

The basic idea of effective interaction may be applied to the investigation of the influence of the third element upon the diffusion process in interstitial alloys. We assume that atoms B substitute for atoms A and atoms C are placed in the interstitial position. The value of the energy of atom C in the interstitial position u_o is

$$u_i = -\sum_{\mathbf{R}}[V_{AC}(\mathbf{R} + \mathbf{h}_i)C(\mathbf{R}) + V_{BC}(\mathbf{R} + \mathbf{h}_i)(1 - C(\mathbf{R}))] \quad (10)$$

In this equation $\mathbf{h}_i$ is the vector of the position of the atom C. V_{AC} and V_{BC} are the values of the interaction potentials between atoms A and C and B and C, respectively. $C(\mathbf{R})$ is the spin-like variable:

$$C(\mathbf{R}) = \begin{cases} 1, & \textit{if an atom in the lattice site } \mathbf{R} \textit{ is of the type A} \\ 0, & \textit{if an atom in the lattice site } \mathbf{R} \textit{ is of the type B} \end{cases}$$

Consequently, the energy u_s of the C atom in the saddle point of the diffusion path (see[3]) is

$$\begin{aligned} u_s = -\sum_{\mathbf{R}}[&V_{AC}(\mathbf{R} + \mathbf{h}_s)C(\mathbf{R}) \\ &+ V_{BC}(\mathbf{R} + \mathbf{h}_s)(1 - C(\mathbf{R}))] \end{aligned} \quad (11)$$

$\mathbf{h}_s$ is the vector of the saddle point position. Eqs. (10) and (11) reproduce exactly the local atomic configuration in the vicinity of the interstitial atom in the diffusion process. Consideration of such local effects is very important in the study of material properties (see, for example,[7]). Calculations of $\bar{u}_i$ and $\bar{u}_s$ may provide the necessary information on the influence of microalloying additives on the diffusion coefficient. These quantities have to be averaged to account for the influence of the crystal media[8].

An alternative way is to introduce the distribution of atoms in A-B substitutional solid solution. This distribution may be described by one occupation probability function n(**R**) that is the probability of finding the atom A at the site **R** of the crystal lattice

$$n(\mathbf{R}) = \langle C(\mathbf{R}) \rangle, \tag{12}$$

where the averaging is done over the Gibbs canonical ensemble. Perfoming such averaging, we may rewrite Eqs. (10) and (11)

$$\bar{u}_i = -\sum_{\mathbf{R}} [V_{AC}(\mathbf{R}+\mathbf{h}_i)n(\mathbf{R}) + V_{BC}(\mathbf{R}+\mathbf{h}_i)(1-n(\mathbf{R}))] \tag{13}$$

$$\bar{u}_s = -\sum_{\mathbf{R}} [V_{AC}(\mathbf{R}+\mathbf{h}_s)n(\mathbf{R}) + V_{BC}(\mathbf{R}+\mathbf{h}_s)(1-n(\mathbf{R}))] \tag{14}$$

Analogous averaging was done in ref.[9] to describe the ordering effects in binary substitutional solid solution. $n(\mathbf{R})$, which determines the distribution of the solute atoms in an ordering phase, can be expanded into Fourier series. It may be represented as a superposition of static concentration waves (SCW)

$$n(\mathbf{R}) = c_a + \frac{1}{2}\sum_j \left[Q(\mathbf{k}_j)e^{i\mathbf{k}_j\mathbf{R}} + Q^*(\mathbf{k}_j)e^{-i\mathbf{k}_j\mathbf{R}}\right] \tag{15}$$

where c_a is a concentration of A-type atoms, $\exp(i\mathbf{k}_j\mathbf{r})$ is a static concentration wave, $\mathbf{k}_j$, is a nonzero wave vector defined in the first Brillouin zone of the disordered binary A-B alloy, index j denotes the wave vectors in the Brillouin zone, $Q(\mathbf{k}_j)$ is a static concentration wave amplitude. As shown in ref.[9], all $Q(\mathbf{k}_j)$ are the linear functions of the long-range-order parameters of the superlattices that may be formed on the basis of the Ising lattice of the disordered solid solution. In the alloy with small concentration of one of the components it is possible to assume the existence of a disordered solid solution. The small concentration of the B atoms immediately leads to the disappearance of the ordering state and all $Q(\mathbf{k}_j)$ become equal to zero.

In the case of fcc lattice that we are studying, there are one octahedral and two tetrahedral interstitial sites per atom. We shall consider only octahedral positions for interstitial atoms. This model follows from the results of the calculations for occupation probabilities for Cu-C which will be discussed in the next section. Making use of the above described averaging it is possible to calculate the influence of the alloying elements upon the diffusion of fiber atoms (we named them C) in the matrix (atoms A). The concentrations of atoms of sort A and B will be c_a and c_b, respectively. Taking into account only the first and the second nearest neighbours we have got from the expressions (Eqs. (13),(14))

$$\begin{aligned}\bar{u}_i &= -\{6\cdot(c_a v_{ac} + c_b v_{bc}) + 8\cdot(c_a v_{ac}''' + c_b v_{bc}''')\}\\ \bar{u}_s &= -\{2\cdot(c_a v_{ac}' + c_b v_{bc}') + 4\cdot(c_a v_{ac}'' + c_b v_{bc}'')\}\end{aligned} \tag{16}$$

Where $v_{ac} = V_{ac}(a/2)$, $v_{ac}' = V_{ac}(a\sqrt{2}/4)$, $v_{ac}'' = V_{ac}(a\sqrt{6}/4)$, $v_{ac}''' = V_{ac}(a\sqrt{3}/2)$, $v_{bc} = V_{bc}(a/2)$, $v_{bc}' = V_{bc}(a\sqrt{2}/4)$, $v_{bc}'' = V_{bc}(a\sqrt{6}/4)$, $v_{bc}''' = V_{bc}(a\sqrt{3}/2)$ are the values of the interatomic interaction energies determined from the values of interatomic interaction potentials V_{ac} and V_{bc} between atoms of sort A and C and B and C, respectively. a is the lattice parameter of the alloy.

The height of the potential barrier is

$$\begin{aligned}\Delta U &= |\bar{u}_s - \bar{u}_i|\\ &= |(6v_{ac} + 8v_{ac}''' - 2v_{ac}' - 4v_{ac}'')c_a\\ &\quad + (6v_{bc} + 8v_{bc}''' - 2v_{bc}' - 4v_{bc}'')c_b|\end{aligned} \tag{17}$$

The values in brackets here have the sense of the heights of barriers ΔU_a and ΔU_b in the diffusion process of C atoms in the metals A and B with fcc lattice. The diffusion coefficient of C atoms that are situated in the octahedral interstices of disordered substitutional solid solution A-B with fcc lattice may be obtained by substituting this expression into Arrhenius-type formulae

$$D \sim \exp\left(-\frac{\Delta U}{kT}\right) \tag{18}$$

That gives

$$D \sim \exp\left(-\frac{c_a\Delta U_a + c_b\Delta U_b}{kT}\right) \tag{19}$$

It is time to study now the changes of pre-exponential factors caused by alloying. The Gibbs free energy of migration includes the entropy term and the pressure-dependent term. The last can be neglected because we are studying the diffusion process at the atmospheric pressure that in our units is approximately zero. The entropy of migration in a binary alloy A-C may be calculated, according to[11], with our values for the migration energy ΔU.

$$S_{mig} = \beta\frac{\Delta U}{T_m} \tag{20}$$

where $\beta \simeq 0.35$[11] and T_m is the melting point. Now we may write the following relation

$$\frac{D_{tern}}{D_{bin}} \simeq \frac{a_{tern}^2}{a_{bin}^2}\exp\left[\frac{(1-\beta\tau)}{kT}\left(\Delta U_{mig}^{bin} - \Delta U_{mig}^{tern}\right)\right] \tag{21}$$

where D_{tern} and D_{bin} are the diffusion coefficients of the carbon in ternary and binary alloys respectively, $\tau = T/T_m$ and ΔU_{mig}^{tern} and ΔU_{mig}^{bin} are the migration energy for ternary and binary alloys.

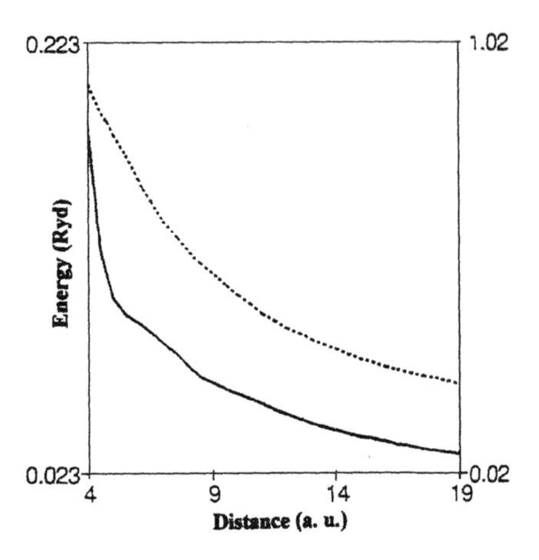

Figure 1 The effective pair potentials of Cu-C, the thick line, and of C-C, the dotted line, in Cu-based metal matrix composite. The left Y-axis denotes energies for Cu-C potential and the right one, the energies for C-C potential. Energies are in Ry. Distances (X-axis) are in atomic units

Table 1 Values of diffusion barriers, ΔU (eV), entropies of migration, S_{mig} ($eV \cdot K^{-1}$), and the ratio D_{tern}/D_{bin} for the interstitial copper-based alloys with carbon

System	Dopant	$S_{mig} \cdot 10^4$	ΔU	D_{tern}/D_{bin}
Cu-C	—	2.554	0.990	—
	Zr	2.657	1.030	0.633
	Al	2.603	1.009	0.805
	Ni	2.575	0.998	0.912
	Si	2.647	1.026	0.662
	Ti	2.652	1.028	0.647

Results and discussion

Our mean-field calculations in the framework of the local density approximation with the semi-empirical pseudo-potentials[10] show the completely different character of C-C interaction in the matrix of diamond and in the Cu matrix (see *Figure 1*). In the Cu matrix the C-C pair potential has a very strong repulsive tendency and in the diamond lattice the behaviour of the pair potential is traditional. The obtained results justify the statement: *the interactions between the same atoms in a different environment are completely different.*

The interatomic potential presented in *Figure 1* was used to calculate the carbon atom energies in different interstitial positions u_o and u_t in Cu matrix. In these calculations we take into account the interaction in the first and the second coordination shell. The energies u_o and u_t were substituted in Eq. (9) and the temperature dependencies of the relative concentrations c_o and c_t were obtained. They are plotted in *Figure 2*.

Our results show clearly that the probability of finding carbon atoms in octahedral positions is much higher than in the tetrahedral ones. This statement will be true up to the temperature $T \sim T_o = 1200$ K. Only with the following increasing of the temperature the probability of the occupation of octahedral positions decreases to 80%. Tetrahedral positions will be occupied by 20% of atoms. Thus one may be convinced to use the model of the interstitial diffusion of carbon atoms in Cu-C solid solution with the carbon atoms randomly situated in octahedral positions. This model is true at least till $T \sim T_o$.

Eqs. (16) and (17) are used to calculate diffusion barriers. We calculated interatomic potentials for the ternary alloys (Cu,Si)-C, (Cu,Zr)-C, (Cu,Ti)-C, (Cu,Al)-C and (Cu,Ni)-C making use of the mean-field theory and form-factors of semi-empirical potentials from Refs.[10]. The concentration of carbon atoms in an alloy was taken to be very small ($\sim 2 \cdot 10^{-2}$ at%) and the concentration of elements, that are added to the copper matrix was taken as 0.1 at%. The values of the

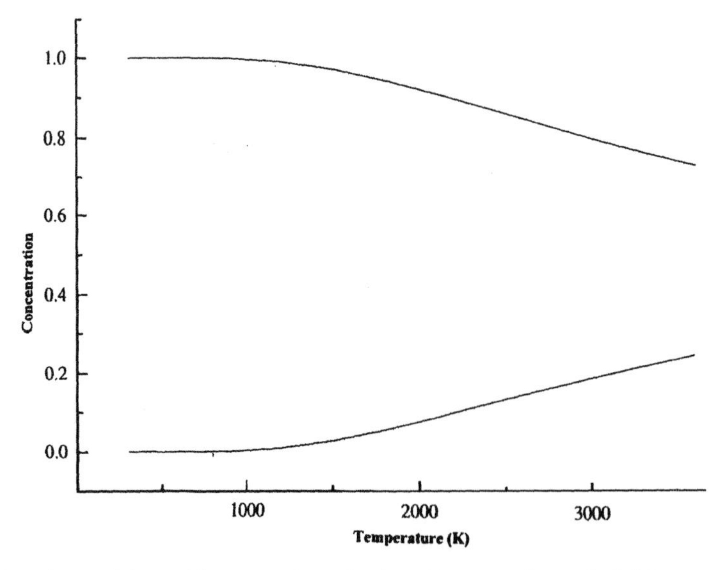

Figure 2 Temperature dependences of the partial concentrations of carbon atoms occupying octahedral (upper curve) and tetrahedral (lower curve) interstitial positions in a copper matrix

potential barrier ΔU for binary alloy Cu-C are given in *Table 1*. We may conclude from the analysis of *Table 1* that alloying of binary system Cu-C by all studied additives increases the height of the potential barrier. The entropy of migration in a binary alloy A-C may be calculated, according to Eq. (20), with the obtained values for the migration energy ΔU. The values of migration entropies, S_{mig}, are presented in *Table 1*:a_{tern}^2/a_{bin}^2, for the (Cu,Zr)-C alloy only slightly deviates from unity and is equal to 1.00065. Thus it is possible to neglect this effect in studying the changes of the diffusion coefficient at small concentrations. Using the data from *Table 1* we estimated for (Cu,Zr)-C alloy at 800 K the ratio $D_{tern}/D_{bin} = 0.633$. The same calculations were performed for copper-carbon alloy with a set of dopants (see *Table 1*). All investigated additives decrease the diffusivity of carbon atoms in a copper matrix. On the basis of the data from *Table 1* it may be predicted that the strongest influence on the diffusion coefficient ratio among the investigated additives may be achieved by Zr alloying of Cu-C system. The predicted reduction of the diffusivity by a factor of two may be a basis for the detailed experimental study of the ways of improvement of carbon fibre interactions with a copper based matrix.

Acknowledgements

We highly appreciate the support of this research by the Grant no. 94-44/2 from the United States-Israel Binational Science Foundation (BSF), Jerusalem, Israel. S.D. acknowledges the Israel Ministry of Absorption for support.

References

1 Kuang, X., Carotenuto, G., Zhu, Z. and Nicolais, L., *Science and Engineering of Composite Materials*, 1996, **5**, 9
2 Gnesin, G.G.and Naidich, Yu. V., *Poroshkovaya Metallurgia*, 1969, **74**, 57
3 Dorfman, S. and Fuks, D., *Composites*, 1996, **27A**, 697
4 Mortimer, D.A. and Nicholas, M.J., *Journal of Materials Science*, 1970, **5**, 149
5 Sun, S.J. and Zhang, M.D., *Journal of Materials Science*,1991, **26**, 5762
6 Stefanik, P. and Stebo, P., *Journal of Materials Science Letters*, 1993, **12**, 1083
7 Krasko, G.L., *ScriptaMetals et Materials*, 1993, **28**, 1543
8 Ellis, D.E., In *MetalCluster Compounds*, ed. L.J. de Jongh. Reidel, Amsterdam, 1994
9 Khachaturyan, A.G., *Theory of Structural Transformations in Solids*. Wiley, New York, 1983
10 Bachelet, G.B., Hamman, D.R. and Schlüter, M., *Physics Review*, 1982, **B26**, 4199
11 Shewmon, P.G., *Diffusion in Solids*. McGraw-Hill, New York, 1962

Materials & Design, Vol. 18, Nos. 4/6, pp. 339–343, 1997
© 1998 Published by Elsevier Science Ltd
Printed in Great Britain. All rights reserved
0261-3069/98 $19.00 + 0.00

PII: S0261-3069(97)00074-5

Microcracking at the change of the deformation path of Al rods

M. Kurowski*, **J. Kuśnierz, A. Grabianowski, E. Bielańska**

Polish Academy of Sciences, Institute of Metallurgy and Materials Science, 25 Reymonta St., 30-059 Kraków, Poland

Received 15 July 1997; accepted 21 July 1997

The investigations were conducted on Al rods pre-deformed using the following technologies: drawing (A); rolling in the mode 'circle-oval-circle' (B); and in the combined mode (C); and next subjected to torsion. Microcracks observed on the surface of samples undergoing torsion occurred in the range of work softening (I) as well as in the range of work hardening (II). Their character was dependent on the pre-treatment of the sample surface. © 1998 Published by Elsevier Science Ltd. All rights reserved.

Keywords: work softening; properties; microcracks

Introduction

The change in the deformation path of rods of circular cross-section which have been subjected to layered work hardening at ambient temperature leads to the reduction of the work of plastic strain[1]. The negative phenomenon here is the occurrence of microcracks[2].

In Cu samples undergoing torsion the occurrence of a reduced range of work hardening (work softening) (I) and the range of increased work hardening (II), as well as the existence of microcracks on the outer surface of rods, following the change in the deformation path, both in the I and II deformation range, have been observed. It has been found earlier that in these samples prepared for torsion in the standard way, the phenomenon of surface microcracks was closely associated, especially in the (I) deformation range, with work hardening originating from the lathe tool[1,2].

In the present study this problem is considered with reference to Al 99.8 rods, which have been shaped by the technologies of drawing (A)[3], rolling in the mode 'circle-oval-circle' (B)[4] and in the combined mode (C)[5], where elongation was the dominating direction of deformation and next subjected to torsion. In the investigations carried out at present, the samples were addi-

*Correspondence to Msc M. Kurowski. Tel.: +48 12 374200/374580; fax: +48 12 372192; e-mail: office@imim-pan.krakow.pl

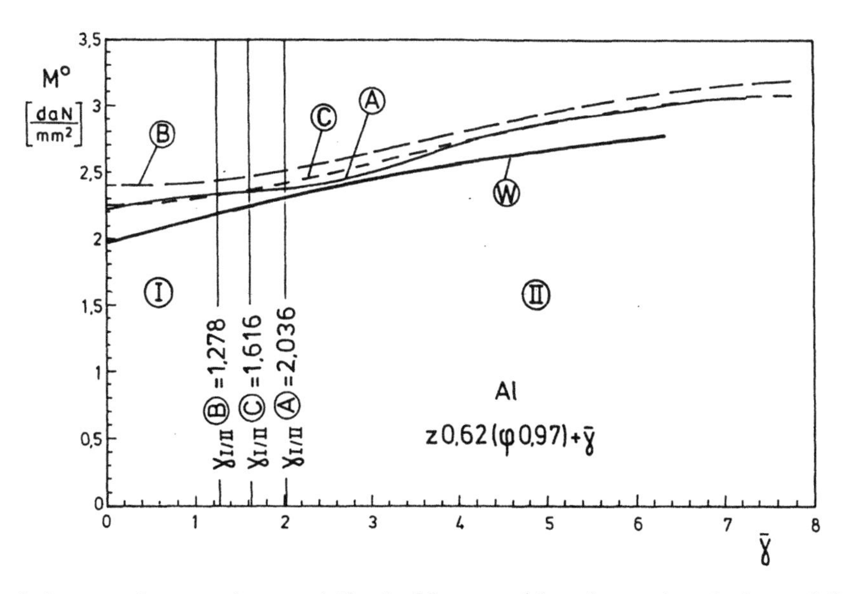

Figure 1 Work hardening curves from a torsion test of Al rods, Φ8 mm pre-deformed up to the reduction $z = 0.62$ ($\varphi = 0.97$) by the technologies: (A) drawing; (B) rolling; and (C) combined technology. The course of work hardening or rods annealed at the recrystallization temperature W is marked

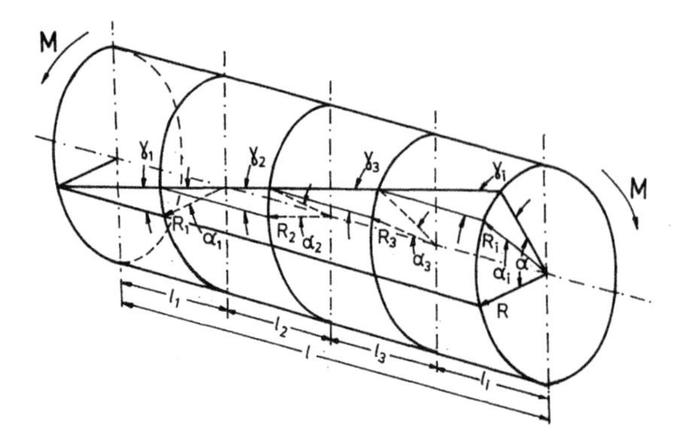

Figure 2 Model of the deformation geometry in the torsion-tested Al rods assumed in the present study

tionally subjected to electrochemical treatment, prior to torsion, to eliminate the results of pre-machining. Consideration has been also given to the dependence between microcracks and the roughness of the outer surface of samples used in the investigations.

Experimental

The investigations were carried out on rods of 13 mm, supplied by a processing plant and annealed at the recrystallization temperature 643 K for 1.5 h. The rods were deformed to obtain the diameter 8 mm by the technology of drawing (A), inducing the layered work hardening distribution with a gradient of a parabolic shape with the vertex directed downwards; by the technology of rolling (B), where the vertex of the parabola of such a distribution is directed upwards; and by an intermediate, combined technology, composed of the two technologies mentioned here (C). The rods were next subjected to torsion in special holders (practically without reducing their diameter) to the limit of uniform strain γ. The course of work hardening is described by the polynomial of IV degree in the system $M^0 = f(\gamma)$ (*Figure 1*), assuming the deformation geometry (*Figure 2*)[6], with:

$$M^0 = \frac{M}{2\pi R^3}; \bar{\gamma} = \frac{R\alpha}{l}; z = 1 - \frac{A}{A_0}; \quad \text{and} \; \varphi = \ln\frac{A_0}{A},$$

where A_0 and A are the rod cross-sections before and after deformation, respectively.

As it is seen, for the technologies (A), (B) and (C), two different ranges can be distinguished: of reduced work hardening (I) and reinforced work hardening (II).

Considering the fact that the occurrence of microcracks may be connected with the rearrangement of work hardening distribution, there have been carried out investigations of hardness distribution HM (according to Meyer) in Al rods before and after the change of the deformation path to the strain 0.5 $\bar{\gamma}_I$ and $\bar{\gamma}_{II}$, corresponding to M^0_{max}. The measurements were conducted using the special spherical penetrator of 442 μm diameter and a load of 17.5 N/30 s and the results were interpreted by means of a graphic computer pro-

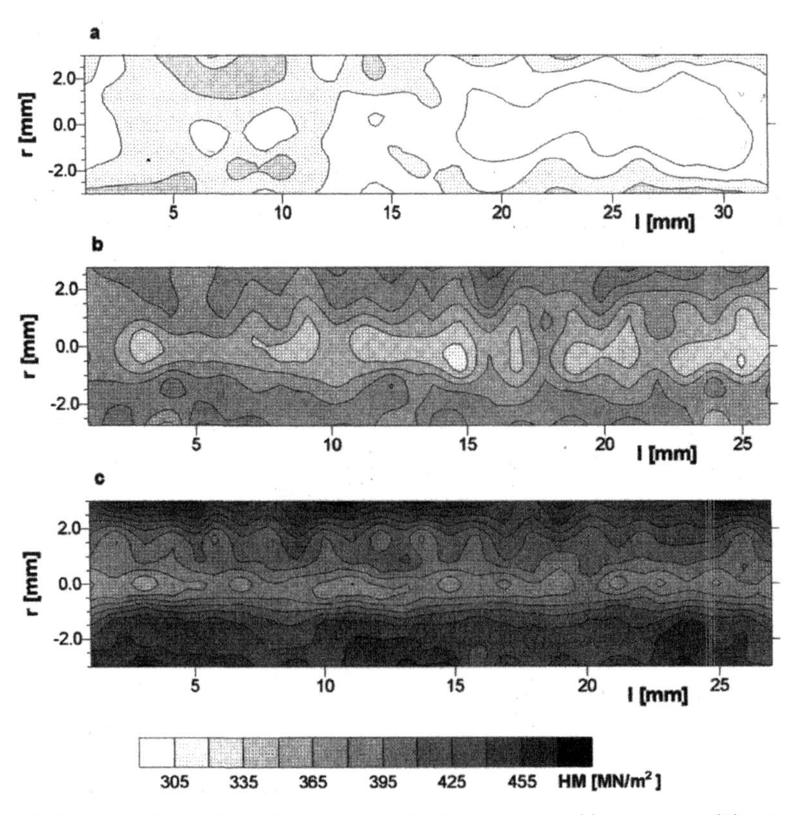

Figure 3 Hardness distribution HM on the longitudinal section of Al rods of 8 mm diameter: (a) after drawing (A) up to $z = 0.62$ ($\varphi = 0.97$); (b) after subsequent deformation by $0.5\bar{\gamma}_I$; and (c) after further deformation up to $\bar{\gamma}_{II}$

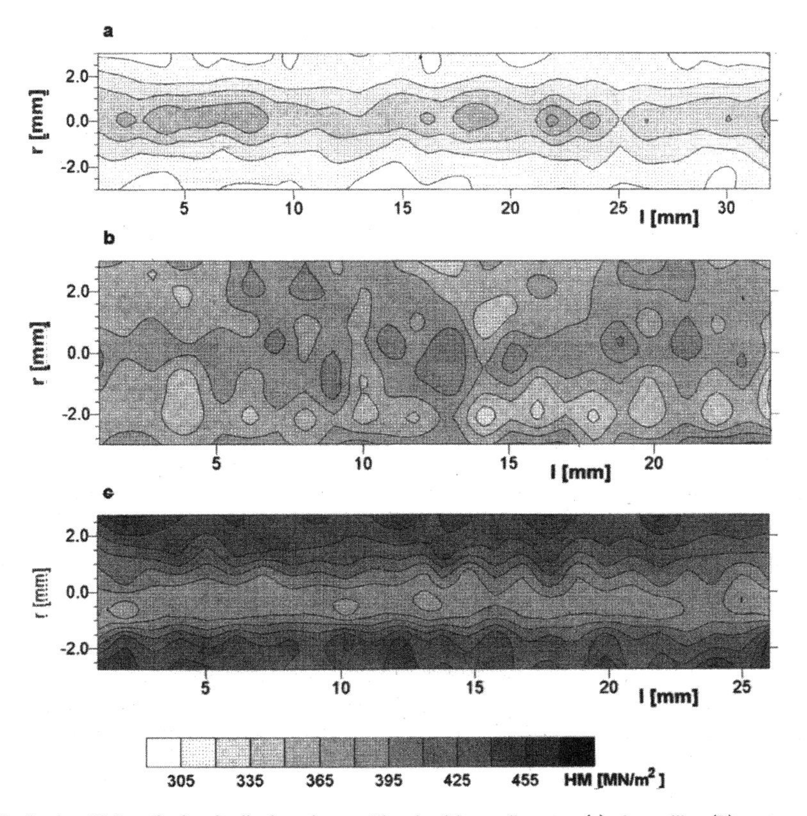

Figure 4 Hardness distribution HM on the longitudinal section on Al rods of 8 mm diameter: (a) after rolling (B) up to $z = 0.62$ ($\varphi = 0.97$); (b) after subsequent deformation by $0.5\bar{\gamma}_I$; and (c) after further deformation up to $\bar{\gamma}_{II}$

gramme (*Figures 3–6*). Such measurements illustrate the work hardening distribution in metal, which is associated with its mechanical properties in the macroscale.

The microcracks of rods were examined, using Philips XL 30 scanning microscope for the samples subjected to the same deformations at which the hardness distributions HM were determined (*Figures 7–9*). It can be noticed that within the range I, in which the unstable metal flow in Al rods occurs to a small extent, there already occur the first microcracks inclined at the angle γ to the axis of the rod undergoing torsion. In the range II of work hardening, on the surface of Al rods, folds of strongly deformed metal are formed, rich in numerous microcracks which change their positions with respect to the rod axis, becoming perpendicular. *Figures 7* and *8* show the observed microcracks on the surface of samples polished electrolytically prior to torsion. For the purpose of comparison, the microcracks formed on the samples surface after careful machining are seen in (*Figure 9*).

The measured surface roughness was only approx.

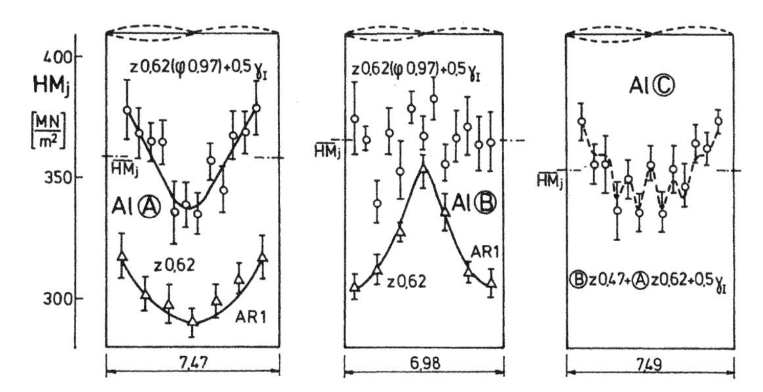

Figure 5 Hardness distribution, HM, at the assumption of a layered work hardening in Al rods (A), (B) and (C) following a change in the deformation path up to the strain $0.5\bar{\gamma}_I$

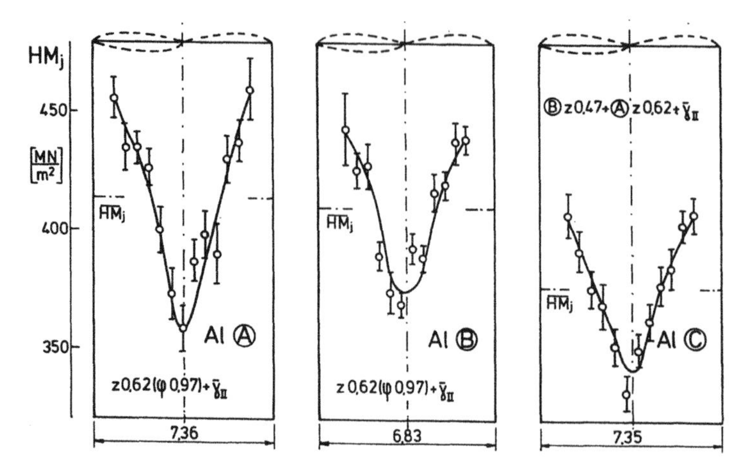

Figure 6 Hardness distribution, HM, at the assumption of a layered work hardening in Al rods (A), (B) and (C) following their torsion up to the strain $\bar{\gamma}_{II}$

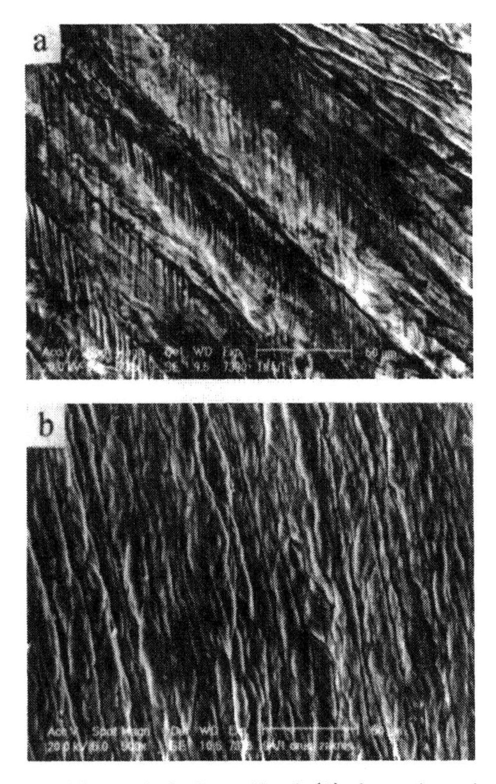

Figure 7 Microcracks in drawn Al rods (A) after a change in the deformation path. Surface sample after electrochemical etching. Photograph taken in a scaning microscope. Magnification 500 ×. The magnification marker is parallel to the sample axis. (a) $z = 0.62$ ($\varphi = 0.97$) + $0.5\bar{\gamma}_I$. (b) $z = 0.62$ ($\varphi = 0.97$) + $\bar{\gamma}_{II}$

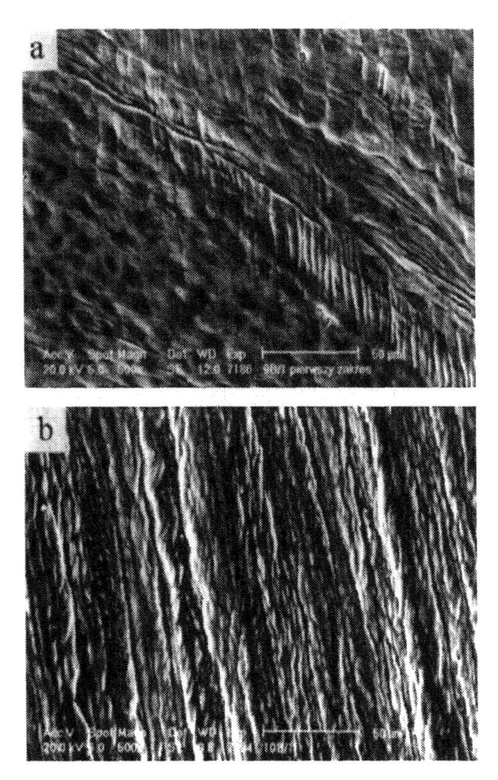

Figure 8 Microcracks in rolled Al rods (B) after a change in the deformation path. Surface sample after electrochemical etching. Photograph taken in a scanning microscope. Magnification 500 ×. The magnification marker is parallel to the sample axis. (a) $z = 0.62$ ($\varphi = 0.97$) + $0.5\bar{\gamma}_I$. (b) $z = 0.62$ ($\varphi = 0.97$) + $\bar{\gamma}_{II}$

30% of the standard value. A considerable effect of the notch caused by the lathe tool on the such microcracks generation can be observed. Both in the range I and the range II of work hardening, they are situated perpendicular to the rod axis, thus they are formed at places where scratches made by the turning tool exist.

Conclusions

Change in the deformation path: elongation-torsion produces microcracks in Al rods, situated mainly at γ angle (*Figure 2*), although microcracks situated in the plane perpendicular to the rod axis have also been observed.

Microcracks in drawn Al rods seem to be greater

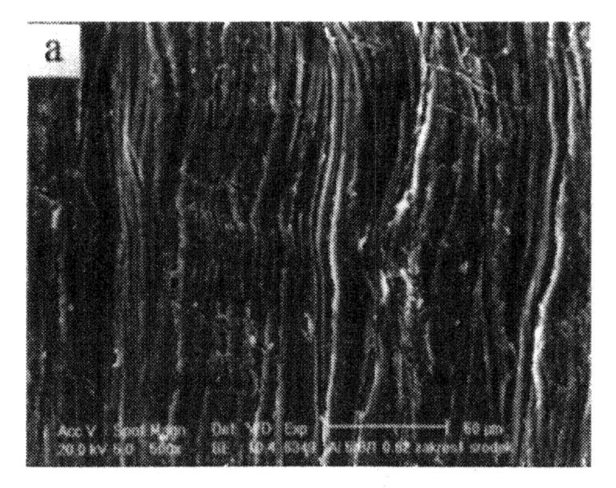

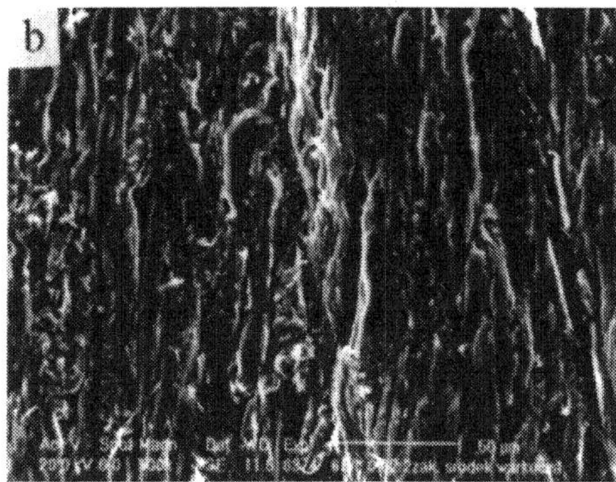

Figure 9 Microcracks on the outer surface of rolled Al rods (B) after a change of the deformation path. Surface sample without electrochemical etching. Photograph taken with a scanning microscope. Magnification 500 ×. The magnification marker is parallel to the sample axis. (a) $z = 0.62$ ($\varphi = 0.97$) + $0.5\bar{\gamma}_{I}$. (b) $z = 0.62$ ($\varphi = 0.97$) + $\bar{\gamma}_{II}$

than in the rolled ones. This phenomenon is associated with the change in the character of work hardening distribution. In rolled rods the gradient of the 'layered' work hardening undergoes a change, whereas in drawn rods the gradient remains the same, only the work hardening distribution shows the 'island-like' form. The 'island-like' work hardening in Al rods should be attributed to the fact that the proportion of the work of friction on the contact metal-tool in the total deformation work is greater when compared with other metals[1].

Change in the deformation path: elongation-torsion induces also the occurrence of microcracks on the outer surface of the rods even at a low value of the shear strain. With high values of γ the effect of microcracks becomes more intensive, contributing to considerable deformations of sample surface.

Great importance of the surface pre-treatment of samples subjected to torsion has been observed. Even a low grade of the sample roughness results in a change of the shape and position of the newly formed microcracks.

References

1 Grabianowski, A., Kurowski, M., Bielańska, E. and Schütz, H. *Zeitschrift für Metallkunde*, 1995, **86**, 12
2 Grabianowski, A., Kurowski, M. and Hamankiewicz, M. *Zeitschrift für Metallkunde*, 1989, **80**, 9
3 Grabianowski, A., Dańda, A. and Ortner, B. *Aluminium*, 1990, **66 J**, 3
4 Grabianowski, A., Dańda, A. and Ortner, B. *Rudy i Metale Nieżelazne*, 1991, **36**, 5
5 Grabianowski, A., Kloch, J. and Ortner, B. *Materials Science and Technology*, 1994, **10**, 227
6 Grabianowski, A. and Kurowski, M. *Zeitschrift für Metallkunde*, 1992, **83**, 1

Materials & Design, Vol. 18, Nos. 4/6, pp. 345–347, 1997
© 1998 Published by Elsevier Science Ltd
Printed in Great Britain. All rights reserved
0261-3069/98 $19.00 + 0.00

PII: S0261-3069(97)00075-7

Structure and mechanical properties of age-hardened directionally solidified AlSiCu alloys

Janusz Król*

A. Krupkowski Institute of Metallurgy and Materials Science, Polish Academy of Sciences, Reymonta 25, 30 059 Cracow, Poland

Received 30 June 1997; accepted 15 July 1997

The structure of directionally solidified Al-Si hypoeutectic alloys are generally composed of Al-matrix and Si-reinforcing phase. The growth direction of the both phases was near ⟨200⟩. The strength properties of an Al-Si alloy with additions of 2 wt.% and 4 wt.% copper have been investigated. These alloys were solution treated, quenched in water and aged at 200°C. Large Al_2Cu precipitates present in D.S. samples dissolved partly, and after ageing, they precipitated as the Θ′ platelets, significantly increasing the mechanical properties of the alloys. Hardness, strength and elongation were measured in the course of ageing. The structure was investigated by means of: XSAS, X-ray phase analysis, lattice parameter measurements and scanning microscopy. © 1998 Published by Elsevier Science Ltd. All rights reserved.

Keywords: directionally solidified Al-Si hypoeutectic alloys; structural; mechanical; properties

Introduction

The structure of directionally solidified eutectic alloys is composed of a matrix and reinforcing phase. At small solidification rates (up to 100 μm s^{-1}), silicon precipitates appear as flakes, but at higher solid fication rates as fibres[1,2]. The flake structure causes a strong increase of ductility[3]. Both phases reveal a sharp crystallographic orientation which is important for the mechanical properties. The crystallographic relationship is dependent on the solidification parameters. The most frequent orientation relationships are as follows: [100]Al||[100]Si[4]; [010]Al||[111]Si[5] and [100]Al||[100]Si[6]. The mechanical properties, which are low in an Al-Si alloy could be increased by an addition generating precipitation strengthening. The highest UTS can be achieved in the aluminium with 10% Si and 4% Cu alloy[7]. The Θ′ precipitates appear in D.S. Al-Si-Cu alloy as disperse, plate-like forms after heat treatment, increasing the alloy strength considerably[7,8]. The object of the presented investigations was to examine the influence of copper addition on the mechanical properties and structure of the heat treated D.S. hypoeutectic aluminium-silicon alloys.

Experimental procedure

Two aluminium alloys of composition AlSi10Cu2 and AlSi10Cu4 (wt.%) were investigated. They were melted and cast (under argon atmosphere), directionally solidified at 28 μm s^{-1}, homogenised at 530°C, quenched in RT water and aged at 200°C. Microhardness was measured on a Zeiss (Neophot 30) microscope at a 20 g load; X-ray phase analysis, lattice parameters and texture were measured on a Philips PW 1710 diffractometer, and XSAS measurements were performed on a Rigaku-Denki diffractometer using a slits camera standardised by Lupolen M1812. The scanning microscope observations were carried out on a Philips XL-30 apparatus. The tensile tests were performed at room temperature and at a strain rate of 5×10^{-5} s^{-1} using an INSTRON testing machine. The diameter of the tensile specimen was 3 mm and the gauge length 15 mm.

Results and discussion

Microhardness measurements

The dependence of microhardness on ageing time is shown in *Figure 1*. The alloy with 2% of copper attained much lower hardness (max. about 65 mHv after 4 h of ageing) than the one with 4% Cu (92 mHv after 8 h). For the longer ageing times (above 10 and 24 h, for the alloys containing 2 and 4% of copper, respectively) microhardness decreased strongly. Hardness of the alloy with higher copper addition measured immediately after quenching was much higher according to the higher content of undissolved Al_2Cu phase.

X-ray phase and texture analysis

According to the results of X-ray phase analysis both alloys were composed of the Al-solid solution, Si-pre-

*Tel.: +48 12 374200; fax: +48 12 372192; e-mail: nmkrol@imim-pan.krakow.pl

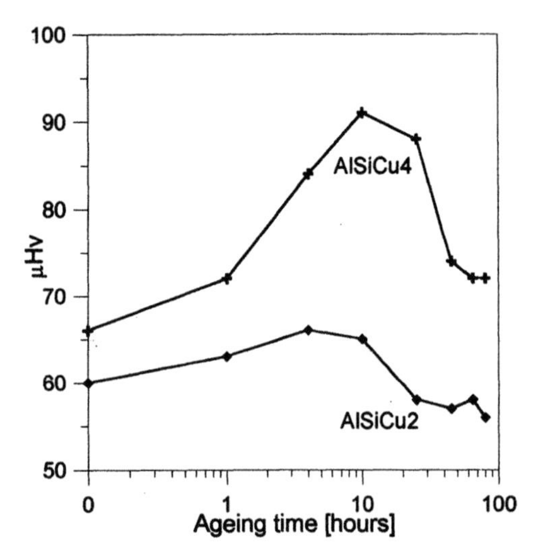

Figure 1 The dependence of microhardness of the investigated alloys on the ageing time

cipitates and Θ′ precipitates. The {200} aluminium diffraction line attained very high intensity, which suggested sharp fibre ⟨200⟩ preferred orientation of the matrix. The changes of copper content in the matrix affected the kinetics of the Θ′-phase precipitation (*Figure 2*), which was calculated from lattice parameter measurements on the basis of the data reported by Elwood and Silcock[9]. Only 1.80 and 3.60% of copper in the alloys with smaller and higher copper content, respectively, dissolved in the matrix after quench. The rate of precipitation was higher in the alloy with 4% of copper, in which after 60 h all copper from the matrix precipitated. In the alloy with 2% Cu, about 0.2% Cu still remained in the matrix after 90 h of ageing.

The crystallographic relationship for the both basic phases of the investigated alloys found from pole figures of the texture was: ⟨200⟩ Al‖⟨200⟩ Si. The matrix preferred orientation was sharp but deviated from the

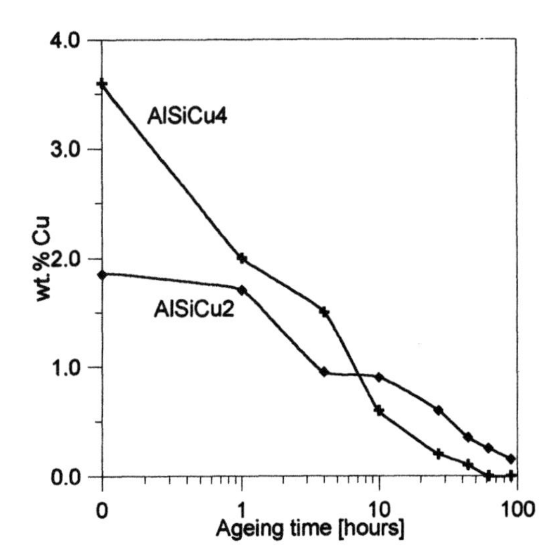

Figure 2 The changes of copper content as a function of ageing time

⟨200⟩ direction of about 10°. The deviations in the silicon phase were smaller but with a small contribution of ⟨111⟩ component of orientation which well corresponded to[5].

X-ray small angle scattering analysis

Guinier radius—related to precipitate size (thickness of the platelets) and integral intensity—related to amount of the precipitates for the both alloys in dependence of ageing time are shown in *Figure 3(a)* and *(b)*, respectively. During ageing two sizes of the precipitates formed in both alloys. The small precipitates (2–3 nm) according to the course of the integral intensity vs ageing time were present in small amounts up to the longest ageing time while the amount of large precipitates (6 to above 9 nm) increase strongly up to slightly above 20 h of ageing. The highest rate of the formation of precipitates between 8 and 16 h of ageing was in a good agreement with the microhardness maximum and the results calculated from the changes of the lattice

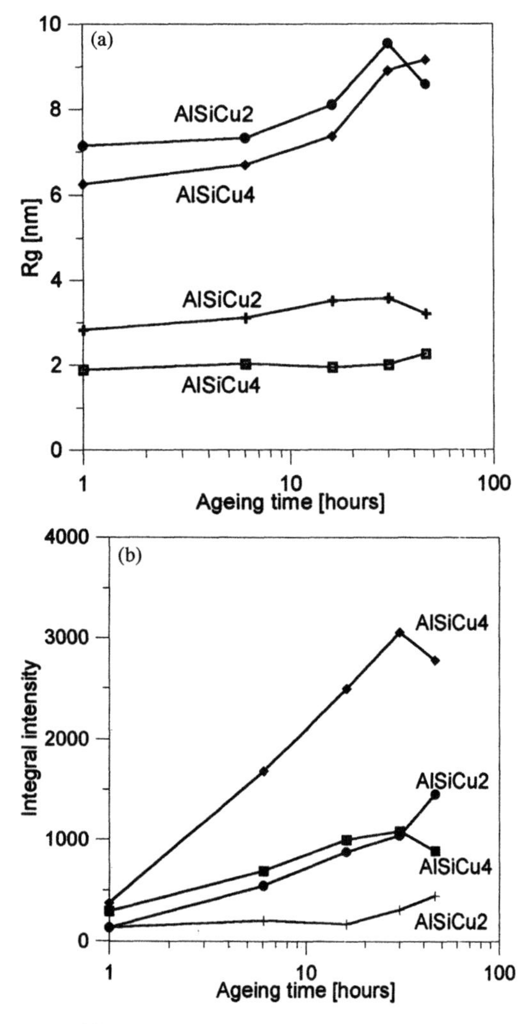

Figure 3 (a) The Guinier radius in dependence of ageing time on the both investigated alloys. (b) The integral intensity as a function of ageing time for the both alloys

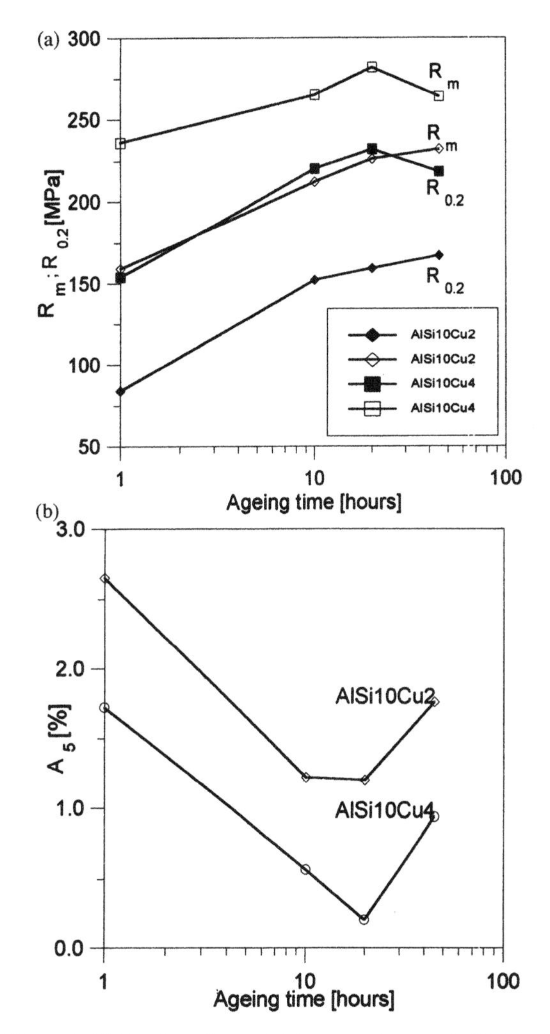

Figure 4 (a) The dependence of ultimate tensile strength (R_m) and yield stress ($R_{0.2}$) on the ageing time of the both alloys. (b) The dependence of elongation (A_5) on the ageing time of the both alloys

parameter (the highest precipitation rate of copper during ageing). The larger thickness of precipitates in the alloy with smaller copper content could be explained by the larger free path between the copper clusters. The large size precipitates (above 8 nm) which formed in large amounts after prolonged ageing, diminished the hardness.

Tensile tests

The tensile tests were carried out on both alloys at room temperature on the test specimens aged at 200°C after 2, 10, 20 and 45 h. The results are shown in the diagrams of the dependences of ultimate tensile strength (R_m), yield stress ($R_{0.2}$) and elongation (A_5) versus ageing time (*Figure 4(a)* and *(b)*). In both alloys the tensile strength and yield stress increased according to the rise of the volume fraction of the large precipitates during ageing. The highest values of R_m and $R_{0.2}$ were found after ageing time between 10 and 20 h. In the alloy with a higher copper addition aged longer as 20 h, these two values diminished according to the results of the structural investigations. The smallest value of elongation was found for the ageing time at which the highest strength occurred and could be bounded with the thickening of the Θ' precipitates.

Scanning microscopy analysis

Fields of matrix, silicon precipitates and large, stable Al_2Cu precipitates were seen in the microstructures of the D.S. alloys in a grown state. After annealing and quenching, the flakes of silicon precipitates remained unchanged, but in the matrix, the Θ' plate-like precipitates occurred. The EDS analysis results corresponded well with the results of the lattice parameter changes. In the quenched state, the alloys with smaller and larger copper additions contained 1.8 and 3.65% Cu, but after 44 h of ageing only about 0.1% or 0% of copper, respectively.

Conclusions

(1) The large, stable Al_2Cu precipitates in the grown state, remained after homogenisation and quenching in small quantities.

(2) The increase of microhardness was in good agreement with the ultimate strength changes and corresponds well to the volume fraction of the copper precipitated from a solid solution of the matrix according to the results of lattice parameter and XSAS measurements.

(3) The decrease of ultimate strength (R_m) and microhardness depends on the growth and transition into the stable, large Al_2Cu precipitates in agreement with the results of the lattice parameter and XSAS measurements.

Acknowledgements

Financial support by grant No. 7 S201 07406 from the State Committee for Scientific Research is gratefully acknowledged.

References

1 Magnin, P., Mason, J.T. and Triverdi, R., *Acta Metall. Mater.*, 1991, **39**, 469
2 Shu-Zu Lu and Hellawell, A., *J. Cryst. Growth*, 1985, **73**, 316
3 Steen, H.A.H. and Hellawell, A., *Acta Met.*, 1975, **23**, 522
4 Shamzuzzoha, M. and Hogan, L.M., *J. Cryst. Growth*, 1987, **82**, 598
5 Eskin, D.G., *Z. Metallkde* 1992, **83**, 762
6 Król, J. and Dytkowicz, A., In *Proc. Conf. Cast Composites '95, Zakopane*, eds. Polish Foundrymen's Technical Associations, Poland, 1995, pp. 126
7 Eskin, D.G., *Z. Metallkde*, 1995, **86**, 60
8 Starink, M.J. and Van Mourik, P., *Met. Trans. A*, 1991, **22A**, 665
9 Elwood, E.C. and Silcock, J.M., *J. Inst. Met.*, 1948, **74**, 457

Materials & Design, Vol. 18, Nos. 4/6, pp. 349–355, 1997

0261-3069/98 $19.00 + 0.00

PII: S0261-3069(97)00076-9

Carbon fibre in automotive applications

H. Adam*

Forschungsgesellschaft Kraftfahrwesen Aachen mbH (fka), Steinbachstr. 10, 52074 Aachen, Germany

Received 24 July 1997; accepted 13 August 1997

Short fiber reinforced composites have proven their substantial potential for automotive application and are state-of-art technology for volume production of non-structural vehicle components. Examples for these components are spread over the vehicle from frontend-reinforcement member made by GMT stamping to intake manifold made by injection moulded thermoplastics. Based on the high potential of advanced composites for both, structural lightweight design and material lightweight design, research and development focussing on innovative engineering solutions is an ongoing process within the automotive industry and related institutions. In this paper, different automotive components made out of advanced composites are presented and discussed with reference to their chances and risks for automotive realization. Furthermore, main restrictions and restraints for the use of advanced composites, e.g. load applications, damage tolerance and high volume production technologies, are explained. Technical solutions as key enablers for industrial realization are shown. © 1998 Published by Elsevier Science Ltd. All rights reserved.

Keywords: carbon fibre; automotive applications; non-structural vehicle components

Introduction

The world-wide competition to reach trade acceptance, cost reduction and to fulfil the legal requirements of the motor vehicle is among other things settled through improved and new materials as well as through optimized construction methods. The request to lower the total expenses as well as the raising of ecological relief can be mentioned as driving force for the usage of new materials. A raising of the ecological relief of future vehicle systems has to go along with a distinct lowering of the automobile pollutant emission — especially the CO_2 — emission. The introduction of marketable innovative products connected with a drastic lowering of the moving masses can contribute decisively to meet the requirements of the future motor-vehicle market (*Figure 1*).

One attempt to solve the problem is found in the introduction of lightweight design by means of the realization of constructive as well as material lightweight measures. Therefore, conventional materials, as e.g. steel, as well as the lightweight materials aluminium alloys and magnesium alloys are used in new developments. In the last few years, more intense effort has been spent on the application of fibre reinforced plastics. While glass fibre reinforced plastics (GRP) are used in several vehicle components, the carbon fibres as reinforcing material are of a relative low significance for the lightweight construction in motor vehicles. In contrast to this especially in the aeronautics and aerospace the CRP components (carbon fibre reinforced plastics) are very significant for the material compound of modern constructions.

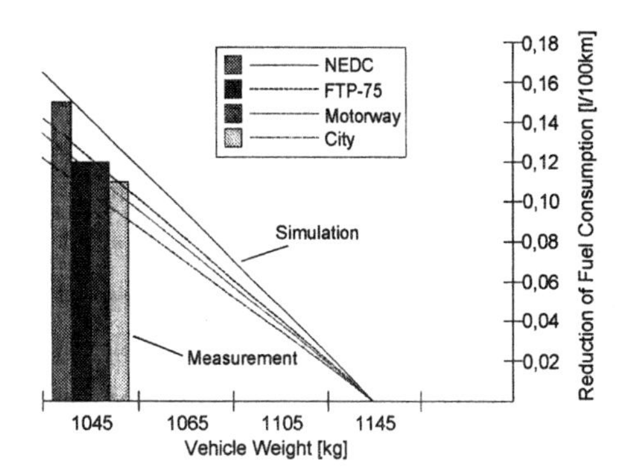

Figure 1 Influence of the weight reduction on fuel consumption for a VW Golf TDI[1]

Characteristics of CRP-materials

CRP-materials belong to the class of fibre compound plastics (FCP). These compound materials are in the main constructed of two main components. These are on the one hand the matrix material and on the other hand the fibre material supporting the reinforcement. According to this dual construction the essential parameters of influence on the characteristics of FCP can be seen. Therefore, e.g. as a result of the choice of fibre type and matrix system, fibre content, fibre diameter, fibre structure and fibre orientation the mechanical

* Tel.: +49 241 8861115; fax: +49 241 8861110; e-mail: adam@fka.de

characteristics of FCP are variable and as a result of the utilization of these design liberties in the material compound, a design that meets all demands can be carried out. The layer construction and the production methods are of a high influence regarding the material characteristics as well.

For CRP-applications, either thermoplasts like PP, PA, PEI, PEEK, PES or duroplasts like UP, EP, VE, phenol-resins (PF) are used as matrix materials. As a result of the improved mechanical characteristics referring to temperature resistance, creeping behaviour and treatment the duroplasts are used preferably. According to the requirements of the material to be used, carbon fibres with particular characteristics can be used. The two fibre types being mainly used are HM (high module) and HT (high tensile) fibres. HM-fibres can be characterized by a high E-module and therefore by a high tensile strength. HT-fibres can be characterized by a higher ductile yield compared with the HM-fibres. Based on this classification further fibre types like IM-fibres (intermediate module) and UHM-fibres (ultra high module) as well as ST-fibres (super tensile) are introduced. High specific rigidities can be seen for CRP-materials at simultaneous low weight compared to usually used metallic materials as steel and aluminium because of the use of reinforced fibres with high tensile strengths and because of the low density of the materials being used for the composition of FCP. In *Figure 2(a)* this fact is illustrated.

In addition to the aspect of the material lightweight construction it can be realized by the usage of variable cross-sections and wall thicknesses. Furthermore, the big design liberty which results from the usage of CRP-materials, offers a huge integration potential as a result of initiating load introduction and function elements. The application of CRP-materials within the scope of the crash management of vehicles is also possible because of the CRP-material's ability to absorb energy extremely well. But the prediction of the CRP's fatigue strength turns out to be difficult. Accord-

(a)

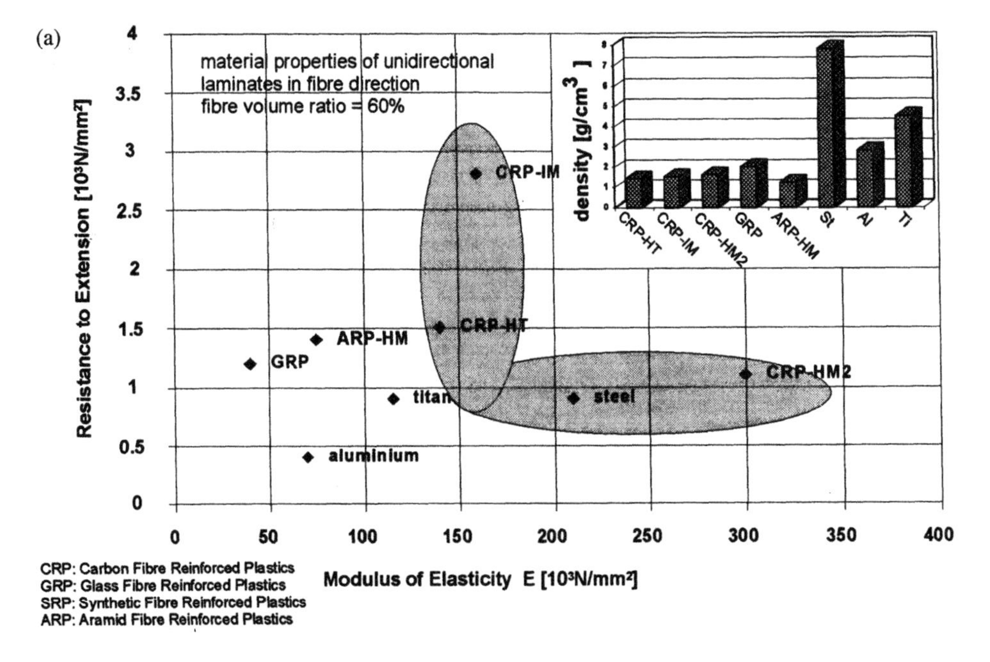

(b)

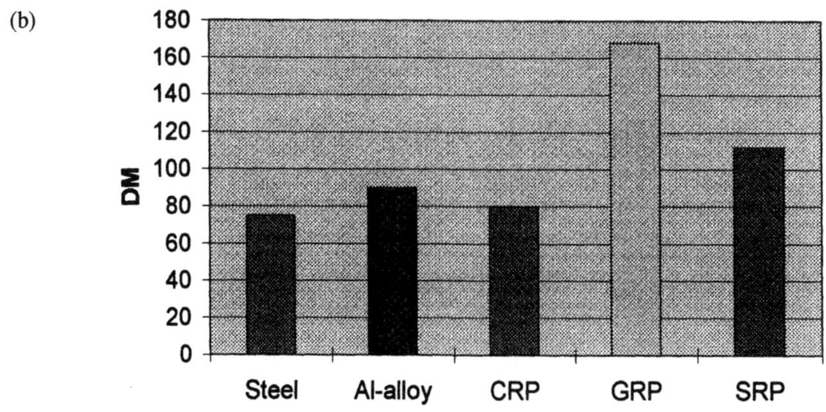

Figure 2 (a) Comparison of specific material characteristics. (b) Comparison of the production costs for one example component (Cabin of MB UX 100 = UNIMOG)[3]

ing to load level and load frequency as well as surrounding temperature a clear stiffness reduction can be seen so that at the moment only the time stiffness can be predicted.

In addition to these aspects of CRP's mechanical characteristics, criteria of economic efficiency and ecological recycling have also to be taken into consideration for the decision on the application of such an alternative material. The recycling of CRP-materials is possible either as mechanical recycling within the frame of the material utilization (down cycling) or as chemical recycling (raw material utilization). Additionally, the possibilities of a thermal utilization by energy recovery through the heat value as well as a refuse dump disposal should be taken into account.

An observation of the economic efficiency is only useful if the excellent mechanical characteristics of CRP-materials are taken into account. Therefore, the question concerning the price of a certain amount of the material to be used does not stand alone in the focal point, but it is more the question concerning the mass of the really used material. A weight reduction potential of 50–70% for CRP-materials results in comparison to metallic materials and other compound materials even if the different demand criteria as buckling, bending, E-module, compression strength and tensile strength are taken into account. Paying attention on this amount factor the result can be seen in *Figure 2(b)*, where a comparison of the production costs is illustrated with reference on the number of components being 4000.

Therefore it can be seen that the production costs for one CRP-application for a small series can be compared with those with a steel structure despite of a higher material price.

(a)

Method	Fibre content	Productivity
Hand lay-up moulding	ca. 50-70 %	low
Tape Winding method	ca. 60 %	medium-high
Braiding method	ca. 50-60 %	high
Vacuum injection method	ca. 30-60 %	high
Pultrusion	ca. 70 %	high
Wet compression moulding	ca. 20-50 %	medium
GMT-SMC	ca. 30 %	high
Spray-up technique	ca. 30 %	medium

(b) **Prototype Tape-Winding of an Axle on a Portal Robot**

Figure 3 (a) Comparison of the reachable fibre volume contents and of production techniques being different in productivity. (b) Production of a commercial vehicle axle by means of the winding method

Production methods

As described in the preceding chapter, the characteristics of FCP are not clearly defined unlike isotropic materials, but they are among other things determined by the used production techniques and their specific restrictions. The choice of one specified production method essentially depends on the resulting costs and on the realizability of the component to be produced in the relevant production method. In order to guarantee an economic production methods with a high throughput are absolutely necessary. Either by means of low clock times or by means of high integrative parts with higher clock times this can be enabled.

The production methods (*Figure 3(a)*), Hand lay-up moulding (wet), Hand lay-up moulding (prepreg) and the tape winding (prepreg) cannot meet the requirements of a high productivity. After all the usage of prepregs is unsuitable as a result of the unfavourable cost situation for automotive applications. From this point of view, the RTM-technology offers a better economic balance. This technology deals with a resin injection method with sized supporting materials. As a result of the application of fibre prepregs and because of the usage of one closed process, favourable clock times are possible in average series. Furthermore standard parts as well as smaller parts can be integrated in the production process and a surface quality corresponding to the automotive standard.

It is true that the pultrusion shows a high productivity, but by means of this method first of all only endless profiles with an easy geometry can be produced. The possibilities referring to the integration of load introduction elements and to the variations of the fibre orientations are admittedly very restricted.

In the methods compression moulding, Spray-up technique, SMC (sheet moulding compound), GMT and BMC (bulk moulding compound) a very high productivity can already be realized. But these production methods, processing short fibres, show too low realizable fibre volume contents as well as insufficient possibilities for the influencing of the fibre orientations for the production of dynamically high loaded space frames. These methods are especially suitable for the production of smaller, shell formed parts.

Components with larger surfaces are produced in injection processes. Fibre mats being bedded in forms are mixed with the resin under pressure in a vacuum.

Those production methods characterized by a high automation level and which help to realize endless fibre reinforcements are on the short-list for the production of rotation symmetrical form bodies: winding method, weaving method and weaving method with rovings. In the case of the production method winding the production speed can be further increased as a result of the application of preconfectioned textile semi-finished products. A commercial vehicle axle produced at the RWTH-Aachen (Technical University of Aachen) by means of the braiding method within the frame of a research project is shown in *Figure 3(b)* as an example for the production of such a body.

Design principles for CRP-applications

For the design of CRP-parts, it is important to reconsider the specific material characteristics; the positive as well as the negative ones. Therefore, for example, because of the relatively expansive material costs, carbon fibres should only be used for those applications, which necessarily demand for the outstanding mechanical qualities of CRP and which can exploit their qualities good enough.

Some other important aspects for the design with CRP are as follows:

- Creeping or relaxation at temperatures near the softening temperature
- Residual stresses at changes in temperature concentration and humidity concentration (because of the differences in material characteristics fibre-matrix)
- Temperature, humidity and chemicals take influence on the stiffness of the material

Particularly the design of load applications for parts made of CRP demands for an intensive care. Basically this aims for a CRP-like solution, which means that the integration of load application must not cause any damage to the fibres and to the matrix. Reconsidering this, a sensible use of the mechanical characteristics offered by CRP can be achieved. Some rules for the design of load applications are as follows:

- The bearing strength of the load application at operating conditions should be higher or equal to the one of the part, which means a possible failure should occur outside the load application
- Differences in thermal extension of joined materials must not cause residual stresses
- Additional masses and additional room have to be minimized
- In order to reach a simple and low-priced design of load applications the special characteristics of the production technology for FRP have to be taken into account
- Existing stiffness demands should be fulfilled which normally means, that there are not any additional pliabilities
- Especially for dynamically loaded parts special damping characteristics can be aimed
- The operating loads should be transferred at the existing working conditions and the bearing strength of a part should not be restricted by the load application
- Load applications for CRP-parts have to be wide spread as possible because locally restricted load applications lead to high stress components and high strain components normal to the laminate layers
- Joining CRP with metals or with metallic joining elements can cause electrical corrosion because the materials' big difference in their electrochemical potential. Especially CRP/alloy combinations and CRP/steel combinations are affected

The proceedings for the substitution of a steel part by a CRP part is shown for an axle body (*Figure 4(a)*), which was developed within a research project at the RWTH Aachen.

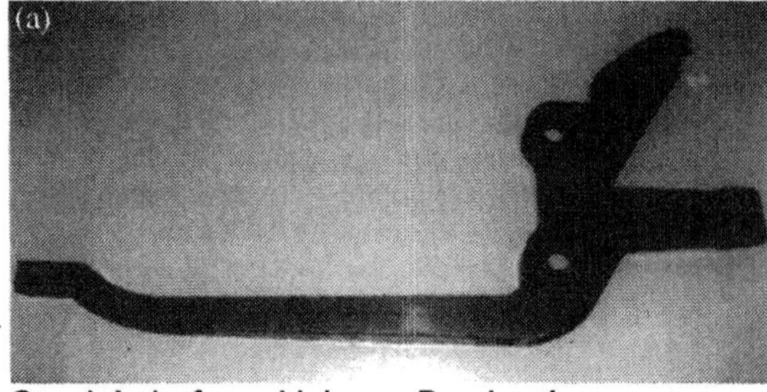

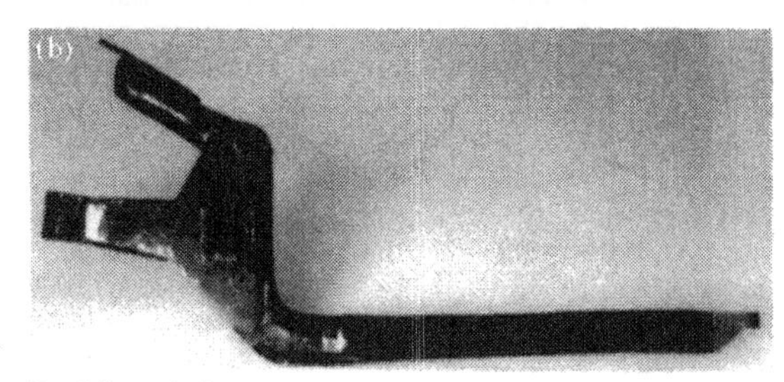

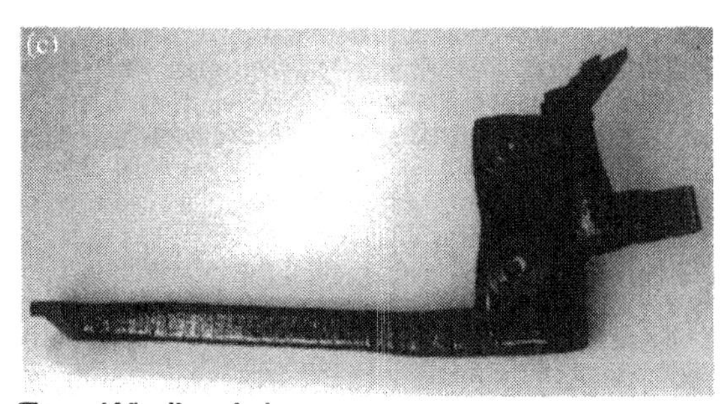

Figure 4 (a) Original axle to be substituted[2]. (b) Axle produced with braiding method[2]. (c) Axle produced with tape winding method[2]

The developmental targets for the substitution of steel parts by parts made of CRP are as follows:

- Reduction of masses
- Increase of comfort
- Reduction of payload stress
- Road saving
- Low production costs for small series
- Possibility of part and function integration
- Increase of maximum payload for commercial vehicles

By using modern CAE-techniques like CAD-design and FE-analysis the development targets could be fulfilled. A weight reduction of more than 50% could be realized. The chosen production technology of tape weaving and winding leads to a economic production as well. The axle bodies are shown in *Figure 4(b)* and *(c)*.

Examples

After having introduced the material characteristics of CRP and the production technologies for FRP this chapter will present possible automotive applications. On the basis of examples of chassis, body and drive-train applications possible uses of CFP are shown.

The use of CRP in body applications

A first example, as shown in *Figure 5(a)*, for the use of CRP in body applications, is the driver cabin of the MB 100 UX (= Unimog).

During the development process the following items were set as targets:

- Keeping the technical demands and requirements
- Target costs as low as possible with reference to a possible later series
- Massive weight reduction compared with a steel cabin
- Good acoustic characteristics
- Integration of additional functions

The material for this application is a duroplastic EP-matrix with carbon fibre prepreg. The different parts of the cabin are joined by structural and functional adhesives. By using this CRP-material, a weight reduction of 63% could be realized. This weight reduction can be reached on the one hand by a lower material mass and on the other hand by effects like loss of corrosion and underfloor protection and loss of noise and warmth insulating materials.

Another example for the use of FRP in automotive engineering is, as shown in *Figure 5(b)*, the Zatol L3 of the Zato Fahrzeugentwicklung GmbH. It is designed as a monocoque sandwich structure with body and chassis made of CRP sandwich elements. Because of the low vehicle weight, it is possible to reach a sporty vehicle behaviour with a relatively low engine power.

Another concept for the use of CRP in body applications is the design of a space frame structure made of CRP. The increasing number of variations on one basic vehicle concept demands for a very variable and flexible body concept. In this context, the space frame technology with a basic chassis made of frameworks connected by gusset structures offers the possibility to economically produce several variations in small series. By using front ends and rear ends made of CRP as well as other CRP-hang-on parts, it is possible to perform a quick and low-priced change of design in terms of a facelift. Furthermore variable cross sections and variable wall thicknesses can increase the wealth of using CRP for a space frame structure.

Within a research study performed by Institut für Kraftfahrwesen Aachen (ika), the characteristics of such a CRP space frame concept were examined. By means of a spaceframe structure, as shown in *Figure 5(c)*, material variations with steel, alloy GRP and CRP were performed. Taking into account that bending and torsion stiffness should be independent of the used material the CRP concept offers the best lightweight potential as shown in *Figure 5(d)*.

Other possible applications in body engineering are flat applications like hoods, hatchbacks and doors.

The use of CRP in chassis applications

The spring rod as shown in *Figure 5(e)* demonstrates the use of CRP for a highly loaded part. This rod represents the connection of the axle rod and the wheel rod and is complexly loaded by bending, torsion and pull/push forces. The chosen production technology for this part is a compression moulding method.

Figure 5(f) shows a carbon fibre brake disk. High tension fibres are put into a matrix made of carbon or SiC (Siliziumkarbit). Because of the specific characteristics of this material (low density, high thermal resistance, low thermal extension), this material can bear the high thermical demands of a high potential brake. However, due to economic reasons this type of brake disk is presently just used in motor racing. Nevertheless there is some effort to try and introduce this type of brake discs into high class production cars.

Other possible applications in chassis parts are springs, rims and rods.

The use of CRP in powertrain applications

Figure 5(h) shows a cardan shaft made by Unicardan produced in a volume of 5000–7000 pieces a year. Because of the high level energy absorption characteristics of CRP a use within the crashmanagement of the vehicle is possible. Furthermore the use of CRP in this application offers economic advantages because the CRP-Shaft can be made of one part, whereas a steel shaft would have to be made of two parts. This is due to the high strain ratio of CRP, which again increases the critical round per minute keeping the same length of the part. This fact is illustrated in *Figure 5(g)*.

Conclusion

In addition to the carrying out of motor improvement measures the application of innovative materials in the body making presents and additional measurement concerning the striving for low consumption and therefore for vehicles being ecological and saving resources. The application of CFP-materials offers the best lightweight potential to realize lightweight concepts by means of applying innovative materials. By means of this material either material lightweight as a result of the outstanding mechanical characteristics at simultaneously low weight or constructive lightweight as a result of functional integration and material equitable construction can be realized.

Besides these aspects of lightweight construction as a result of the high energy absorption ability the material can also contribute to an improved crash management and therefore to an improved passive safety of the vehicle by, e.g. the application of crash elements of CFP.

The application of production methods being suitable according to requirements on the component and to characteristics of the component as well as the observation of process engineering restrictions enables an economic production in one production segment up to ca. 4000 units.

Conclusively it must be said that today's application of CFP-materials stands at the very beginning of an exploitable developmental potential. Finally all concerned partners—vehicle manufacturers and feeder plants, manufacturer of production machines for the making of fibre compound materials and material

(a) **MB Unimog UX100**

Reference: Mercedes Benz

(b) **Body made of FRP of the Zato L3**

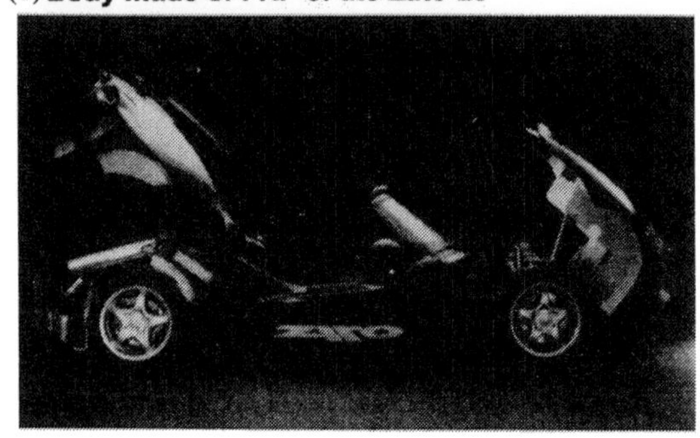

(c)

(d)

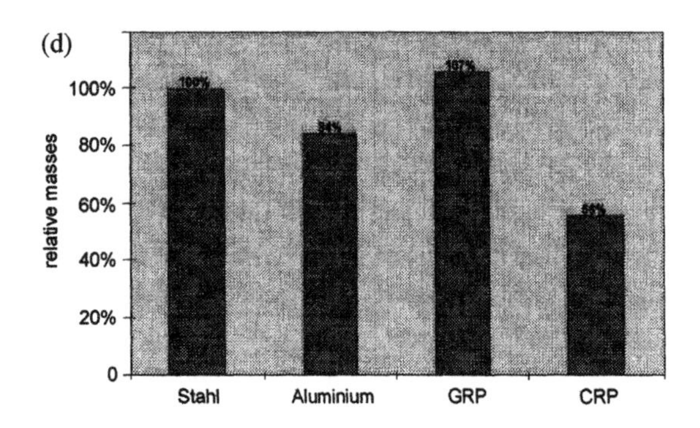

(e) **FRP Spring Rod**

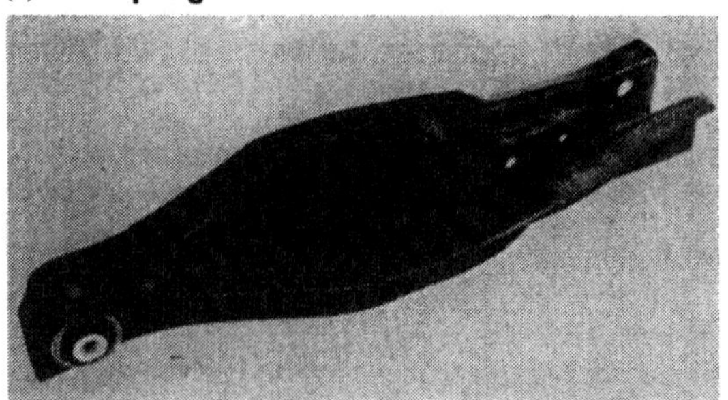

(f) **C/C Brake Disc**

Reference: Mercedes Benz

(g)

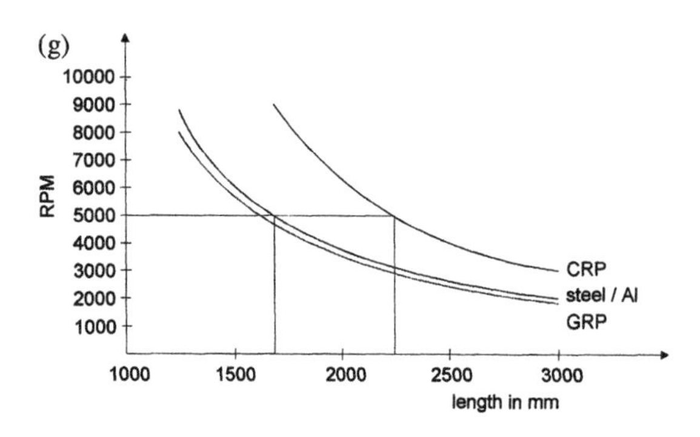

(h) **Technology of the UNICARDAN Cardan Shaft**

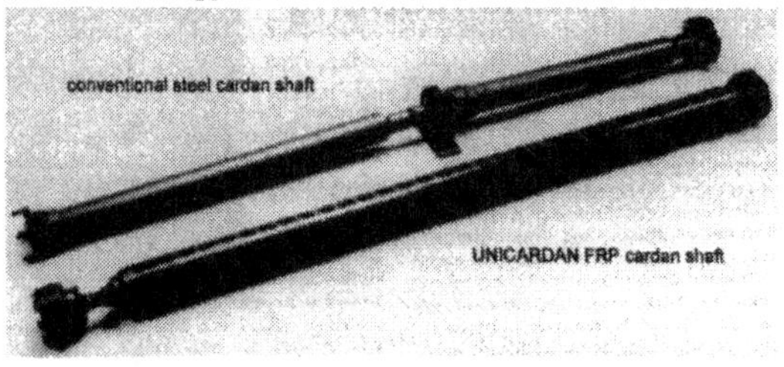

Reference: UNICARDAN

feeder plants—have to contribute actively that this innovative material is used increasingly in automotive technologies. Perhaps the peripheral conditions described here may lead to a change of the topical situation and contribute to the solution of vehicle specific problems by means of CFP.

Figure 5 (a) Driver cabin of the MB UX 100 (Unimog)[4]. (b) Zato L3[5]. (c) Spaceframe structure[6]. (d) Comparison of spaceframe masses[6]. (e) Spring rod of CRP by Mercedes Benz[7]. (f) Carbon-brake disc. (g) Comparison of cardan shaft lengths. (h) Cardan Shaft by UNI-CARDAN[8]

References

1 Wallentowitz, H., Rappen, J. and Gossen, F., *Fuel Saving Potential by Weight Reduction and Optimizing of Ancillary Components: Simulation and Bench Test*. VDI Berichte Nr. 1307, 1996
2 Steinacker, T., *Nutzfahrzeugachse in Faserverbundbauweise Schriftenreihe Automobiltechnik*. Dissertation, Aachen, 1996
3 N.N UX 100, *The New Unimog, the New Dimension*. Mercedes Benz AG, 1996
4 Roth, Sigolotto, Keller, *Kümmerlein Fahrerhausstrukturen in CFK-Technologie, Chance oder Utopie*. Kunststoffe im Automobilbau, 1995
5 N.N ZATO L3, Zato Gesellschaft für Fahrzeugentwicklung, 1997
6 Wallentowitz, H., Adam, H. and Bröcking, J., *Einsatzpotential von Space frame Strukturen aus FVK im Automobilbau*. VDI-Tagung 1996
7 Schreiber, W., Beckmann, H. D. and Rathje, V., *Faserverbundwerkstoffe im Automobilbau Congress-Zentrum*. Würzburg, 1989
8 Hoffmann, W. and Schafferus, T., *Die zweite Generation von Faserverbund-Kardanwellen*. Automobiltechnische Zeitschrift 96, 1994

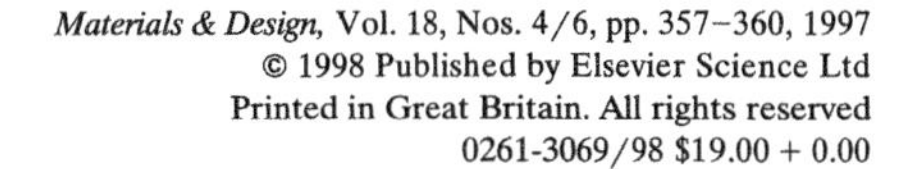

Materials & Design, Vol. 18, Nos. 4/6, pp. 357–360, 1997
© 1998 Published by Elsevier Science Ltd
Printed in Great Britain. All rights reserved
0261-3069/98 $19.00 + 0.00

PII: S0261-3069(97)00077-0

Tensile properties and microstructures of NiAl–20TiB$_2$ and NiAl–20TiC in situ composites

J. T. Guo[a,*], D. T. Jiang[a], Z. P. Xing[b], G. S. Li[a]

[a]*Institute of Metal Research, Chinese Academy of Sciences, 72 Wenhua Road, Shenyang 110015, China*
[b]*Beijing Institute of Aeronautical Materials, Beijing 100095, China*

Received 25 August 1997; accepted 28 August 1997

A hot-pressing aided exothermic synthesis (HPES) technique was developed to fabricate NiAl matrix composites reinforced with TiB_2 and TiC particles which were in situ reaction synthesized from elemental powders. These particles were uniformly dispersed in the matrix. The resulting products were hot isostatically pressed to nearly complete densification. It was found that the tensile yield strengths of the composites at 900°C were about two times stronger than that of unreinforced NiAl and were approximately three times stronger at 980°C. The interfaces between NiAl and TiC or TiB_2 were atomically flat, sharp and free from any interfacial phases in most cases, however, a thin interfacial amorphous layer or overlapped interfacial layer was observed at the interfaces in some cases. This type of interfacial structure may be beneficial to the strength of the composites. © 1998 Published by Elsevier Science Ltd. All rights reserved.

Keywords: tensile properties; microstructures; NiAl-20TiB$_2$; NiAl-20TiC; in situ composites

Introduction

The search for new high-temperature structural materials has stimulated much interest in ordered intermetallics. The intermetallic compound NiAl displays a number of favorable properties for use at high temperatures. These include high melting temperature (1640°C), low density (5.86 g cm^{-3}) and excellent oxidation resistance. However, limited ductility and toughness as well as poor impact resistance continue to be critical issues which will impede near term production implementation[1].

One potential means of improving performances of NiAl has been the addition of either continuous or discontinuous reinforcements[2–7]. Alman and Stoloff have provided some data on the compressive and tensile behavior of NiAl–20 vol.% TiB_2 produced by a reactive hot isostatic pressing (RHIP) technique[5]. The drawback of this approach is that the particle size is limited by what is commercially available and the surfaces are not very clean, which may badly influence the interfacial bonding. Furthermore, the spatial distribution of the particles is not uniform. The high-temperature strength improvements in the RHIP-processed composites were not obvious. XD™ process has been employed to fabricate NiAl dispersed with 2.7–30 vol.% TiB_2 particles[2–4]. These particles significantly improved the high-temperature compressive strength of NiAl. But the fracture toughness improvements in these composites were only marginal[8].

*Correspondence to Dr J. T. Guo. Tel.: +86 24 3843531; fax: +86 24 3891320; e-mail: jtguo@imr.ac.cn

In this paper, NiAl matrix composites containing 0, 20 vol.% TiB_2 and 20 vol.% TiC particles were fabricated using HPES technique. The microstructure, interfaces and tensile properties of the composites were studied in detail.

Experimental procedures

Elemental powders of Ni (98% ≤ 1 μm), Al (98% ≤ 13 μm), Ti (99% ≤ 75 μm) and B (95% ≤ 1 μm) or C (98% ≤ 1 μm) were blended to obtain the desired compositions. Details of HPES process were reported elsewhere[7,13]. Some HPES samples were hot iso-statically pressed (HIP) at 1165°C/150 MPa for 4 h. Transmission electron microscopy (TEM) and high-resolution electron microscopy (HREM) were used to characterize the microstructure and interfaces of NiAl-TiB_2 and NiAl-TiC. The TEM or HREM foils were prepared by a conventional procedure which involved in cutting disks with a diameter of 3.0 mm, mechanically polishing them to 50 μm, dimpling to 20 μm and finally ion milling. TEM and HREM observations were carried out on Phillips EM420 and JEM 2000EX II, respectively. The tensile specimens were electro-discharge machined from the HIP-processed compacts with gauge dimensions of $3 \times 3 \times 18$ mm, and tested in uniaxial tension at 900 and 980°C at a strain rate of 1.67×10^{-4} s^{-1}. Fracture surfaces from the tests were characterized using a scanning electron microscopy (SEM).

Figure 1 Dislocation arrangement in as-fabricated NiAl-20 vol.% TiB_2

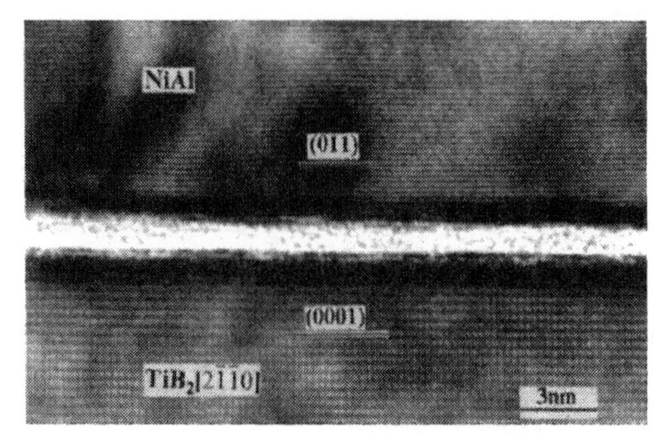

Figure 3 HREM image of the NiAl/TiB_2 interface showing an amorphous layer between TiB_2 and NiAl

Results and discussion

Microstructure of NiAl–TiB_2

The relative densities of the monolithic NiAl and NiAl–20 vol.% TiB_2 composite were 100% and 98.4%, respectively. In the composite, the TiB_2 particles (0.5 ~ 3 μm) uniformly dispersed in the matrix. TEM examination of as-fabricated composite indicates that the TiB_2 particles with brighter contrast were generally polygonal and faceted[9,10]. The grain size (~ 20 μm) of NiAl in the composite has been reduced about one order of magnitude compared to that of monolithic NiAl. There were few dislocations in the TiB_2 particles, while a lot of uniformly arranged dislocations existed in the NiAl matrix, as shown in *Figure 1*. Via HIP processing, the relative density of the NiAl–20 vol.% TiB_2 reached 99.7%, near fully densified. The microstructure of the composite didn't show obvious changes. Dislocation arrangement and density in the HIP processed NiAl–TiB_2 composites were also similar to those in the as-fabricated composites.

There was no consistent orientation relationship between TiB_2 particles and NiAl matrix, as concluded by Wang and Arsenault's result[11]. *Figure 2* gives an example showing HREM image of the NiAl/TiB_2 interface, however, only one-dimensional lattice fringes appeared in the NiAl side. From a number of such observations, it shows that the interfaces in the composite are parallel to the low index and densely packed $\{01\bar{1}0\}$ and {0001} planes of the TiB_2 in most cases. These two surfaces are the expected low-energy surfaces in hexagonal crystals and lead to high thermal stability. Furthermore, TiB_2 particles bonded well to the NiAl matrix and the TiB_2/NiAl interfaces were atomically flat, sharp and free from any interfacial phases in most cases. This may be beneficial to the strength of the composite.

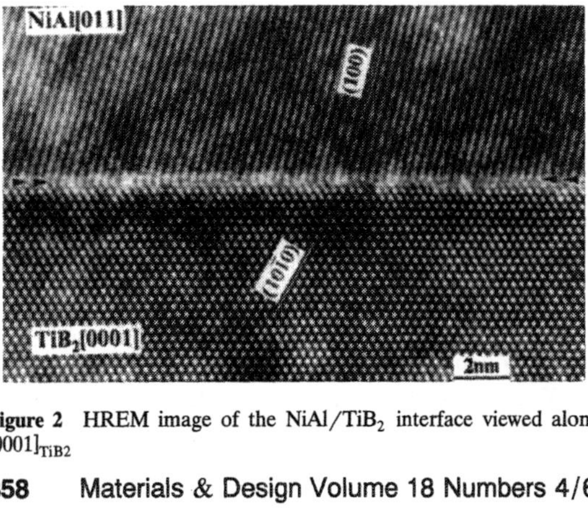

Figure 2 HREM image of the NiAl/TiB_2 interface viewed along $[0001]_{TiB2}$

The HREM images of these relationships had been given elsewhere[9], and semi-coherent interfaces between TiB_2 and NiAl were observed. Most of the interfaces investigated by HREM were smooth and free from any interfacial phases. However, thin interfacial amorphous layers were occasionally observed at TiB_2/NiAl interfaces. *Figure 3* shows such a HREM image viewed along $[2110]_{TiB_2}$. As fabricated, the thickness of this amorphous layer is about 1.5 nm. Perhaps such layers can absorb the thermal residual stress at the interfaces and might beneficially influence the composite toughness.

Microstructure of NiAl–TiC

After HPES process, NiAl–20 vol.% TiB_2 was successfully reaction synthesized from elemental powders. No other phases were found from X-ray diffraction pattern. The relative density of the product changed from 98.9% to 99.4%. The TiC particles are uniformly dispersed in the matrix, however its dispersion uniformity isn't as good as TiB_2 particles in the matrix. In the as-fabricated material, TiC particles (0.2–1.0 μm) are mostly polygonal and the NiAl/TiC interfaces were perfectly smooth, sharp and free of any interfacial phase in most cases. After HIP, the TiC particles tended to become round (see *Figure 4*) and a narrow region with some modification of the NiAl/TiC interface was frequently observed (*Figure 5*). The interfacial layer with the visible different image could be either an overlapped image of NiAl and TiC, or alternatively a reaction product between NiAl and TiC. If it was an overlapped image of NiAl and TiC, the Moiré fringe spacing (as indicated by the arrow) could be calculated using the following expression[12]:

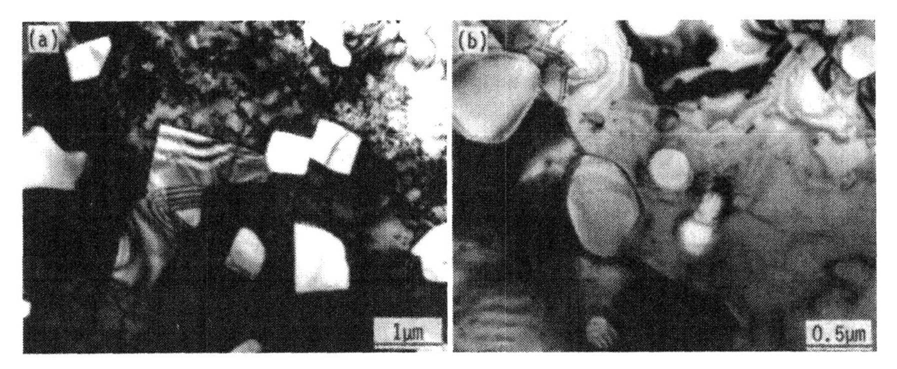

Figure 4 TEM photographs of NiAl-20 vol.% TiC. (a) HPES and (b) HPES + HIP

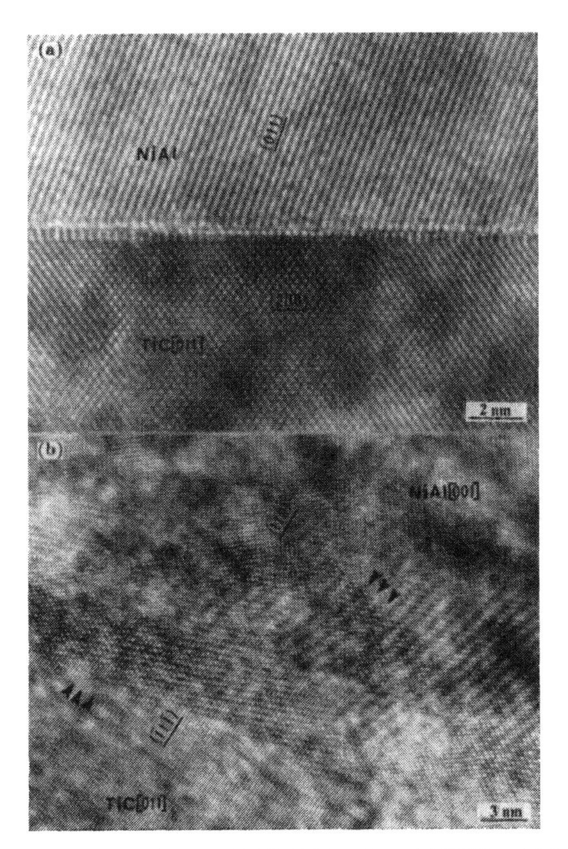

Figure 5 HREM images of NiAl/TiC interfaces. (a) HPES and (b) HPES + HIP

$$d_M = \frac{d_1 d_2}{\left[(d_1 - d_2)^2 + d_1 d_2 \beta^2\right]^{1/2}}.$$

In the equation, $d_1 = d^{\mathrm{TiC}}_{(1\bar{1}\bar{1})} = 0.249$ nm, $d_2 = d^{\mathrm{NiAl}}_{(110)} = 0.204$ nm, $\beta \approx 0°$. The calculated $d_M = 1.1$ nm, which is 3.8 times the interplanar spacing of $(110)_{\mathrm{NiAl}}$. This is perfectly consisted with the Moiré fringe spacing observed in *Figure 5*. So the thin interfacial layer is an overlapped image between NiAl and TiC. Such an overlapped interface increases the interfacial strain fields, and plays an important role in the strength enhancement of the composite.

Elevated temperature tensile behavior

The results obtained from the uniaxial tension tests at high temperatures are shown in *Table 1*. The strengths of the two composites were approximately twice that of the unreinforced NiAl at 900°C and about three times that of NiAl at 980°C. The strength effect of TiC particles is better than that of TiB_2. Furthermore, the tensile strengths at 900°C of the two composites were about twice that of NiAl–20 vol.% TiB_2 fabricated by RHIP technique[6]. TiB_2 particles were in situ formed in the matrix by HPES technique. As to RHIP process, the original TiB_2 particles were directly mixed with Ni and Al powders. The in situ formed TiB_2 particles can be well bonded with the matrix with flat and sharp interfaces, so the composite showed higher tensile strength than that fabricated by RHIP process.

The tensile fracture surfaces at 900°C indicates that monolithic NiAl exhibited typical ductile fracture showing dimples and evidence of microvoid coalescence. The two composites exhibited dimple rupture in the matrix with matrix–particulate debonding, however, remnants of the matrix were often observed on the exposed TiB_2 fracture surfaces, again suggesting a good matrix/particulate bonding in the composite. Similar fracture surfaces were observed on both materials at 980°C and on the composite at 800°C.

4. Conclusions

The conversion to product was complete after HPES process from the NiAl matrix composites reinforced with TiC and TiB_2 particles. In both composites, the strengthening particles uniformly dispersed in the matrix, specifically, the TiB_2 particles showed relatively better distribution characteristic. HIPing didn't significantly influence the two composites' microstructure, except for improved densification and that the TiC particles tended to become round. The tensile strength of both materials evidently increased compared to monolithic NiAl. The strengthened composite fabricated by HPES showed higher strength than that fabricated by RHIP process. The good strengthening effect of the two composites can be partially attributed to the well-bonded interfaces between NiAl and particles, which is illustrated by HPES to be smooth and sharp.

Table 1 Tensile properties of NiAl matrix composites

Materials	Conditions	Temp (°C)	YS (MPa)	UTS (MPa)	EL (%)	R.A. (%)
Ni-50Al	HPES + HIP	900	89	93	14.7	28.4
Ni-49.5Al–20 vol.% TiB_2	HPES + HIP	800	241	262	7.1	14.1
		900	162	169	8.5	14.8
		980	152	169	11.1	16.9
Ni-49Al–20 vol.% TiC	HPES + HIP	900	178	201	5.3	9.0
		980	156	173	7.7	16.3
Ni-49.5Al–20 vol.% TiB_2	RHIP	800	190	207		12.5
		900	86	99		9.4

In addition, an interfacial amorphous lay observed between NiAl and TiB_2 and overlapped interfacial layer existed between NiAl and TiC are also probably beneficial to the strength enhancement of the composites.

Acknowledgements

This work was partially supported by the National Natural Foundation of China (No. 59331012) and National Advanced Materials Committee of China, to whom we are very grateful.

References

1 Darolia, R., Walston, W.S. and Nathal, M.V., In *Superalloys 1996*, ed. R.D. Kissinger, D.J. Deye, D.L. Anton, A.D. Cetel, M.V. Nathal, T.M. Pollock and D.A. Woodford. The Minerals, Metals and Materials Society, 1996, p. 561
2 Whittenberger, J.D., Viswanadham, R.K., Mannan, S.K. and Sprissler,B., *Journal of Materials Science*, 1990, **25**, 35
3 Wang, L. and Arsenault, R.J., *Materials Science and Engineering*, 1990, **A127**, 91
4 Whittenberger, J.D., Kumar, K.S. and Mannan, S.K., *Materials at Higher Temperatures*, 1991, **9**, 3
5 Alman, D.E. and Stoloff, N.S., *International Journal of Powder, Metallurgical*, 1991, **27**, 29
6 Xing, Z.P., Dai, J.Y., Guo, J.T., An, G.Y. and Hu, Z.Q., *Screen Metallurgical Materials*, 1994, **31**, 1141
7 Xing, Z.P., Yu, L.G., Guo, J.T., Dai, J.Y., An, G.Y. and Hu, Z.Q., *Journal of Materials Science Letters*, 1995, **14**, 443
8 Kumar, K.S., Mannan, S.K. and Viswanadham, R.K., *Acta Metallurgical*, 1992, **40**, 1201
9 Dai, J.Y., Xing, Z.P., Li, D.X., Guo, J.T. and Ye, H.Q., *Materials Letters*, 1994, **20**, 23
10 Ramberg, J.R. and Williams, W.S., *Journalof Materials Science*, 1987, **22**, 1815
11 Wang, L. and Arsenault, R.J., *Metallurgical Transactions A*, 1991, **22A**, 3013
12 Li, D.X., Pirouze, P. and Heuer, A., *Phil Mag*, 1992, **A65**, 403
13 Guo, J.T. and Xing, Z.P., *Acta Metallurgical Sinica (English Letters)*, 1995, **8**, 455.

Materials & Design, Vol. 18, Nos. 4/6, pp. 361–363, 1997
© 1998 Published by Elsevier Science Ltd
Printed in Great Britain. All rights reserved
0261-3069/98 $19.00 + 0.00

PII: S0261-3069(97)00087-3

The effect of microstructure on the mechanical properties of two-phase titanium alloys

Jan Sieniawski*, Ryszard Filip, Waldemar Ziaja

Rzeszów University of Technology, W.Pola 2, 35-959 Rzeszów, Poland

Received 4 September 1997; accepted 18 September 1997

This article presents the results of investigations of microstructure and mechanical properties of two-phase $\alpha+\beta$ titanium alloys with different volume fraction of the β-phase. Microstructure of the specimens was examined using an optical microscope. Fracture surfaces were observed by SEM technique. The influence of the microstructure and phase composition on the mechanical properties of the alloys was studied. Static tensile tests, hardness tests and fatigue investigations were performed. It was noticed that the volume fraction and chemical composition of the β-phase has a significant effect on mechanical properties and cracking process during fatigue. © 1998 Published by Elsevier Science Ltd. All rights reserved.

Keywords: titanium alloys; microstructure; fatigue strength

Introduction

Use of two-phase titanium alloys in industry is limited by economic factors. Characteristic conditions of the metallurgical processes and manufacturing of the product imposed by the special properties of the titanium are no longer a problem for production engineering. Thus properties make it possible to use titanium alloys in conditions calling for certain special properties, such as corrosion resistance, heat resistance and paramagnetism. The main factors determining functional characteristics of these materials are diffusion and diffusionless transformations taking place during heat treatment. The control of these processes by means of selection of heat treatment conditions and chemical composition of the phases that are present in these alloys enables advancement in operational properties[1,2].

Mechanical properties of titanium alloys are an important criteria of material service capabilities both in aerospace and industrial applications. Microstructure of the alloy is one of the important factors controlling both the tensile strength and the fatigue strength[3–5]. The aim of the study was to determine the effect of the alloy microstructure, the volume fraction and chemical composition of the β-phase on its mechanical properties (including fatigue behaviour). Therefore to determine the influence of the α and β phases on the strength two alloys were selected for the test: Ti-6Al-2Mo-2Cr and Ti-6Al-5Mo-5V-1Cr-1Fe for which the coefficient of the β phase stability $K_\beta = 0.6$ and 1.2, respectively. The mechanical tests were made using specimens produced of material with globular and lamellar microstructure. The presence of lamellar α phase in titanium alloys guarantees fair mechanical properties[6–9]. The analysis of the effect of heat treatment conditions on the geometrical parameters of the microstructure was necessary in the course of the study in order to define their optimum values.

*Correspondence to Jan Sieniawski. Tel.: +48 17 625406

Material and research methodology

The materials tested were high strength, two-phase $\alpha+\beta$ titanium alloys Ti-6Al-2Mo-2Cr and Ti-6Al-5Mo-5V-1Cr-1Fe. The chemical composition of the alloys is presented in *Table 1*. The materials tested were vacuum melted and rolled in the $\alpha+\beta \rightarrow \beta$ temperature range in order to obtain fine grained, equiaxial microstructure after stabilising annealing which provides the best plasticity and high phase stability. Lamellar microstructure of α phase was acquired in both alloys by means of proper cooling rates selection, which were specified on the basis of CCT diagrams.

Dilatometric test were performed employing absolute dilatometer LS-4 in order to determine the influence of the cooling rate on the temperature of $\alpha+\beta \rightarrow \beta$ phase transformations, kinetics of phase transformations during continuous cooling and morphology of the two-phase $\alpha+\beta$ microstructure. In dilatometric tests the protective argon atmosphere was used. Rounded specimens 4 mm in diameter and 15 mm long were applied for the tests.

Microstructure of the specimens was investigated by optical microscope Neophot 2. Specimens were etched at 270 K, two etchants with the following chemical composition: 50% HNO_3 + 40% HF + 10% H_2O and 10% HNO_3 + 2% HF + 88% H_2O were used. The

Table 1 Chemical composition of the alloys tested

Alloy	Alloy elements contents (wt.%)							
	Al	Mo	V	Cr	Fe	C	Si	Ti
Ti-6Al-2Mo-2Cr	6.3	2.6	—	2.1	0.40	0.05	0.20	Balance
Ti-6Al-5Mo-5V-1Cr-1Fe	5.8	5.3	5.1	0.9	0.80	0.05	0.15	Balance

basic geometrical parameter of lamellar microstructure of the alloys tested (thickness of the α-phase lamellae (t)) was determined using also the optical microscopic examination.

The fracture surfaces of the specimens after fatigue tests were observed using scanning electron microscope Novascan 30 with a 7-nm resolution at 15 kV acceleration voltage. Fractographic examinations were indispensable for analysis of the fatigue damage process.

Mechanical testing of the specimens with equiaxial as well as lamellar microstructure were performed. The hardness test was carried out using Vickers hardness tester with a load of 980 N. The static tensile test was carried out on the UTS-100 testing machine. The specimens were of cylindrical shape. During the test the strain rate $\dot{\varepsilon} = 0.005\ s^{-1}$ was applied. Four specimens in each series were examined and average values of the mechanical properties were determined. Ultimate tensile strength — R_m, 0.2% proof stress — $R_{0.2}$, reduction in area — Z and elongation — A_5 were defined.

Fatigue strength was examined in a rotational bending test using the UMBS-4 machine. The step method was applied for fatigue strength determination. From 15 to 18 specimens were tested with the stress step of 20 MPa and terminal number of cycles $N_g = 2 \cdot 10^7$ in order to determine the average fatigue limit.

Results and their analysis

Microstructure of the Ti-6Al-2Mo-2Cr alloy obtained after cooling at a controlled cooling rate (lamellar) — is presented in *Figure 1*. It was noticed that thickness and length of the α-phase decrease with increasing cooling rate and with increasing content of the β stabilising elements.

As a result of dilatometric examination the $\beta \rightarrow \alpha + \beta$ phase transformation temperature was determined, particularly the start temperature, $T_{s\beta_M}$ and the finish

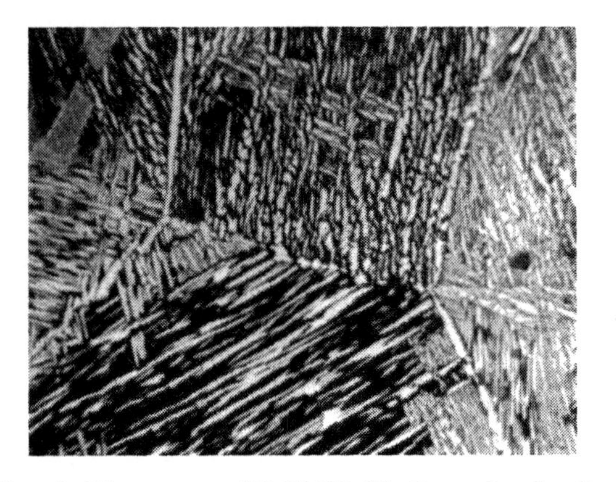

Figure 1 Microstructure of Ti-6Al-2Mo-2Cr alloy, α-phase lamellae in β-phase matrix; (500 ×)

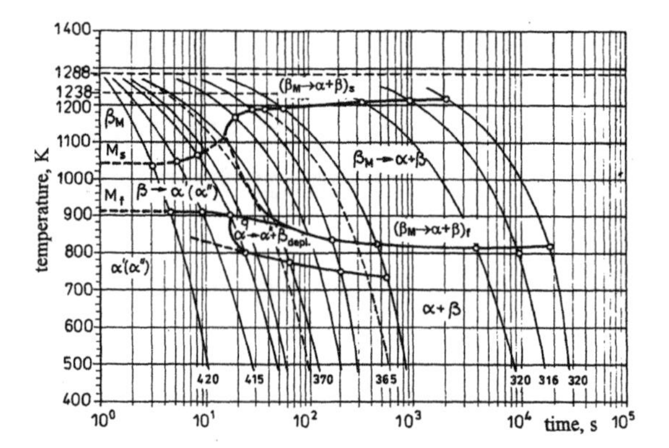

Figure 2 CCT diagram for Ti-6Al-2Mo-2Cr alloy

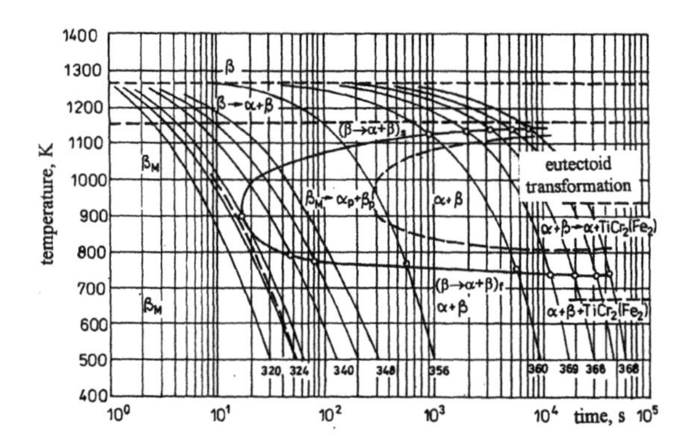

Figure 3 CCT diagram for Ti-6Al-5Mo-5V-1Cr-1Fe alloy

temperature, $T_{f\beta_M}$, of the metastable β_M phase decomposition and temperature ranges of martensitic phases α' or α'' presence. The microstructure obtained while cooling at different rates was identified. The range of the cooling rate for development of lamellar α-phase was additionally determined (*Table 2*). Dilatometric examination, microstructure observation and X-ray structural analysis enabled the construction of CCT diagrams of the alloys studied (*Figures 2* and *3*).

Continuous cooling of Ti-6Al-2Mo-2Cr alloy from the β-phase range leads to development of:

- microstructure composed both of martensitic phases α' or α'' (denoted as $\alpha'(\alpha'')$ in *Figure 2*) and stable α and β-phases at the cooling rate $v_c = 2 - 40\ K\ s^{-1}$;
- stable lamellar microstructure $\alpha + \beta$ at cooling rate $v_c < 2\ K\ s^{-1}$ and for Ti-6Al-5Mo-5V-1Cr-1Fe alloy:
- microstructure composed of metastable β_M phase and stable α and β-phases at the cooling rate $v_c = 1.2 - 40\ K\ s^{-1}$;
- stable lamellar microstructure $\alpha + \beta$ at the cooling rate $v_c < 1.2\ K\ s^{-1}$.

Table 2 Phase composition of the alloys after cooling with controlled cooling rates

Alloy	Average cooling rate ($K\ s^{-1}$)				
	48	9	1.2	0.08	0.004
Ti-6Al-2Mo-2Cr	α'	$\alpha + \beta + \alpha'(\alpha'')$	$\alpha + \beta$	$\alpha + \beta$	$\alpha + \beta, TiCr_2$
Ti-6Al-5Mo-5V-1Cr-1Fe	β_M	$\beta_M + \alpha$	$\beta + \alpha$	$\beta + \alpha$	$\beta + \alpha, TiCr_2, Fe_2$

Table 3 Mechanical properties of the two-phase titanium alloys tested

Alloy	Microstructure	$R_{0.2}$ (MPa)	R_m (MPa)	Z (%)	A_5 (%)	HV	Z_{go} (MPa)
Ti-6Al-2Mo-2Cr	Equiaxial	940	1008	45	14	310	506
	Lamellar, $t = 2.0$ μm	1043	1207	42	12	338	540
Ti-6Al-5Mo-5V-1Cr-1Fe	Equiaxial	1043	1113	41	12	310	519
	Lamellar, $t = 1.5$μm	1225	1285	40	7	340	552

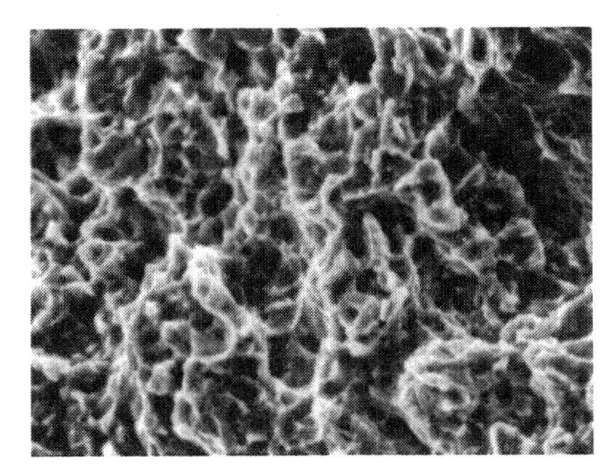

Figure 4 Fracture surfaces of titanium alloy Ti-6Al-2Mo-2Cr (400 ×)

The mechanical properties of the alloys are presented in *Table 3*. Presence of lamellar α-phase after heat treatment in the β-range improves both tensile and fatigue strength. Fractographic examination revealed a similar fracture mechanism in both alloys. In the Ti-6Al-2Mo-2Cr alloy the zones of large plastic deformation can be seen in β-phase regions and propagating crack passes round the colonies of the parallel α-phase lamellae (*Figure 4*). In Ti-6Al-2Mo-2Cr the plastic deformation zones fraction on the fracture surface is smaller because of smaller amount of the β-phase.

The increase in fatigue strength of the alloys with lamellar microstructure is the result of change in crack direction and secondary crack branching. When β-phase regions are small, they cannot absorb large amounts of energy and retard the crack propagation. Sufficient thickness of β-phase areas enables absorption of energy in the process of plastic deformation in regions ahead of the crack tip and therefore slows down the rate of crack propagation.

Summary

The studies carried out proved that presence of the stable β-phase in the alloy microstructure leads to improvement in its mechanical properties (Table 3). Lamellar precipitates of the α-phase that are present in both alloys behave similarly in fatigue cracking conditions. They stand in the way of the crack tip forcing it to propagate in the plane parallel to the lamella surface. In two-phase alloys the α-phase lamellae are separated by the β-phase precipitates of high strength and plasticity. Cracking process is similar in two-phase alloys, i.e. the crack is blocked on the α-phase lamella and propagates along its surface and crack propagation is hindered through the energy absorption in the process of plastic deformation. It is to be noted that increase in the strength of the alloy by solution hardening is possible by means of chemical composition modification (addition of β-stabilisers). Moreover, in a two-phase alloy with the bigger value of the coefficient of the β-phase stability K_β, significantly smaller phase precipitates were obtained after cooling which in turn caused frequent changes of the crack propagation direction and absorption of considerable amount of energy in the process of fatigue crack reinitiation.

References

1 Eylon, D., Fujishiro, S., Postans, P. J. and Froes, F. H., *Journal of Metals*, 1984, **36**, 55–61
2 Sastry, S., Peng, T. C., Mechter, P. J. and O'Neal, J. E., *Journal of Metals*, 1984, **36**, 21–28
3 Colagelo, V. J. and Heizer, F. A., *Analysis of Metallurgical Failures*. John Wiley, New York — Toronto, 1987
4 Emburg, J. D. and Zok, F., Micromechanisms of Fracture. In *Proceedings of the 26th Conference of Metallurgists — Winnipeg 1987*. Pergamon Press, New York, 1988
5 Sieniawski, J., *Scientific Papers of the Rzeszow University of Technology* — Mechanics No. 10, Rzeszow, 1985 (in Polish)
6 Kahveci, A. I. and Welsch, G. E., *Scripta Metals et Materials*, 1991, **25**, 1957–1962
7 Müller, C., Correlation between microstructure and fracture toughness of titanium alloys. In *Proceedings of Strength of Metals And Alloys — Montreal 1985*. Pergamon Press, Toronto, 1985
8 Starke, E. A. and Lütjering, G., *Transactions ASM*, 1978, **3**, 225–231
9 Sieniawski, J., *Inzynieria Materialowa*, 1993, **14**, 64–68 (in Polish)

Materials & Design, Vol. 18, Nos. 4/6, pp. 365–367, 1997

0261-3069/98 $19.00 + 0.00

PII: S0261-3069(97)00088-5

Effect of forging conditions and annealing temperature on fatigue strength of two-phase titanium alloys

K. Kubiak*, J. Sieniawski

Department of Materials Science, Rzeszów University of Technology, W. Pola 2, 35-959 Rzeszów, Poland

Received 4 September 1997; accepted 18 September 1997

This paper reports on the results of studies on the influence of deformation degree and temperature in the die forging process and annealing temperature on fatigue strength of forgings made of two-phase, martensitic titanium alloys (Ti–6Al–4V and Ti–6Al–2Mo–2Cr). Dilatometric and metallographic studies have been carried out along with fatigue tests. The influence of phase composition and microstructure obtained after cooling at different rates on fatigue strength has been determined. © 1998 Published by Elsevier Science Ltd. All rights reserved.

Keywords: deformation; die forging process; alloy

Introduction

A large number of elements made of two-phase, martensitic titanium alloys are fabricated in the process of plastic working. Service life and reliability of these alloys depend on the microstructure obtained after working and heat treatment and on the surface layer condition[1]. The development of different microstructure and phase compositions in various regions of forging is the result of the high sensitivity of two-phase titanium alloys to rate, degree and temperature of plastic deformation[2,3]. Proper selection of these parameters allows one to control the phase composition of the alloy and hence its mechanical properties[4]. The dynamic strength is one of the basic criteria determining the possibility of using the alloy in structural elements[5]. Therefore, an effort has been made in the paper to determine the effect of plastic working conditions (strain degree, forging temperature) and annealing temperature on fatigue strength of two-phase, martensitic titanium alloys (Ti–6Al–4V and Ti–6Al–2Mo–2Cr).

Materials and experimental

The materials tested were two-phase $\alpha + \beta$, martensitic titanium alloys, Ti–6Al–4V and Ti–6Al–2Mo–2Cr with the chemical composition as shown in *Table 1*.

Dilatometric examination was applied to determine the start and the end temperature of the $\alpha + \beta \rightleftarrows \beta$ phase transformation during alloy heating at the rate $v_c = 0.08$ K s^{-1}. On the basis of the phase transformation temperature the conditions of plastic working of the alloys were defined. Forging was accomplished in the β-range (1320 K) and in the $\alpha + \beta$-range (1170 K). Semi-finished products (∅ 65 × 120) were forged on the crank operated power press (LKM 4000) and then annealed at 950, 1060 and 1250 K over 1 or 3 h. The microstructure of the specimens was examined using an optical microscope (Neophot 2).

The fatigue behaviour was examined in a rotational bending test according to the PN-76/H-04326 standard on the UMBS-4 machine with a loading frequency of 50 Hz. Between six and 10 specimens were tested on each stress level and the terminal number of cycles was set to $N = 2 \times 10^7$. Specimens, 7 and 3.5 mm in diameter were made by forging in both as the forged and heat-treated condition. Results of fatigue tests were statistically analysed in order to determine the average life N_{av}, standard deviation s of the average life and confidence intervals for the fatigue limit and fatigue strength.

Results and analysis

Forging temperature, deformation degree and annealing temperature of the Ti–6Al–4V and Ti–6Al–2Mo–2Cr alloys have a very pronounced effect on fatigue strength of the studied alloys (*Table 2*). It applies especially to the Ti–6Al–2Mo–2Cr alloy as the presence of chromium raises its susceptibility to phase composition changes (martensitic transformation $\beta \rightarrow \alpha'$).

Metallographic examination revealed a significant effect of forging temperature on the microstructure of

*Correspondence to K. Kubiak. Tel.: +92 48 17625406; e-mail: krkub@ewa.prz.rzeszow.pl

Table 1 Chemical composition of the Ti–6Al–4V and Ti–6Al–2Mo–2Cr alloys

Alloy	Stability factor K_β	Alloying elements and impurities content (wt.%)										
		Al	Mo	V	Cr	Fe	C	Si	H	N	O	Ti
Ti–4Al–4V	0.3	6.1	—	4.3	—	0.16	0.01	—	0.015	0.06	0.12	Bal.
Ti–6Al–2Mo–2Cr	0.6	6.3	2.6	—	2.1	0.40	0.05	0.2	0.016	0.016	0.09	Bal.

Table 2 Fatigue strength of the Ti–6Al–4V and Ti–6Al–2Mo–2Cr alloys as a function of temperature and degree of deformation and annealing temperature of forgings

Alloy	Forging temp. K	Fatigue strength, mPa						
		Plastic deformation $\epsilon_1 = 0,35$			Plastic deformation $\epsilon_2 = 2,0$			
		Annealing temperature (K)			Annealing temperature			
		—	950	1250	—	950	1060	1250
Ti–6Al–2 Mo–2Cr	1170	546	605	—	261	285	232	—
	1320	482	—	465	351	372	—	326
Ti–6Al–4V	1170	511	558	—	262	283	221	—
	1320	426	—	417	340	362	—	322

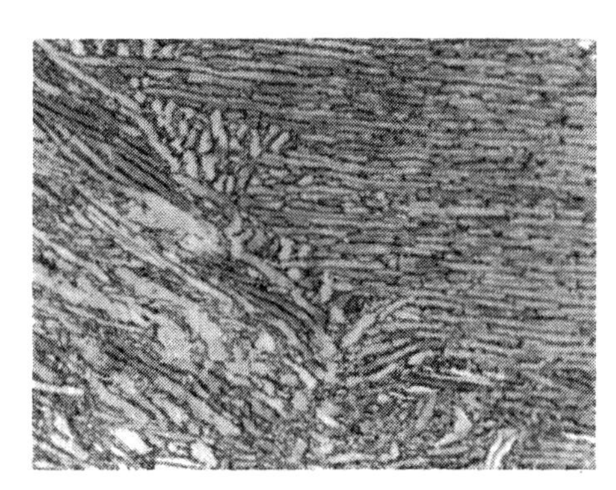

Figure 1 Microstructure of the Ti–6Al–2Mo–2Cr alloy after forging at 1170 K (globular microstructure after plastic deformation), 400 ×

Figure 2 Microstructure of the Ti–6Al–2Mo–2Cr alloy after forging at 1320 K (lamellar microstructure), 400 ×

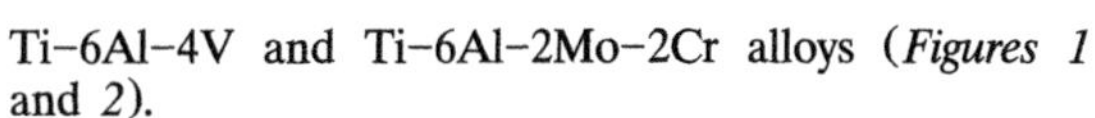

Ti–6Al–4V and Ti–6Al–2Mo–2Cr alloys (*Figures 1* and *2*).

After forging in the $\alpha + \beta$-range (1170 K) a globular microstructure was formed with grain deformation depending on the degree of plastic strain. A characteristic feature of this microstructure was significant irregularity, i.e. various distortion of α-phase grains (*Figure 1*). The microstructure irregularity is related to the strain degree. The strain degree was calculated by numeric simulation using FORGE – 2 software[6]. The higher the deformation degree the lower the microstructure irregularity and larger α-phase grains distortion.

Forging in the β-range (1320 K) generated the lamellar microstructure (*Figure 2*). Parallel lamellae of the α-phase formed colonies in primary β-phase grains. Forging in the β-range led to primary β-phase grains refinement that followed the dynamic recrystallization processes.

The annealing temperature of forgings after plastic working has an influence on fatigue strength. A maximum fatigue strength of Ti–6Al–4V and Ti–6Al–2Mo–2Cr alloys was obtained after annealing at 950 K for 3 h. An increase in fatigue strength after annealing (950 K/3 h) compared with the strength as the forged condition was:

- Ti–6Al–4V alloy
  — forging temperature 1170 K, 9%;
  — forging temperature 1320 K, 4%.
- Ti–6Al–2Mo–2Cr alloy
  — forging temperature 1170 K, 10%;
  — forging temperature 1320 K, 6%.

Annealing at 1250 K for 1 h led to a decrease in fatigue strength when compared with that of the alloys both in as the forged condition and annealed at 950 K for 3 h. At the forging temperature of 1320 K the reduction in fatigue strength was approximately 2–5% for the Ti–6Al–4V alloy and 4–8% for the Ti–6Al–2Mo–2Cr alloy. A decrease in fatigue strength is related to the volume fraction of the recrystallized microstructure which depends on deformation degree.

Hardness measurements of alloys forged at 1170 K and 1320 K showed that annealing temperature is a

Table 3 Recrystallized microstructure fraction for Ti–6Al–4V and Ti–6Al–2Mo–2Cr alloys for various deformation conditions and heat treatment

Alloy	Recrystallized microstructure fraction (%)			
	Deformation degree and heat treatment			
	$\epsilon = 2,0$ 950°C/3 h	$\epsilon = 2,0$ 1250°C/1 h	$\epsilon = 0,35$ 950°C/3 h	$\epsilon = 0,35$ 1250°C/1 h
Ti–6Al–4V	34	66	26	63
Ti–6Al–2Mo–2Cr	36	67	27	60

critical factor influencing the fraction of recrystallized microstructure and yet deformation degree has the minor effect (*Table 3*).

Annealing of Ti–6Al–4V and Ti–6Al–2Mo–2Cr alloys at 1250 K for 1 h promoted the formation of new recrystallized grains and lamellae of the α-phase and increased the fraction of recrystallized regions in the microstructure to 66 and 67%, respectively. This was also confirmed by microscopic examination.

Summary

Analysis of the results from the Ti–6Al–4V and Ti–6Al–2Mo–2Cr titanium alloys examination has shown that:

- Continuous cooling of the Ti–6Al–4V and Ti–6Al–2Mo–2Cr alloy forgings from the β-range produces microstructure in the shape of colonies of parallel α-phase lamellae in primary β-phase grains. Cooling from the $\alpha + \beta$ range develops globular microstructure of α and β phases.
- The fatigue strength of the alloys studied depends on forging temperature. The maximum fatigue strength was obtained by forging at 1170 K ($\alpha + \beta$-range). Forging at 1320 K (β-range) led to a decrease in fatigue strength (approx. 13 and 20% for Ti–6Al–2Mo–2Cr and Ti–6Al–4V, respectively);
- An increased plastic deformation degree lowers fatigue strength. The decrease in fatigue strength is smaller in the case of forging in the β-range than in $\alpha + \beta$-range.
- The fatigue strength of the alloys depends on the heat treatment of forgings. Annealing at 950 K for 3 h led to an increase in fatigue strength and annealing at 1250 K for 1 h caused a decrease in fatigue strength.

References

1 Bylica, A., Sieniawski, J., *Tytan i jego stopy* (*Titanium and its alloys*). PWN, Warszawa, 1985 (in Polish)
2 Sieniawski, J., Scientific Papers of the Rzeszow Univ. of Technology. Mechanics No. 10, 1985 (in Polish)
3 Buhl, H., *Advanced Aerospace Materials*. Springer Verlag, Berlin, 1992
4 Katcher, M., *Engineering Fracture Mechanics*, 1973, p. 4
5 Sieniawski, J., *Kryteria i sposoby oceny materiałów na elementy lotniczych silników turbinowych* (*Criteria of materials evaluation for turbine aeroengines applications*). Oficyna Wydawnicza Politechniki Rzeszowskiej, Rzeszów, 1995 (in Polish)
6 Program FORGE – 2, CMFM, Ecole Nationale Superieure des Mines, de Paris rue C. Daunesse, Sophia Antipolis 06560 Valbonne cedex, France

ELSEVIER

Materials & Design, Vol. 18, Nos. 4/6, pp. 369–372, 1997

0261-3069/98 $19.00 + 0.00

PII: S0261-3069(97)00078-2

Metallurgical problems associated with the production of aluminium-tin alloys

Tomasz Stuczyński*

Institute Non-Ferrous Metals, Light Metals Division, Department of Foundry Technology, Pilsudskiego 19, 32-050 Skawina, Poland

Received 26 August 1997; accepted 28 August 1997

Of all the plain bearing alloys available, aluminium alloys have a better combination of ideal bearing characteristics than any other single material. Conventional Babbit alloys, especially the lead bronzes are increasingly being replaced by aluminium alloys, particularly in the automobile industry. It has been known, that the addition of elements, like tin or lead, improves the antiscoring and antifrictional properties of aluminium. Aluminium tin alloys have a wide miscibility gap in the molten state and are virtually insoluble in each other during solidification. Further difficulties arise from the large freezing range of the alloys, which together with the wide density difference between the two components greatly increase the tin segregation during alloy preparation. It has, therefore, been difficult to introduce and uniformly disperse tin in aluminium to the desired extent by conventional melting and casting techniques. The present authors have developed a simple foundry technique which has been used successfully to disperse and retain tin at contents up to 20 wt/ in aluminium ingots, which is the subject of the present paper. © 1998 Published by Elsevier Science Ltd. All rights reserved.

Keywords: metallurgical problems; production; aluminium-tin alloys

Introduction

The use of aluminum alloys for slide bearings had aroused interest already back in the thirties. Rolls-Royce then successfully applied aluminum-tin alloys for the production of sliding elements in aircraft engines. During the forties similar alloys were applied in the automotive industry in the USA.

The main reason that sparked off the general interest in aluminum alloys was in fact the quest for a material having appropriately high sliding features, capable of withstanding substantially higher loads than the babbits and thus allowing to avoid lots of problems commonly encountered in both the manufacturing process and the subsequent utilization of the copper-lead and copper-tin-lead alloys.

As a result of the pursued research the aluminum alloys comprising 6–7% of tin were eventually obtained (e.g. the ALCOA 750 alloy used for monolithic bushing, or alloys with 20% of tin used in manufacturing bimetallic bushings on the steel sub-base). In this way, a truly universal bushing material was found, on the one hand characterized by good sliding features, on the other by its capability to withstand substantial loads.

Due to the underdeveloped infrastructure of the transport industry in Poland, no products using aluminum-tin alloys, especially the ones of high tin content (i.e. up to 20%) were manufactured for this specific use until the 1990s.

The dynamic growth of Polish economy in the recent years, especially the rapid development of automotive industry has inspired a lot of interest in the possibility of designing an effective manufacturing technology for the aluminum-tin alloys based monolithic and bimetallic slide bearings that would conform to the world standards in this domain.

The present paper aims to present the issues regarding the metallurgical aspects of the aluminum-tin alloys production. It is based on part of the research project carried out at the IMN-OML Skawina, co-financed by the National Committee for Scientific Research (Komitet Badañ Naukowych)

Basic metallurgical problems

The manufacturing process of aluminum-tin bearings comprises the following stages: melting—casting ingots—ingots rolling (inter-stage annealing)—integration with steel sub-base–creating of bi-metals. One of the conditions for obtaining a final product of required specification is making an aluminum-tin alloy ingot with tin evenly distributed throughout its volume.

Fulfilment of this condition is a difficult issue; the reason laying in the very structure of the tin equilibrium system in conjunction with the substantial difference of the respective specific weights of aluminum and tin.

Aluminum-tin alloys are characterized by very high tribological features and may be regarded as the most

* Tel.: +12 764088/776672; fax: +12 764776; e-mail: zwstuczy@cyf-kr.edu.pl

suitable bearing materials conforming to all modern technical requirements, provided that the technology of their manufacture has been mastered. These specific traits along with the difficulties entailed in the production technology result directly from the two component aluminum-tin equilibrium system, being strongly connected with the heterogenic structure of the weakly enriched alloy component in the α solution and the small tin-rich phase.

According to the aluminum-tin equilibrium system (*Figure 1*) the tin solubility in solid solution at 900 K is 0.1%, while at the eutectic temperature it goes down to the 0.005–0.07%. Tin with aluminum forms an eutectic system at low eutectic temperature of 501–502 K and reveals a strong deviation of the eutectic point towards tin (99.5% Sn). Creating the bearing alloys structure on the aluminum-tin basis starts with the appearance of first crystals of solid solution out of super cooled liquid.

As a result of a low value of the phase distribution coefficient $k = C_s/C_L$ a diffusion of high quantities of tin admixture appears on a crystallization front. As the effect of crystallization the resultant appearance of concentration supercooling becomes non-durable and tends to form cellular or dendritic surface. After the formation of cellular or dendritic crystals, the highly segregated eutectic liquid solidifies. As a result a net of developed dendritic crystallites with precipitates of the eutectic tin phase within inter-dendritic spaces appears in the structure of the alloy.

It follows from the above description of the aluminum-tin alloys solidifying process that an even distribution of the tin-rich eutectics in the cast material structure depends on the α solution growth process. It means that all parameters causing the refining of grains as well as the decrease of inter-dendritic distances will also be instrumental in making a more even distribution of tin.

Effective methods of grain refinement should be applied in the process of ingots production—original material for bimetallic bearing manufacturing—as well as such casting techniques that would allow for a faster solidifying process, so as to create the appropriate conditions for obtaining the correct structure.

The existence of substantial differences in the respective specific weights of aluminum and tin poses the danger of gravity segregation in the metal bath during the melting stage and the preparation for casting. This phenomenon could effectively be minimized by the application of such melting method that would allow for the shortest possible melting time, while at the same time securing constant motion of the metal bath by way of moving the melting process into the induction crucible furnace.

It is also very important to make sure that the ingot casting time is the shortest possible, so as to preclude the incidence of gravity segregation of the metal bath remaining in the crucible.

Aluminum-tin alloys production technology

As a result of the conducted research and the pertinent experiments, the method of aluminum-tin alloy production for sliding elements (i.e. monolithic and bimetallic bearing on the metal sub-basis) was defined. This technology is based on the two essential elements.

The first one advocates melting the metal in the induction crucible furnace. As a result the alloy is intensively stirred during the melting process to preclude segregation. One should bear in mind, however, that mixing of the liquid in the induction furnace can only be effective when the metal is being heated, and the attendant technological requirements provide for maintaining a stable temperature of metal bath at a certain level, so the furnace must be switched off periodically.

In order to keep up the mixing of the bath in the idle periods, a special device was constructed, joining both the mixer and the device for refining the bath through flotation and neutral (argon) gas blowing. The application of this method provides for keeping the metal bath

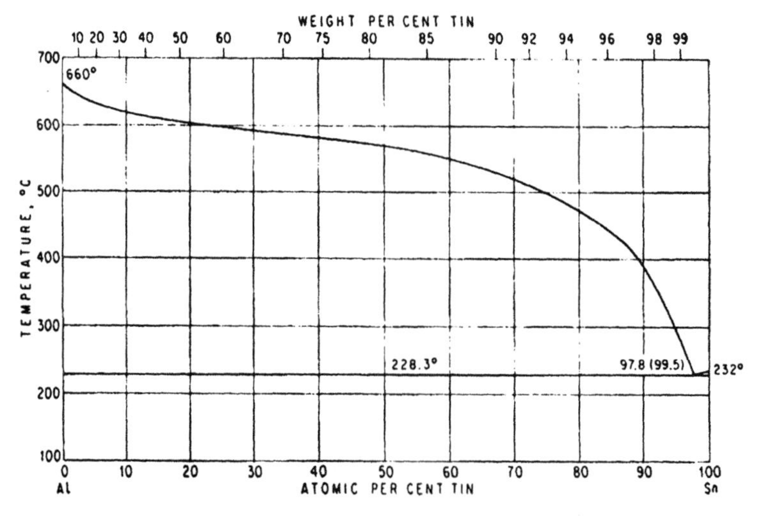

Figure 1 Al-Sn equilibrium system [1–4]

in constant motion while having it effectively degasified at the same time, let alone the elimination of any solid non-metallic inclusions. This precludes the danger of gravity segregation during the melting of aluminum-tin alloys and ensures that the gasification level of a metal bath does not exceed 0.15 ppm H_2.

The issue of even distribution of tin eutectics diffusions in the microstructure of the cast alloys has been solved in two ways. Firstly, the high efficiency grain refiner AlTi5B1 has been applied. It is being introduced as a wire into the liquid metal, in the proportion of 1 kg ton^{-1} of the alloy to be produced.

Secondly, a great emphasis has been put upon the selection of the most suitable method of ingots casting. There are two basic methods of ingots casting in the common foundry practice; the ingots being semi-products for further processing. The first method provides for the casting of ingots into a chill mould or sand mould, although is practically not applied to aluminum or any aluminum alloys processing nowadays.

The second method enables ingots casting of ingots in a continuous or semi continuous process. The basic advantage of the second method, apart from the better efficiency, is the achievement of higher intensity in the metal solidifying process.

In the case of die castings, the speed of solidification expressed by temperature drop within time unit is of several degrees per second, whereas in continuous or semi continuous casting the solidifying speed is approx. several dozens of degrees per second, so it is substantially higher.

There are following benefits: cast metal is of higher uniformity, both with respect to its chemical content as well as to the size of grains, the very structure of an ingot showing much finer grains, without the columnar crystals zone so characteristic for the die castings.

The ingots obtained by continuous or semi continuous casting have more consistent structure—without any visible porosity in both macro and micro scale.

As a consequence the material cast by way of the continuous or semi continuous method is far better suited for further processing than the one obtained through die casting. Also, the economic factors should be taken into account before deciding whether the aluminum and aluminum ingots should be produced through the semi continuous or continuous systems. The continuous casting lines are characterized by high capacity, with a yield of at least several thousand tons a year. At the same time the installation of such lines requires a lot of space. The vertical semi continuous casting, however, is fully justified economically with the annual yield of just several hundred tons. It should be born in mind, that a device for semi continuous casting is much simpler in terms of its construction and operational requirements, so the necessary investment is much lower.

As the volume of the planned production does not exceed 2.000 tons, the technology for the production of aluminum tin alloys put forward in the present paper is the of the semi continuous casting system, using the low crystalizators of the 'hot-top' type. For this purpose a device for semi continuous casting has been designed and installed. It has been equipped with a fully integrated computerized control system using General Electric microprocessor and coupled with a PC.

Summary

As a result of the application of the above described technology for aluminum-tin alloy production the appropriate conditions for obtaining ingots were established; the ingots to be prospectively used as an input material for the manufacture of bi-metallic bearings characterized by the microstructure, as illustrated in *Figure 2*.

Following the rolling with 50% cold work in one pass and annealing in the temperature of 623 K for 4 h, the material has received a microstructure, where the tin-rich eutectics appears in the fine and evenly distributed grains (*Figure 3*).

A band rolled down to 2 mm thickness and prepared for the subsequent integration with steel sub-base manifests a microstructure where the similarly fine and evenly distributed eutectics in the structure of metal (*Figure 4*).

The lack of stripe-like distribution of aluminum-tin eutectics and directional marks of plastic processing proves that the processes of casting, rolling and annealing were carried out faultlessly in terms of the attendant technological constraints. This particular type of

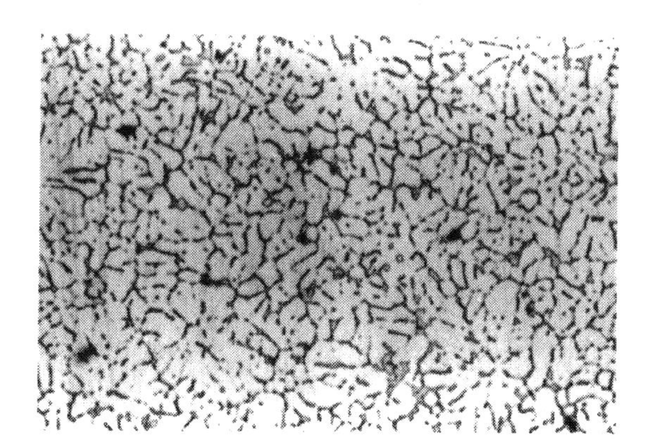

Figure 2 AlSn20Cu1 alloy mictrostructure in as-cast condition (semi-continouous casting). Magnification 100 ×. Unetched

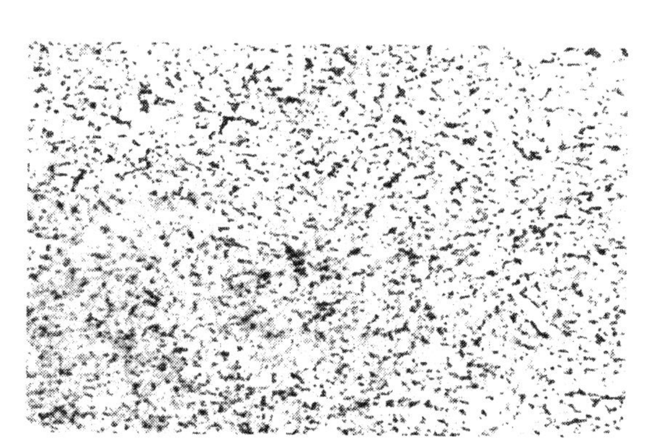

Figure 3 AlSn20Cu1 alloy mictrostructure: 50% cold worked in one pass and annealed (623 K for 4 h). Magnification 100 × X. Unetched

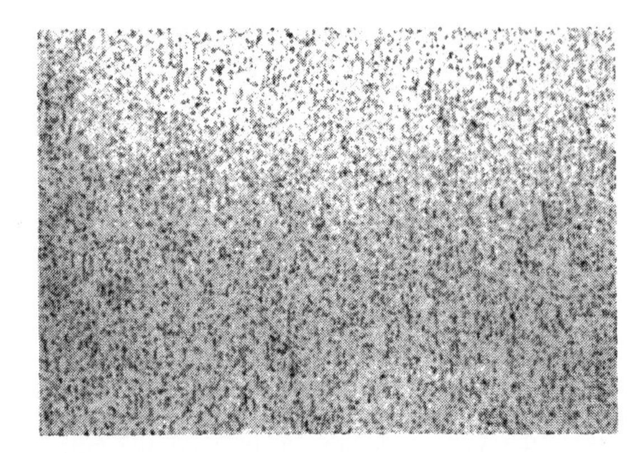

Figure 4 AlSn20Cu1 alloy mictrostructure. Band rolled down to 2 mm thickness and prepared for the integration with steel. Magnification 100 ×. Unetched

microstructure in the aluminum-tin alloy is by far the most suitable for the manufacture of bimetallic bands, which has been further corroborated by the results of the tribological tests.

Acknowledgements

The author of the present paper would like to take this opportunity to express his gratitude to the National Committee for Scientific Research (Komitet Badañ Naukowych) as well as to the Management of the Slide Bearings Factory 'BIMET' for having been granted the opportunity to work on this Project. Special thanks are also due to Lucjan Pierewicz, MSc (Eng) and to Wladyslaw Wezyk, MSc (Eng) for their inspiring and creative contributions.

References

1 Pratt, G.C., *International Metallurgical Reviews*, 1973, **18**, 62
2 Underwood, A., *The Selection of Bearing Materials, Sleeve Bearing Materials*. ASM, Metals Park, OH, 1949
3 Hunsicker, H., *Aluminium Alloy Bearnings: Metallurgy, Design and Service Characteristics, Sleeve Bearing Materials*. ASM, Metals Park, OH, 1949
4 Hansen, M., *Constitution of Binary Alloys*. McGraw-Hill, New York, 1958
5 Stuczyński, T., *Report no. 5047 / 95*, unpublished

Materials & Design, Vol. 18, Nos. 4/6, pp. 373–377, 1997
© 1998 Published by Elsevier Science Ltd
Printed in Great Britain. All rights reserved
0261-3069/98 $19.00 + 0.00

PII: S0261–3069(97)00079–4

Investigation of annealing behavior of nanocrystalline NiAl

L. Z. Zhou[a], J. T. Guo[a,*], G. S. Li[a], L. Y. Xiong[a], S. H. Wang[a], C. G. Li[b]

[a]Department of Superalloy and Intermetallics, Institute of Metal Research, Chinese Academy of Sciences, 72 Wenhua Road, Shenyang 110015, China
[b]Beijing Institute of Aeronautical Materials, Beijing 100095, China

Received 16 June 1997; accepted 10 July 1997

Nanocrystalline NiAl prepared by mechanical alloying and the vacuum heat pressing method has been studied to investigate its thermal stability. Isothermal annealing was conducted at 1000°C. The results show that a significant grain growth takes place from 30 nm at the initial state to 56 nm by annealing for 30 h, then maintains this value regardless of annealing time up to 100 h. Microhardness test results show that the microhardness of the annealed specimens reduces significantly with annealing time up to 5 h and then increases slightly with further annealing due to off-stoichiometry. In addition, positron annihilation spectroscopy has been used to investigate the interfacial defects in the nanocrystalline NiAl specimens. The results show that there are three kinds of defects in the interfacial region of the nano-NiAl alloy, which are monovacancy-sized free volumes, microvoids and large voids. During the present annealing process, the sizes of the three defects have no significant change, but the amount of the defects, especially for monovacancy-sized free volumes and microvoids varies clearly with annealing time. © 1998 Published by Elsevier Science Ltd. All rights reserved.

Keywords: nanocrystalline NiAl; annealing behavior; thermal stability

Introduction

The demand for new structural materials in high performance aerospace applications has been the driving force behind the development of intermetallics, ceramic and composite materials. These efforts have been necessitated by the need for greater engine operating efficiencies. The B2 cubic crystal structure intermetallic NiAl possesses many attractive properties for structural use at elevated temperatures in hostile environments, e.g. high melting point, low density, high thermal conductivity and excellent oxidation resistance. However, it suffers from poor fracture resistance at low temperature, which indicates that use of the monolithic materials is improbable[1–4]. A number of attempts have been made to overcome this drawback[5–9]. One of possible routes is the synthesis of nanocrystalline materials which may transform nominally brittle compounds into ductile materials[10,11]. Indeed, recent work shows that room temperature ductility can be improved by grain size refinement[12,13].

Nowadays, nanocrystalline (NC) materials have received much interest because of their unusual properties and special microstructure[14–17]. Several potential applications have already been identified, but the widespread industrial implementation of these NC materials will ultimately depend upon their long-term resistance to environmental degradation, i.e. corrosion resistance and thermal resistance. According to well-known Gibbs–Thomson equation, the driving force for grain growth increases with the reduction of grain size and it might be extremely large for the nm-sized grains, even at room temperature. However, experimental observations indicate that most NC materials synthesized by various methods exhibit inherent grain size stability up to reasonably high temperature, sometimes as high as approx. 0.5 Tm, which is comparable with that of the grain growth in conventional coarse-grained polycrystals[15,17]. Investigations on the grain size stability have been reported in various NC materials, including pure metals, oxides, compounds and composites. But there is still lack of work on consolidated NC intermetallics, especially on NiAl. Besides, except for grain size, interfacial characteristics (e.g. defects) should also be important as far as the properties of NC materials are considered, since there is a large amount of interface in NC materials. Thus the stability of interface microstructure in NC materials needs further investigation.

In this work, we present an investigation on annealing behavior of NC NiAl. Grain size stability and mechanical properties stability have been studied. Interface defects as well as their variation with annealing time have been investigated by positron annihilation spectroscopy, with an effort to explore the relationship between mechanical property and microstructure.

*Correspondence to J. T. Guo. Fax: +86 24 3891320; e-mail: jtguo@imr.ac.cn

Experimental materials and procedures

NC NiAl compacts were prepared by mechanical alloying and vacuum heat pressing method. Mechanical alloying was carried out in a SPEX 8000 mixer/mill. Ni and Al elemental powders, mixed at a composition of $Ni_{50}Al_{50}$, were milled under Ar atmosphere at room temperature. The ball-to-powder weight ratio is 10:1. After 30 h of milling, the elemental powders were completely transformed to NiAl phase with an average grain size of 10 nm. After degassing in vacuum at 400°C for 1 h, the NC powder was vacuum hot pressed at 900°C/62.5 MPa for 1 h. The average grain size of the compact (91% density of bulk material) was 30 nm.

Samples taken from such a compact were subjected to isothermal annealing at 1000°C in air, using a muffle furnace and then air-cooled after annealing. The grain sizes of the annealed samples were determined by conventional Scherrer equation. Microhardness was conducted on a MVK-H3 microhardness meter with a load of 100 mg and load time of 10 s.

Positron lifetime experiments were carried out by using a fast–fast coincidence ORTEC system with a time resolution of 250 ps (FWHM) under experimental conditions. A 3.7×10^5 Bq source of ^{22}Na was sandwiched between two pieces of the nanocrystalline samples. The measurement of positron annihilation was performed at room temperature (18°C) and the spectra of all the samples were analyzed by applying the computer program PATFIT[18].

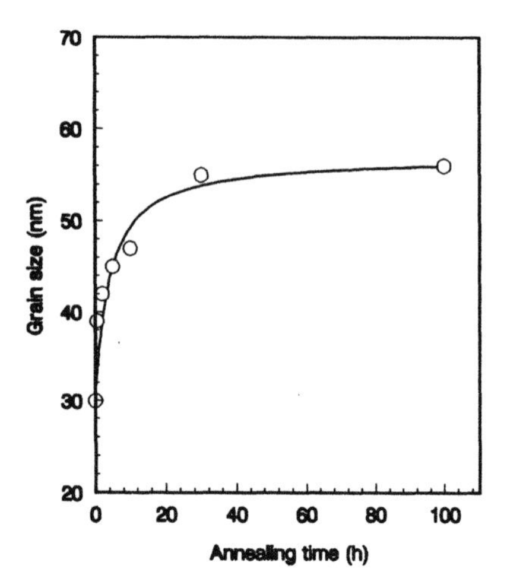

Figure 2 Grain size as a function of annealing time for the nanocrystalline NiAl

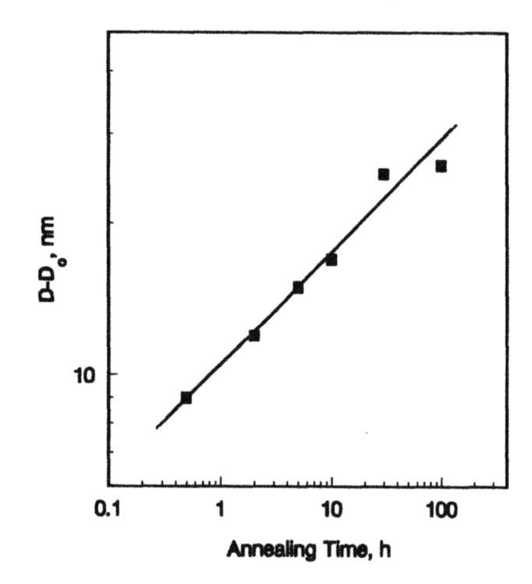

Figure 3 A plot of log $(D - D_0)$ vs. log (t)

Results and discussion

Grain size stability

In order to study the grain growth kinetics in the NC NiAl material, isothermal annealing was conducted at 1000°C for different times. *Figure 1* shows the XRD patterns for the samples unannealed and/or annealed for 100 h. The dominant Bragg peaks are corresponding to NiAl phase, the smaller peaks are corresponding to α-Al_2O_3 phase, i.e. a small amount of α-Al_2O_3 exists in the NiAl samples. There was no phase transformation occurring in these samples during a long time annealing. The grain size as a function of annealing time is shown in *Figure 2*. It can be seen that a significant grain growth takes place from 30 nm at the initial state to approx. 55 nm by annealing for 30 h, then maintains this value regardless of annealing time up to 100 h. *Figure 3* shows a plot of log $(D - D_0)$ vs. log (t) for the NC NiAl. It was found that a regression line with a slope of 1/5 fits well to these data points. Thus, the grain growth kinetics can be expressed as $D - D_0 = kt^{1/n}$, where k is a constant and $n = 5$. The exponent n is slightly lower than that ($n = 6$) in conventional coarse-grained NiAl[19], but much higher ($n = 2$) in pure metals[20] and is also higher ($n = 3$) in mechanically alloyed NC Al_3Nb[21] and that ($n = 4$) in NC Al_3Nb-5.5 wt.% Ti[22].

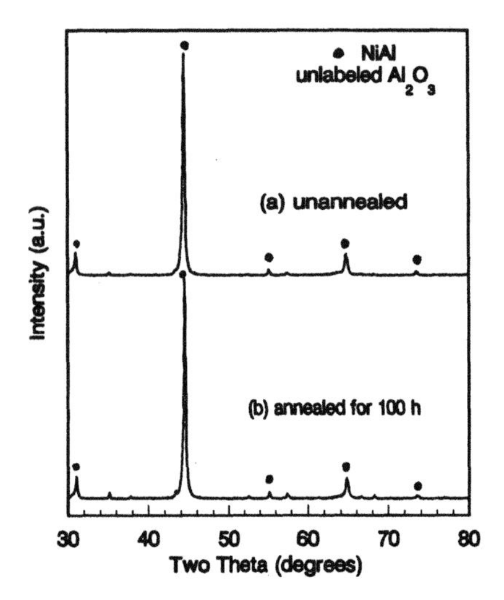

Figure 1 XRD patterns for the nanocrystalline NiAl unannealed (a) and annealed for 100 h (b)

The results presented above suggest that the NC NiAl material exhibits inherent grain size stability and therefore provides considerable promise for its industrial application as high-temperature materials. The inherent grain size stability in NC materials is usually attributed to structural characteristics of the materials, such as the grain size and its distribution, grain mor-

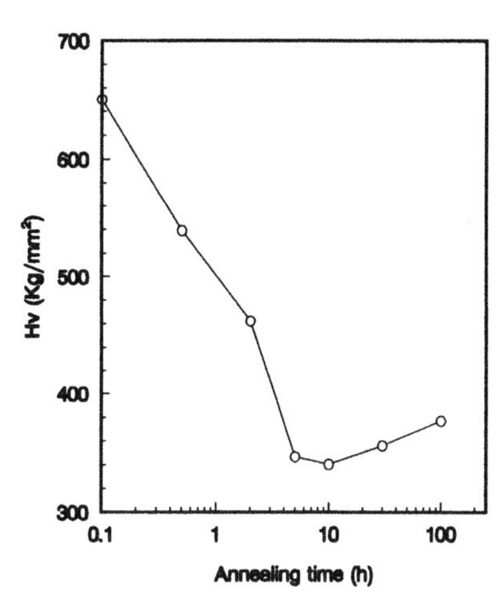

Figure 4 Microhardness as a function of annealing time for the nanocrystalline NiAl

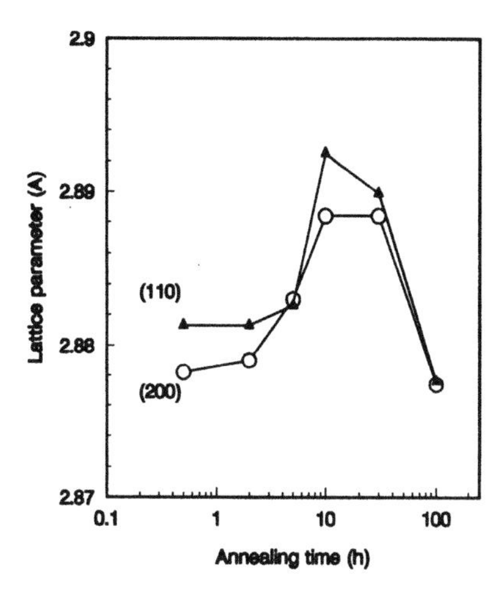

Figure 5 Lattice parameter as a function of annealing time for the nanocrystalline NiAl

phologies, the nature of interfaces and triple junctions, porosity in the sample and so on[17]. In this work, the following factors might play an important part in the inherent grain size stability: (a) second-phase particles (α-Al_2O_3) drag which results in the pinning of grain boundaries; (b) the drag of solute, such as Fe and C atoms which come from contamination during ball milling; and (c) porosity drag, 9% porosity present in the NC NiAl is expected to inhibit the grain growth considerably.

Microhardness stability

Microhardness Hv has been measured on these annealed samples and the result is shown in *Figure 4*. It is seen that microhardness of the annealed samples reduces significantly with annealing time up to 5 h and then increases slightly with further annealing. For the unannealed sample, the value is 650 kg mm^{-2}, which is significantly higher than Hv 330 of stoichiometric coarse-grained NiAl[23], but lower than Hv 887 for NiAl (grain size = 15 nm, 80% dense) reported by Hanbold *et al.*[24] and is comparable with Hv 650–697 (grain size = 70 nm, 92–94% dense) reported by Smith *et al.*[25]. Thus, the hardness value appears to be reasonable.

Several factors (e.g. Hall–Petch effect, deviation from stoichiometry) may be responsible for the variation of microhardness Hv in the annealed samples. A quantitative estimation of Hall–Petch effects (i.e. $\sigma = \sigma_0 + Kd^{-1/2}$) can be made based on the value of $K = 0.522$ MPa $m^{1/2}$ for powder extruded NiAl material (d = 10–200 μm) reported by Bowman *et al.*[26]. Assuming σ_0 to be constant for all these annealed samples, the Hall–Petch effect (from 30 nm to 46 nm) is $\Delta\sigma = -583$ MPa. Using the relationship that Vickers hardness is three times the compressive strength[27], $\Delta Hv = -175$ is obtained.

In NiAl, deviation from the stoichiometric composition has a significant effect on mechanical behavior. It was reported by Hahn *et al.* that the average hardening rate for Al-rich NiAl was 40 MPa/at.% and that for Ni-rich alloys it was approx. 70 MPa/at.%[28]. In this work, the chemical analysis indicates that the Ni:Al ratio after milling is 48:52, then we have a enhanced strength of 80 MPa corresponding to a hardness of 24 kg mm^{-2}.

The above two factors give a total value of approx. 200 kg mm^{-2}, considering that (i) weakening mechanisms (e.g. viscous type flow) operate in NC materials, the Hall–Petch slope K is much smaller in nanometer range than is observed at more normal grain sizes[16], which is also verified by the lower hardness in the sample annealed for 5 h (Hv = 346), though it has a nanometer range of grain size; and (ii) since a small amount of α-Al_2O_3 has been formed in the consolidated material, the off-stoichiometry may be decreased, therefore the above value is overestimated. From this analysis, it is seen that grain growth and off-stoichiometry cannot account for all the markedly decrease of microhardness. There must exist another mechanism which also affects the mechanical properties of the annealed samples.

With an effort to explore the reason for the slight increase of microhardness in the long-time annealed samples, the lattice parameter has been carried out and the results are shown in *Figure 5*. It is seen that the lattice parameter varies with annealing time, suggesting

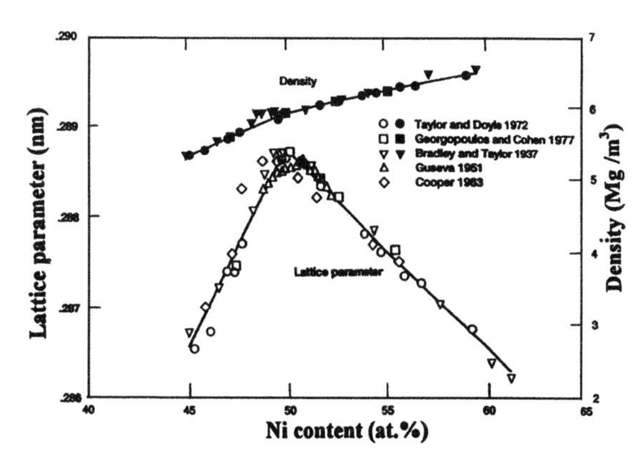

Figure 6 Room-temperature lattice parameter of NiAl as a function of stoichiometry[29]

Table 1 Positron lifetime results of the nanocrystalline NiAl material annealed for different times

Sample no.	τ_1 (ps)	τ_2 (ps)	τ_3 (ps)	I_1 (%)	I_2 (%)	I_1/I_2	I_3 (%)
1	185.3 ± 8.1	334.7 ± 10.7	2265 ± 292	55.29 ± 7.34	43.93 ± 7.27	1.26	0.78 ± 0.07
2	171.8 ± 3.6	332.1 ± 8.5	2086 ± 109	65.90 ± 2.44	33.10 ± 2.42	1.99	1.00 ± 0.06
3	174.0 ± 4.0	342.0 ± 9.4	2134 ± 137	70.25 ± 2.79	28.82 ± 2.73	2.44	0.94 ± 0.06

that Al content in the NC NiAl materials varies with annealing time due to oxidation, since lattice parameter is significantly dependent on Ni:Al ratio (*Figure 6*). The peak value of lattice parameter found here is very close to $a_0 = 0.2887$ nm for the stoichiometric NiAl at room temperature[30], which indicates that the Ni:Al ratio in the corresponding samples is very close to the stoichiometry (1:1), that is also confirmed by the microhardness. Therefore it is understood that the slight increase of microhardness is caused by off-stoichiometry due to oxidation in the materials.

Interface defects stability

In NC materials, interface characteristics is very important as mentioned above, while positron annihilation spectroscopy has been used as a powerful method to investigate the interface structured of NC materials. In this work, positron lifetime spectroscopy has been carried out on several samples (i.e. no. 1 unannealed; no. 2 annealed for 2 h; no. 3 annealed for 5 h). *Table 1* lists the positron lifetime results of the NC NiAl. It can been seen that three lifetimes (τ_1, τ_2, τ_3) exist in these samples, i.e. three kinds of defects exist in the interface regions of the NC NiAl, which are monovacancy-sized free volumes, microvoids and large voids, with I_1, I_2 and I_3 corresponding to their intensities. During the present annealing, the three lifetimes have no obvious variation, which indicates that the sizes of the three defects show no significant change, but the amount of the defects, especially for monovacancy-sized free volumes and microvoids, varies clearly with annealing time. As is well known, the amount of the defects (especially in interfaces) will obviously influence the deformation behavior, such as fracture toughness and microhardness. Thus, in the present work, the rapid decrease of Hv with the increase of annealing time should also be attributed to the variation of the interface defects. It is worth noting that the amount of microvoids in the samples is decreasing with annealing time, while the amount of monovacancy-sized free volume is increasing with annealing time, which may suggest that microhardness is seriously dependent on the microvoids, but not the monovacancy-sized free volume.

Conclusion

The grain growth kinetics can be expressed as $D - D_0 = kt^{1/n}$, where k is a constant and $n = 5$. The nanocrystalline NiAl material exhibits intrinsic grain size stability, its grain size can maintain nanometer range during long-time annealing at 1000°C.

The microhardness is significantly reduced at the early annealing stage, then slightly increased with annealing time. The reduction of microhardness is mainly attributed to grain growth and interface defects and the increase is to off-stoichiometry.

Three kinds of interface defects exist in the nanocrystalline NiAl, which are monovacancy-sized free volumes, microvoids and large voids. During the present annealing, the sizes of the three defects do not significantly change, but the amount of monovacancy-sized free volumes in the samples is clearly increased with annealing time, while the amount of microvoids is decreased with annealing time.

Acknowledgements

This research is supported by the National Natural Science Foundation of China (Project 59331012) and the National Advanced Material Committee of China.

References

1 Miracle, D. B. *Acta Metallurgica*, 1993, **41**, 649
2 Neobe, R. D., Bowman, R. R. and Nathal, M. V. *International Materials Review*, 1993, **38**, 193
3 George, E. P., Yamaguchi, M., Kumar, K. S. and Liu, C. T. *Annual Review of Materials Science*, 1994, **24**, 409
4 Darolia, R. *Journal of Materials Science and Technology*, 1994, **10**, 157
5 George, E. P. and Liu, C. T. *Journal of Materials Research*, 1990, **5**, 754
6 Enami, K., Martynov, V. V., Tomie, T., Khandros, L. G. and Nenno, S. *Transactions Japanese Institution on Metals*, 1981, **22**, 357
7 Darolia, R. *Journal of Metals*, 1991, **43**, 44
8 Darolia, R., Lahrman, D. and Field, R. *Scripta Metallurgica*, 1992, **26**, 1007
9 Lshida, K., Kainuma, R., Ueno, N. and Tinishizawa. *Metallurgical Transactions*, 1991, **22A**, 441
10 Karch, J., Birringer, R. and Gleiter, H. *Nature*, 1987, **330**, 536
11 Bohn, R., Hanbold, T., Birringer, R. and Gleiter, H. *Scripta Metallurgica*, 1991, **25**, 811
12 Dollar, M., Dymek, S., Hwang, S. J. and Nash, P. *Scripta Metallurgica*, 1992, **26**, 29
13 Whittenberger, J. D., Viswnadham, R. K., Mannan, S. K. and Kumar, K. S. *Materials Research Society Symposium Proceedings*, 1989, **122**, 621
14 Gleiter, H. *Progress in Materials Science*, 1989, **33**, 223
15 Andrievski, R. A. *Journal of Materials Science*, 1994, **29**, 614
16 Suryanarayana, C. *International Materials Review*, 1995, **40(2)**, 41
17 Lu, K. *Materials Science and Engineering*, 1996, **R16**, 161
18 Kirkegard, P., Eldrap, M., Mogensen, O. E. and Peterson, N. *Computer Physics Communications*, 1981, **23**, 307
19 Haff, G. R. and Schulson, E. M. *Metallurgical Transactions*, 1982, **13A**, 1563
20 Ralph, B. *Materials Science and Technology*, 1990, **6**, 1139
21 Isonishi, K. and Okazaki, K. *Journal of Materials Science*, 1993, **28**, 3829
22 Kawanishi, S., Isonishi, K. and Okazaki, K. *Materials Transactions JIM*, 1993, **34**, 49
23 Hayashi, K. and Kihara, H. In *Sintering '87*, vol. 1. eds. S. Somiya, M. Shimada, M. Yoshimura and R. Watanabe. Elsevier Science Publishers, New York, 1987, p. 255

24 Haubold, T., Bohn, R., Birringer, R. and Gleiter, H. *Materials Science and Engineering*, 1992, **A153**, 679
25 Smith, T. R. and Vecchio, K. S. *Nanostructure Materials* 1995, **5**, 11
26 Bowman, R. R., Noebe, R. d., Raj, S. V. and Locci, I. E. *Metallurgical Transactions*, 1992, **23A**, 1493
27 Tabor, D. *Review of Physics and Technology*, 1970, **1**, 145
28 Hahn, K. H. and Vedula, K. *Scripta Metallurgica*, 1989, **23**, 7
29 Noebe, R. D., Bowman, R. R. and Nathal, M. V. *NASA Technical Paper 3398*, 1994
30 Taylor, A. and Doyle, N. J. *Journal of Applied Crystallography*, 1972, **5**, 201

Materials & Design, Vol. 18, Nos. 4/6, pp. 379–383, 1997

0261-3069/98 $19.00 + 0.00

PII: S0261-3069(97)00089-7

Characteristic features of silumin alloys crystallization

Stanislaw Pietrowski*

Technical University of Łódź, Institute of Materials Engineering and Chipless Technology, Stefanowskiego no. 1/15, 90-924 Łódź, Poland

Received 23 September 1997; accepted 26 September 1997

The results differential thermal analysis (DTA) of crystallization of multicomponent silumin alloys with 8 ÷ 22 wt %Si and Mg, Cu, Ni and Fe minor admixtures are presented in the paper. It has been proved that to any one of the solid phases crystallizing from liquid (α, β, Mg_2Si, Al_6Cu_3Ni, Al_2Cu, Al_9Fe_2Si) a particular thermal effect can be deduced and identified. A new hypothesis concerning hypereutectic silumin alloys crystallization has been proposed in the work. © 1998 Published by Elsevier Science Ltd. All rights reserved.

Keywords: silumin alloys; crystallization; DTA analysis

Introduction

The aim of the work is to present the characteristics of silumins crystallization using thermal-derivative analysis (ATD). Examination of Al alloys using ATD is presented in several works[1–6]. Theoretical data about ATD method for silumins are described by Jura and Jura[7]. The maximum of the crystallization curve and its intersections with zero-axis are the most typical points of this curve. The point are the characteristic temperatures of crystallization and phases transformations in the solid state[7–12]. This paper is, to some extent, a generalization and development of research carried on in the Institute of Materials Engineering of Technical University Łódź and partially presented elsewhere[8–12].

Experimental details

The representative chemical composition of the examined silumins is shown in *Table 1*. Crystallization of silumins was examined with the use of the CRYSTALDIGRAPH PC\AT equipment by produced Z-TECH company and the ATD sampler with solidification module M = 0.54 cm.

Results

The ATD curves of model hypo-, hyper- and eutectic unmodified silumins (Al–Si) are shown in *Figures 1(a)–(c)*. Crystallization of the hypoeutectic silumin begins after undercooling to the lowest temperature of liquidus $t_{Lmin} = t_{pk} = 599°C$ (point P_k on the crystallization curve *Figure 1a*). It starts with precipitations of α-phase dendrites. The heat of their crystallization increases the alloy temperature up to the maximum liquidus $t_{Lmax} = t_B = 601°C$ (point B). The maximum thermal effect of α-phase dendrites' crystallization occurs at point A. The process of crystallization of α-phase dendrites continues until the lowest solidus temperature $t_{Smin} = t_C = 567°C$ (point C) is reached. Crystallization of $\alpha + \beta$ (Al + Si) eutectic begins at that temperature because of supersaturation of the rest of the liquid by silicon. The eutectic heat of crystallization causes its maximum thermal effect to occur at point D on the derivative curve and the temperature of the alloy increases up to the maximum solidus $t_{smax} = t_E = 570°C$ (point E). The crystallization process ends at the temperature $t_F = 551°C$.

Crystallization of eutectic silumin (*Figure 1b*) begins after undercooling to the lowest solidus temperature $t_{Smin} = t_C = 570°C$ (point C). It begins with precipitations of α-phase dendrites. The liquid is supersaturated

Table 1 Chemical composition of examined silumins

Alloy number	Chemical composition (wt%)				
	Si	Mg	Ni	Cu	Fe
1	7.52	—	—	—	—
2	12.48	—	—	—	—
3	17.65	—	—	—	—
4	8.62	—	—	—	0.51
5	8.60	—	—	—	0.87
6	8.58	—	—	—	1.32
7	8.63	—	—	—	1.74
8	12.98	1.24	1.36	1.31	0.28
9	4.09	1.28	—	1.80	0.20

*Tel.: +48 42 312275; fax: +48 42 366790

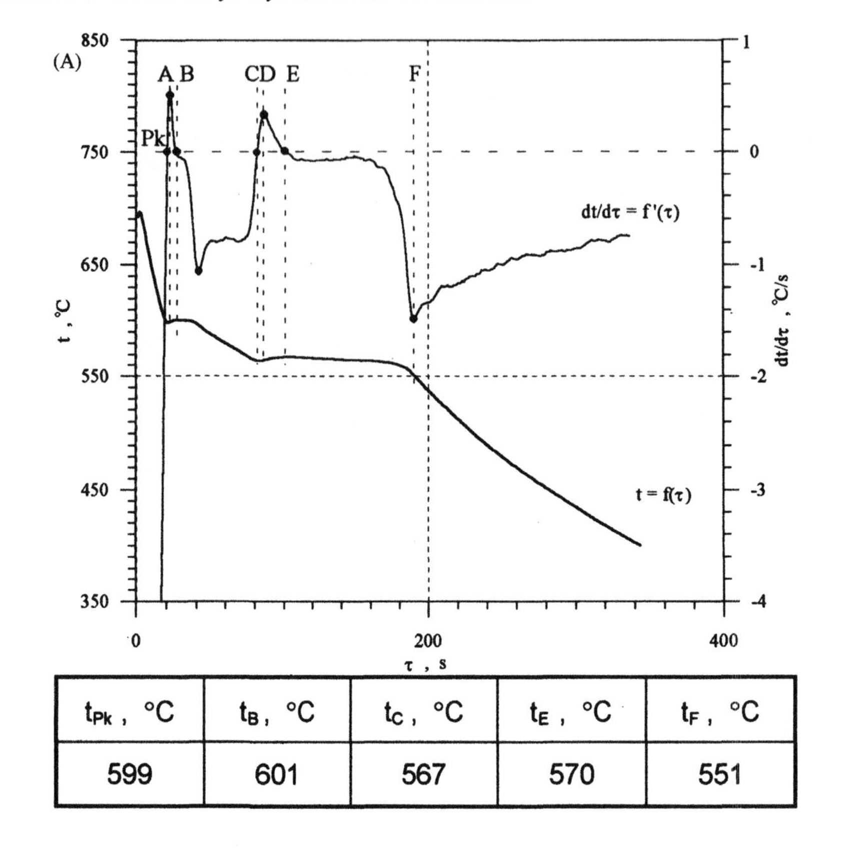

t_{Pk}, °C	t_B, °C	t_C, °C	t_E, °C	t_F, °C
599	601	567	570	551

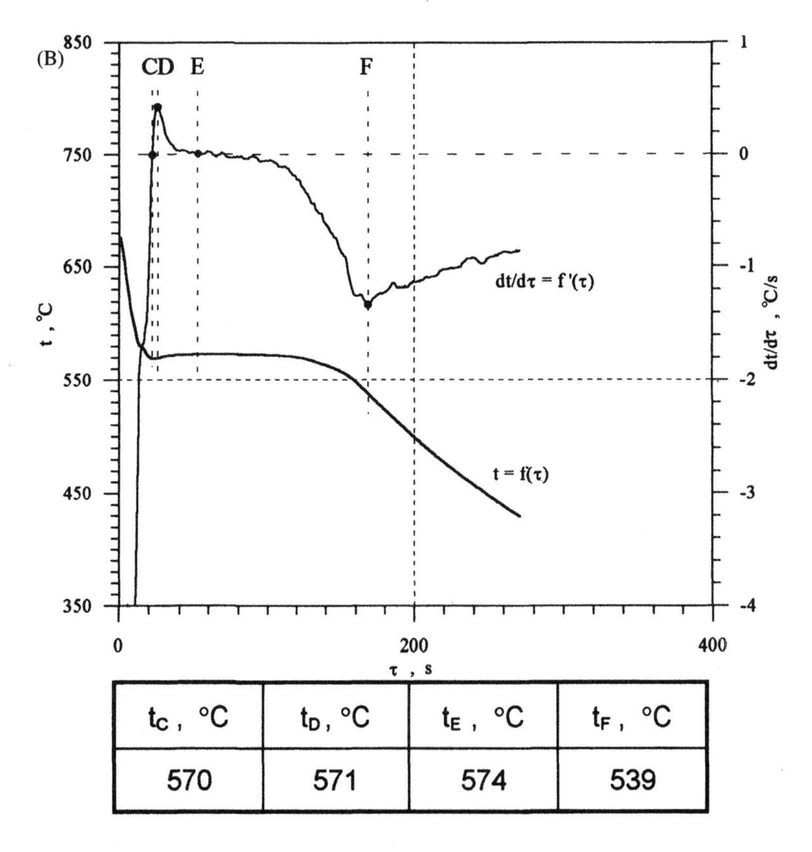

t_C, °C	t_D, °C	t_E, °C	t_F, °C
570	571	574	539

in silicon and enters upon a zone of associate eutectic growth that is asymmetric according to[13]. The crystallization of β-phase occurs and then the α-phase crystallizes on it. The alloy crystallization ends of the temperature $t_F = 539°C$ (point F).

Crystallization of hypereutectic silumin (*Figure 1c*) begins after maximum undercooling of the liquid to the lowest liquidus temperature $t_{Lmin} = t_{Pk'} = 612°C$ (point $P_{k'}$). It begins with the precipitation of large β-phase crystals. The maximum thermal effect of their crystallization occurs at the A point. The liquid surrounding the extended β-phase crystals becomes considerably depleted in silicon. According to the Gibbs–Thomson effect, the free energy of the crystal with a small radius of curvature is higher than the free energy of the crystal with a high radius of curvature. Therefore the silicon concentration in liquid around these latter decreases. Because of that the α phase becomes nucleated and grows around the large primary β crystals. The occurrence of α phase causes the formation of an additional thermal effect B at temperature $t_B = 594°C$. The examinations have shown that α phase crystallizes on β phase crystals if the silicon content in silumin is higher than 15 wt%. The presented crystallization proceeds to the lowest solidus temperature $t_{Smin} = t_C = 574°C$ (C point) when the crystallization of the $\alpha + \beta$ eutectic begins. Crystallization of the eutectic and the whole silumin ends at temperature $t_F = 554°C$ (F point).

Addition of Fe (up to approximately 0.50 wt%) to silumins causes the crystallization of $\alpha + Al_9Fe_2Si + \beta$ ternary eutectic without any change of ATD curves. The increase of Fe addition up to 0.87 wt% brings about the change of silumin crystallization and hence a change of ATD curves as it is shown in *Figure 2*. Crystallization of α-phase dendrites (zone P_k-A-B-J on the crystallization curve) generates the increase of Fe and Si concentration in front of the solidification front up to the content high enough for peritectic crystallization ($L + Al_{12}Fe_3Si \rightarrow \alpha + Al_9Fe_2Si$). The peritectic reaction begins by nucleation and growth at temperature $t_J = 575°C$ (J point). The peritectic transformation ends at temperature $t_{Smin} = t = 567°C$ (C point) and the crystallization of ternary eutectic ($\alpha + Al_9Fe_2Si + \beta$) begins. Crystallization of silumin ends at the temperature $t_F = 556°C$ (F point).

Additions of Mg, Ni and Cu (separately or together) cause the following ternary eutectics to crystallize successively after completion of the binary or ternary eutectic crystallization: with Mg_2Si or Al_6Cu_3Ni or Al_3Ni silumins without Cu admixture or with Al_2Cu in the other case. Heat of their crystallization causes thermal effects on the derivative curve as shown in *Figure 3*. The crystallization process of hypoeutectic silumin proceeds up to the point F in the same way as in the former case. Crystallization of binary eutectic ends at $t_F = 537°C$ (F point) and the ternary eutectic with Mg_2Si phase begins to crystallize. That crystallization continues up to the t_H temperature; beginning with this temperature the eutectic with Al_6Cu_3Ni phase crystallizes. That phase crystallizes up to temperature

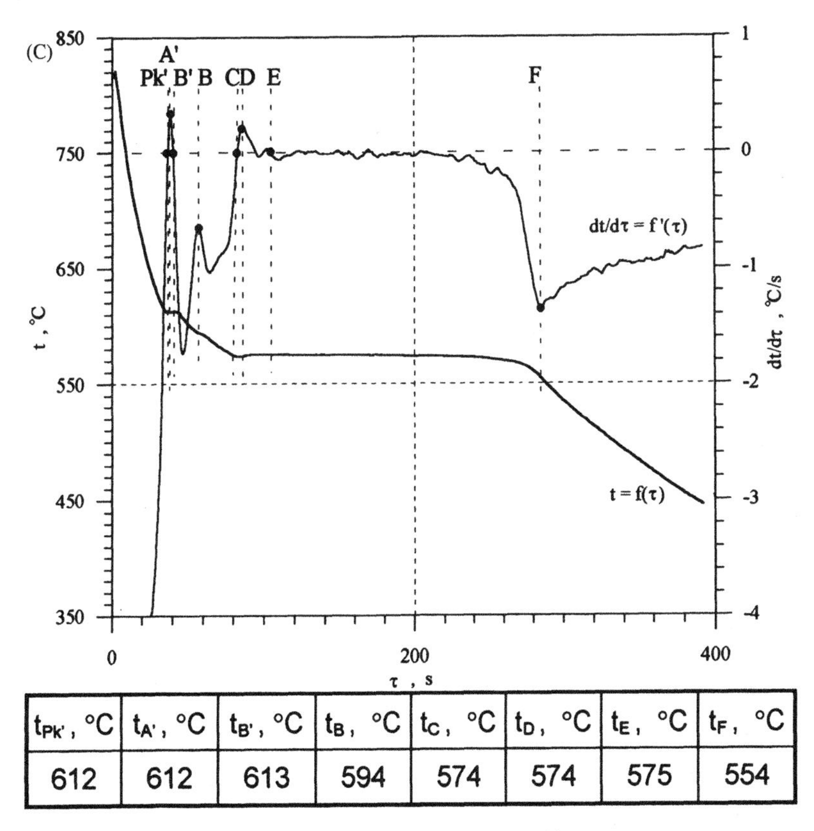

$t_{Pk'}$, °C	$t_{A'}$, °C	$t_{B'}$, °C	t_B, °C	t_C, °C	t_D, °C	t_E, °C	t_F, °C
612	612	613	594	574	574	575	554

Figure 1 (a) Representative ATD curves of hypoeutectic silumin with 7.50% Si content. (b) Representative ATD curves of eutectic silumin with 12.48% Si content. (c) Representative ATD curves of hypereutectic silumin with 17.61% Si content

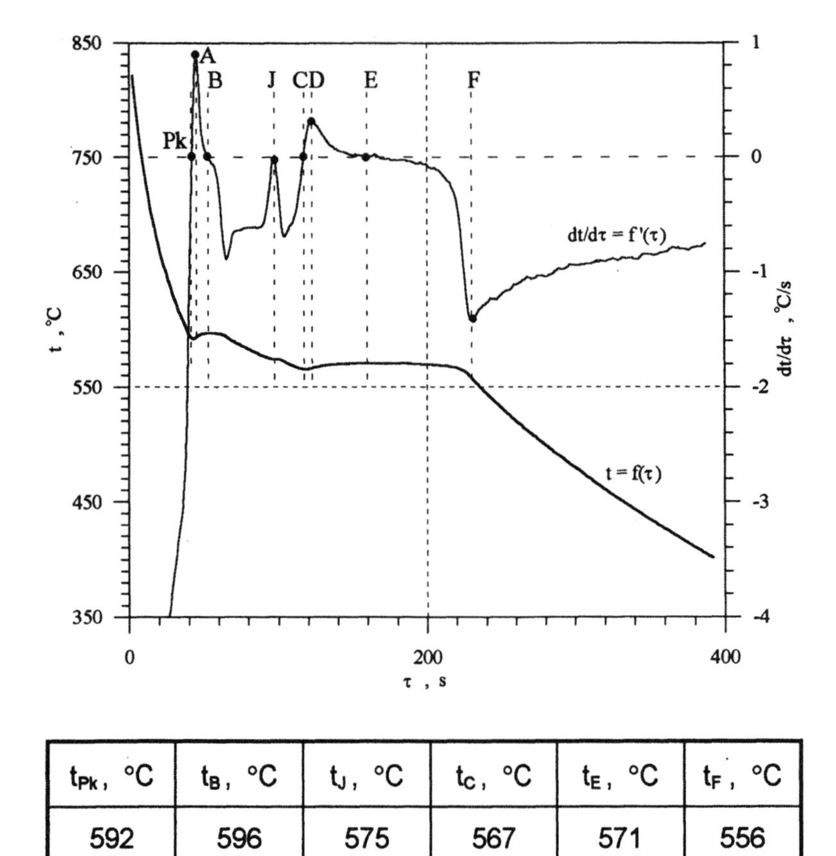

t_{Pk}, °C	t_B, °C	t_J, °C	t_C, °C	t_E, °C	$t_{F'}$, °C
592	596	575	567	571	556

Figure 2 ATD curves of silumin with 8.60% Si and 0.87% Fe content

$t_{F'} = 506$°C (F point) when the eutectic with Al_2Cu phase crystallizes. Crystallization of this eutectic and of the whole silumin ends at the temperature t_G.

Conclusions

The following conclusions result from the described experiments:

- the crystallization process of multicomponent silumins with addition of Mg, Ni, Cu, Fe is complex because;
- in the hypereutectic unmodified silumins containing more than 15 wt% Si, the α phase is growing hypoeutectically on big primary β crystals; and
- the knowledge of the lowest temperature of eutectics' crystallization enables the proper choice of the temperature for solution heat treatment.

References

1 Argyropoulos, S., *Quantitative kontrolle der veredelung von Al–Si gusslegierungen unter anwendungeiner thermoanalysetechnik. Giesserei-Praxis*, 1985, **11**, 173

2 Roučka, J., Kotlaba, K., *Použiti termické analýze k posouzeni chemickěcho složeni a struktury slitin typu Al–Si. Slěvárenstvi*, 1986, **5**, 198

3 Sakwa, J., *Anwendung der termischen analyse und der derivativen diferential-thermoanalisezur abschaetzung des einlusses von Mg, P und Cu sowie Na und Ti auf den kristallisationsprozess bei aluminium-silicium-legierungen. Giessereiforschung*, 1986, **3**, 112

4 Sakwa, J., *Zur kristallisation der aluminiumguslegierung AlSi7Mg. Giessereiforschung*, 1988, **4**, 134

5 Roučka, J., Janová, D., Hrubý, J., *Thermal Analysis as the Means of Al–Si–X Phase Analysis. Solidification of Metals and Alloys. Polish Academy of Sciences. Foundry Commission*, 1992, **17**, 124

6 Mulazimoglu, M. H., *Studies on the Minor Reactions ond Phoses in Strontium-treated Aluminium — Silicon Casting Alloys. Cast Metals*, 1993, **1**, 16

7 Jura, S., Jura, Z., *Solidification of Metals and Alloys. Polish Academy of Sciences, Foundry Commission*, 1996, **28**, 57

8 Pietrowski, S., Władysiak, R., *Estimation of Modification and Crystallization Effect of Hypereutectoid Silumin by ATD Method. Solidification of Metals and Alloys. Polish Academy of Sciences, Foundry Commission*, 1992, **17**, 114

9 Pietrowski, S., An assessment of the silumins crystallization with the ATD method. *Inżynieria Materiałowa*, 1994, **1**, 3

10 Pietrowski, S., Władysiak, R., *The Crystallization of Synthetic Near-Eutectic Silumins Assessment with ATD Method. Solidification of Metals and Alloys. Polish Academy of Sciences, Foundry Commission*, 1994, **20**, 65

11 Pietrowski, S., Władysiak, R., *The crystallization of synthetic hypoeutectic silumins assessment with ATD method. Solidification of Metals and Alloys. Polish Academy of Sciences, Foundry Commis-*

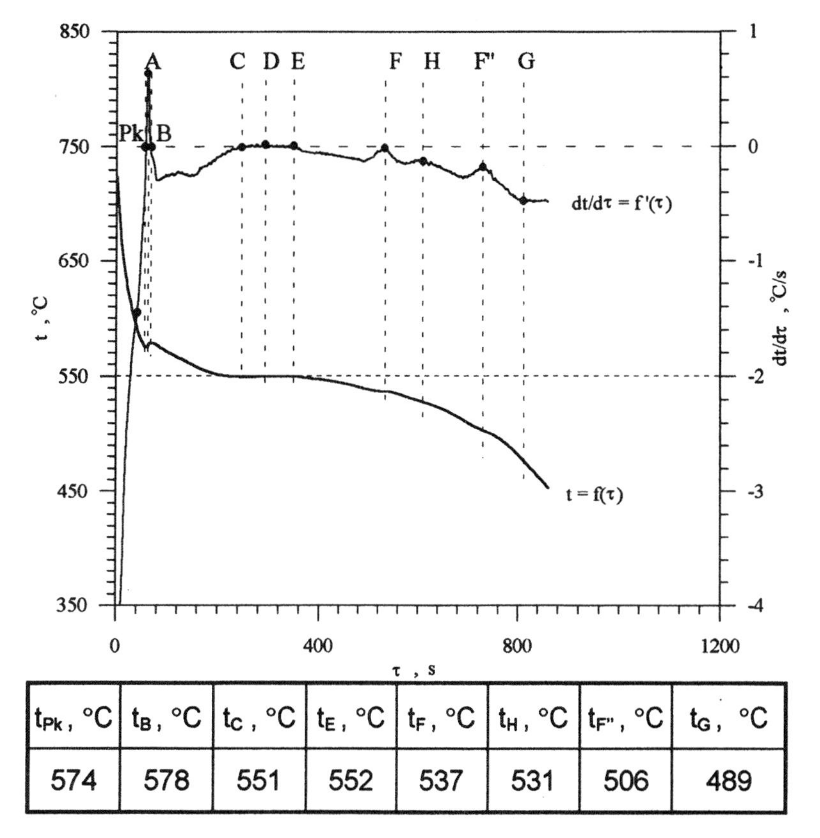

t_{Pk}, °C	t_B, °C	t_C, °C	t_E, °C	t_F, °C	t_H, °C	$t_{F''}$, °C	t_G, °C
574	578	551	552	537	531	506	489

Figure 3 ATD curves of piston silumin AK102

sion, 1995, **22**, 46

12 Pietrowski, S., Władysiak, R., *Control of piston silumins with ATD method. Solidification of Metals and Alloys. Polish Academy of Sciences, Foundry Commission*, 1996, **28**, 160

13 Hamani, M. S., Pikunow, M. W., *Osobiennosti struktury spławow Al–(7 ÷ 17)% Si Al–(25 ÷ 39)% Cu w swiazi s rozlicznymi usłowiami kristalizacji. Cziornaja Metałłurgia*, 1992, **3**, 53

Materials & Design, Vol. 18, Nos. 4/6, pp. 385–388, 1997
© 1998 Published by Elsevier Science Ltd
Printed in Great Britain. All rights reserved
0261-3069/98 $19.00 + 0.00

PII: S0261-3069(97)00080-0

The effect of copper addition on the structure and strength of an Al–Li alloy

O. Kabisch[a], W. Gille[a], J. Król[b,*]

[a]*Martin-Luther-Universität-Halle-Wittenberg, Hoher Weg 7, D-06120 Halle, Germany*
[b]*A. Krupkowski Institute of Metallurgy and Materials Science, Polish Academy of Sciences, Reymonta 25, Cracow, 30-059, Poland*

Received 30 June 1997; accepted 15 July 1997

In this study the effect of copper addition on the structure, precipitation kinetics and hardness in the Al–Li and Al–Li–Cu alloys aged at 200°C was investigated. The structures of precipitates were studied using X-ray-small-angle-scattering (XSAS) and transmission electron microscopy (TEM) methods. The changes in the structure parameter (Rg) of both alloys was calculated using two methods, the Guinier approximation and correlation function $\gamma(r)$. By use of a plot of $r\,\gamma(r)$ the distribution law of the T_1 disc thickness was obtained and the coexisting spherical particles of δ' were estimated. Two types of δ' precipitates of approximately 2 nm size and above 8 nm and the T_1 precipitates of thickness between 3 and 4 nm were observed. © 1998 Published by Elsevier Science Ltd. All rights reserved.

Keywords: copper; AL–Li; structure; strength

Introduction

The formation of δ' precipitates within a modulated structure due to spinodal decomposition was found during early stages of ageing[1–3]. The diagram calculated by Gomiero *et al.*[4] predicted a miscibility gap which was metastable with regard to the α–δ' equilibrium. After ageing of the Al–Li alloy at low temperatures two different sizes of δ' precipitates appear[5]. To improve the mechanical properties of the Al–Li alloys (which attain a moderate strength) more complex alloy system were investigated. It is necessary to incorporate the elements which modify the misfit of δ' precipitates[6]. In the alloys of the ternary Al–Li–Cu system the disc shaped, hexagonal T_1 (Al_2LiCu) forms[7,8] which increases the strength of aged Al–Li alloys significantly[9]. Caponetti[10] found copper addition causes an elipsoidal shape deformation of the δ' precipitates. The aim of the present article is to show the effect of copper on the mechanical properties, structure and evolution of the precipitates in Al–Li alloys aged at 200°C.

Experimental procedure

The alloys were cast under a pure argon atmosphere (<1 ppm N_2) in a special chamber. The nominal compositions were (in wt.%) AlLi 2.5, AlLi 2.4, Cu 2.0. The alloys were homogenised at 550°C, quenched in RT water and aged at 200°C. Hardness was measured using Vickers method at 5 kg load. X-ray small angle scattering (XSAS) measurements were performed on a Rigaku–Denki camera with vertical slits. For the evaluation of the size (Rg-Guinier radius) of the precipitates, Guinier approximation of a scattering function of a single particle [$j_p(s)$] was applied according to equation

$$j_p(s) = \exp[-\mathrm{Rg}^2 h^2/3]$$

where
$h = 4\pi/\lambda \sin(\Theta)$, λ is the wavelength of Cu Kα radiation and Θ is the scattering angle, The γ–correlation function (described later) follows from the recorded intensity $I(h)$ and was calculated according to the equation given in the next chapter.

The TEM examinations were carried out on a PHILIPS CM-20 transmission electron microscope operating at 200 kV. Thin foils were obtained by jet electropolishing.

Results and discussion

Hardness test

In *Figure 1* the changes of hardness for both investigated alloys are shown. Copper addition increases significantly the hardness of the aged alloy and shifts its maximum to longer ageing times. The highest hardness of the binary alloy was 67 Hv after 10 h of ageing but in the ternary alloy hardness grow to 131 Hv after 24 h. After longer ageing times hardness slightly decreases as the size of the precipitates increases.

*Correspondence to J. Król. Tel.: +48 12 374200; fax: +48 112 372192; e-mail: nmkrol@imim-pan.krakow.pl

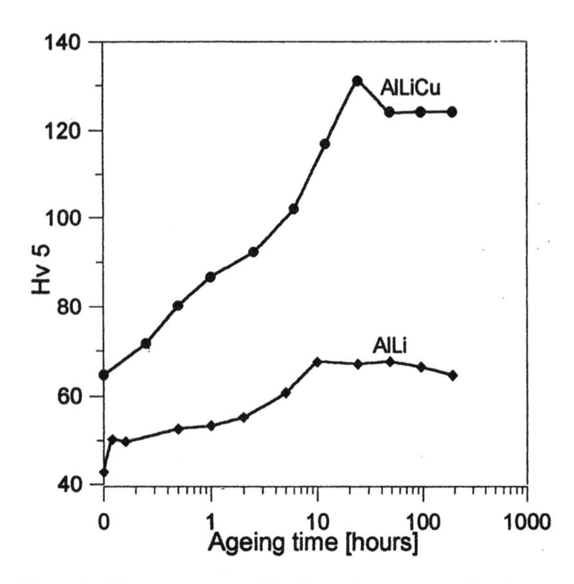

Figure 1 The dependence of hardness changes vs. ageing time at 200°C of AlLi and AlLiCu alloys

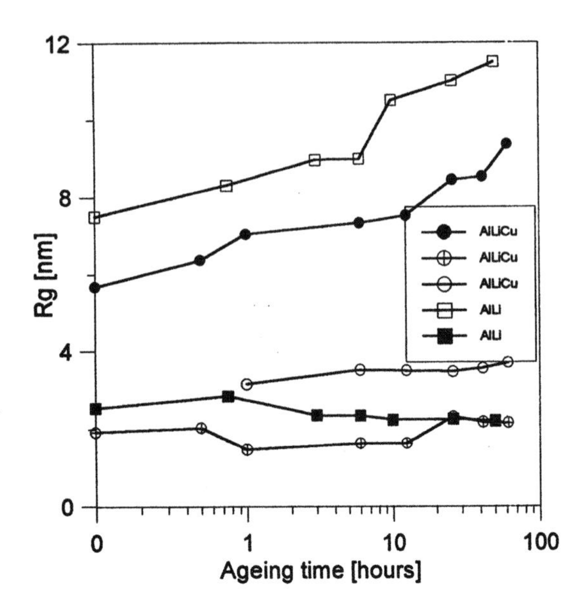

Figure 2 The dependence of Guinier radii vs. ageing time of AlLi and AlLiCu alloys

X-ray small angle scattering investigations

The results of XSAS are shown, for the both alloys in the form of the diagram in *Figure 2*. In the binary alloy Guiner radius (Rg) of the small precipitates of δ' phase decrease their size but the large one grew very fast in the time up to 6 h (to 9.4 nm) and slower for the longer ageing time, reaching 11.5 nm after 50 h. In the ternary alloy a small amount of δ' (Al_3Li) precipitates were observed immediately after quenching as the spherical particles of two sizes. The large particles grew fast up to 7.6 nm after 12 h and for the longer time of ageing slowly reaching 9.4 nm after 60 h. After 1-h ageing, plate-like particles approximately 3 nm thick were additionally formed, apart from the two types of globular ones and grew up to 3.7 nm after 60 h. These values were estimated from the cross-section of the disc-like Al_2Cu and Al_2LiCu type precipitates. The size of the large δ' precipitates in the binary alloy was much larger than in ternary one. In the Al–Li–Cu alloy T_1 particles acted as nucleation sites for the δ' precipitates. In the ternary alloy it can be assumed that the volume fraction of Al_3Li particles was smaller by the formation of the Al_2LiCu phase and the nucleation sites were more numerous than in the Al–Li alloy.

Transmission electron microscope investigations

In the binary alloy the rapid growth of δ' precipitates, like spherical particles (as white circles in the dark field image, taking the 010 δ' spot), occurred during ageing and after 4 h attained the diameter of approximately 17 nm. For the ternary alloy the plate-like T_1 phase was visible on the micrograph. The plates run across the $\langle 110\rangle$ and $\langle 100\rangle$ directions.

Disc-like precipitates (d.l.p.)

In this section a special consideration of the correlation function of XSAS at one ageing-time-state is carried out. Under the condition $T_a = 200°C$, $t_a = 41$ h, disc-like precipitates (d.l.p.) exist, together with two types of spherical particles.

The scattering intensity of the d.l.p. dominates. The reason for this is the higher electron density (copper content in phase Al_2Cu) on one hand and the relatively large volume of the d.l.p. on the other. The amount of the globular particles is much smaller than that of the plate-like ones. The lateral dimension of the d.l.p. is clearly longer than 70 nm. This value is therefore outside the resolution interval of the experiment. A connection between the lateral dimension and H (thickness of the d.l.p.) is not considered here.

The procedure for the determination of the thickness distribution V(H) of the T_1-phase. The correlation function follows from the recorded intensity $I(h)$ (see *Figure 3*). Here h is defined by $h = 4\pi/\lambda \sin(\Theta)$ and 2Θ is the scattering angle. $\gamma(0) = 1$ holds true. In a first step, for the study of the d.l.p. the function $r\gamma(r)$ is tailor-made. $\gamma(r)$ of an infinitely extended plate can be represented in two r-intervals. If $0 < r < H$, γ is linear, but for $H < r, \gamma(r) \sim 1/r$ holds true. From the latter it follows that $r\gamma(r)$ has a maximum value at the position $r = H$.

$$\gamma(r) = \frac{\int_0^\infty h^2 \cdot I(h) \cdot \dfrac{\sin(hr)}{hr} \mathrm{d}r}{\int_0^\infty h^2 \cdot I(h) \mathrm{d}h}.$$

The value of this maximum is $H\gamma(H) = H/2$. In *Figure 4* these connections are demonstrated in the case of a plate of height $H = 1$. This theoretical behaviour coincides with the correlation function, calculated from the experiment (see *Figure 3*). In detail $H \approx 4$ nm holds, a value which coincides with the result obtained from the Guinier-approximation Rg = 3.6 nm (*Figure 2*). Furthermore, by use of this thickness-value

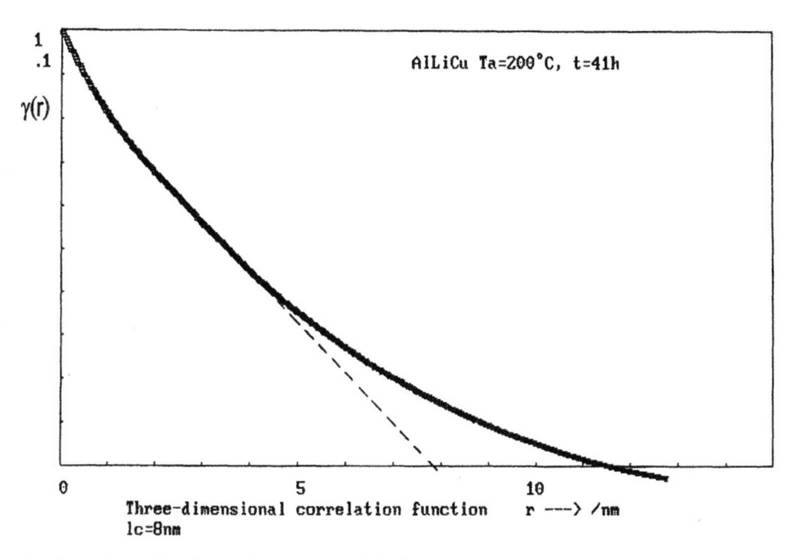

Figure 3 Three-dimensional correlation function in dependence of the length r

and by use of the first derivative of the linear part of γ(r), the volume fraction f of the d.l.p. can be estimated. The basic relation is[11]

$$\left|\frac{1}{\gamma'(r_1)}\right| = \bar{l}\cdot(1-f),\ 2\ \text{nm} <_{r1} < \text{H}.$$

Here l is the mean chord length of the d.l.p. For a single plate of thickness H, $l = 2H$ holds. By use of the experimental results (see *Figure 3*), 7.5 nm < $-1/\gamma'(r_1) < 8$ nm and $l = 8$ nm, follows $0 < f < 0.06$. More specific results follow from the consideration of the function V, $V \sim [\gamma''(r)\ r^3\ V(H)$ is the assumed distribution density of H. Because of $\gamma(r) \sim 1/r$, if $H < r$, it follows $\gamma''(r) \sim 1/r^3$ if $H < r$. Thus the product $r^3\gamma''(r)$ should be a constant for $H < r$ and therefore $V(H)$ is given by

$$V(H) = k\cdot[\gamma''(H)\cdot H^3]'.$$

Here k is a normalization constant for the relation

$$\int_{H_1}^{H_2} V(H)\mathrm{d}H = 1.$$

H_1 and H_2 are the corresponding limits with respect to the dimension of the δ'-phases and the resolution limit of the experiment. In this connection, it must be emphasized that the point function $V_k(H_k)$ is not smoothed at the end of the procedure. All the smoothing operations necessary work within the LAPA-program[11].

The interpretation of the V(H) result. The detailed steps of data evaluation can be reconstructed by means of the series of the functions: $I(h) \rightarrow h^4\cdot I(h) \rightarrow \gamma(r) \rightarrow r\gamma(r) \rightarrow V(H)$. We are perfectly aware that the calculation of V(H) contains some noise-terms. Therefore, on principle it is not possible to obtain a stable solution at all. Nevertheless, a relatively smooth function $V(H)$ was obtained. The main peak, 3 nm $< H <$ 4 nm, corresponds to the distribution law of the thickness H of the d.l.p. From *Figure 5* it is clear that the whole $V(H)$-distribution stretches from $H_1 = 2$ nm to $H_2 = 5$ nm. This means, we are concerned with a relatively extensive distribution interval. The first moment of $V(H)$ is 4 ± 0.5 nm, which is completely in agreement with the discussion of the model behaviour of the function $r\gamma(r)$. This value also agrees with the corresponding TEM-micrographs and classical XSAS-data evaluation.

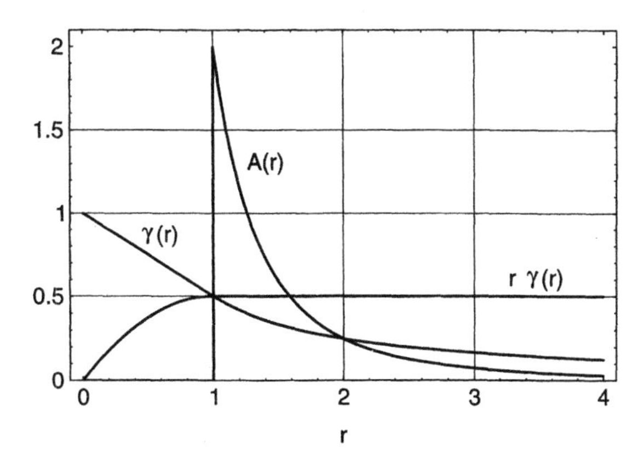

Figure 4 Structure functions (correlation function; $r\gamma(r)$ and chord distribution density $A(r)$ of an infinitely large plate with the high (thickness) H = 1

Of course, a lot of other peaks occur. It can be assumed that the prepeak A at $H = 2$–3 nm corresponds to the small spheres of the δ'-phase. Probably the peak B at approximately $H = 6$–8 nm is connected with the other, large δ'-phase-particle (see *Figure 5*). These assumptions are supported by the TEM micrographs and particularly by the corresponding Rg-values (*Figure 2*) obtained directly from the scattering curve.

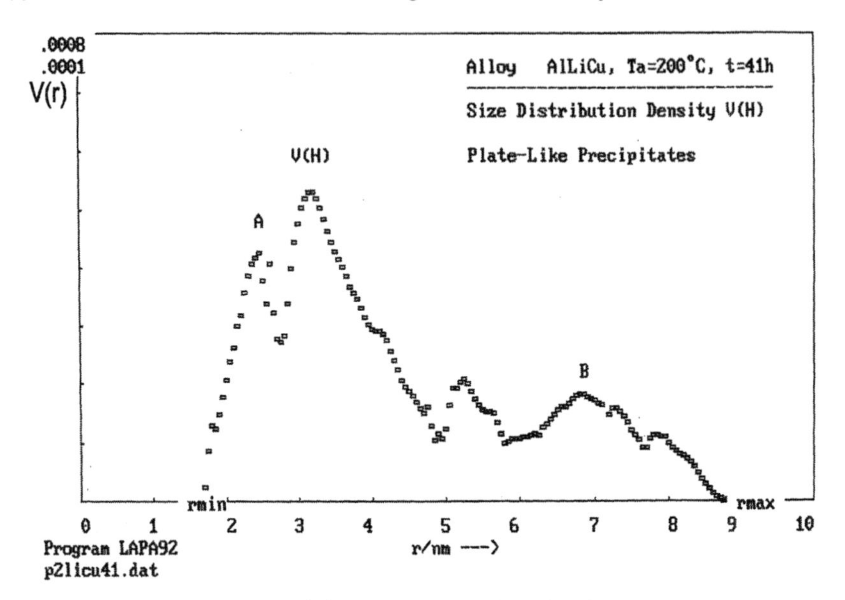

Figure 5 The dependence of size distribution density $V(H)$ for plate-like particles vs. length r

Acknowledgements

This work was supported by the Foundation of the Polish–German Cooperation in project No 1185/93/LN and by Polish–German Cooperation in project No X082.21 (Kernforschungs-zentrum Karlsruhe) of the mutual cooperation.

References

1 Radmilovic, V., Fox, A. G. and Thomas, G., *Acta Metallurgica*, 1989, **37**, 2385
2 Khachaturyan, A. G., Lindsey, T. F. and Morris, J. W., *Metallurgical Transactions*, 1988, **19A**, 249
3 Lee, H. Y., *Materials Chemistry and Physics*, 1992, **32**, 336
4 Gomiero, P., Livet, F., Simon, J. P. and Lyon, O., In *Proceedings of the Conference on Aluminium-Lithium*, ed. M. Peters and P.-J. Winkler. DGM Informationsges. Oberursel 1992. p. 69
5 Dutkiewicz, J. and Król, J., *Physica Status Solidi*, 1994, **41**, 317
6 Spooner, S., Williams, D. B. and Sung, C. M., In *Aluminium-Lithium Alloys III*, ed. C. Baker. The Institute of Metals, London, 1986, p. 329
7 Ardell, A. J. and Huang, J. C., *Materials Science and Technology*, 1987, **3**, 176
8 Vecchio, K. S. and Williams, D. B., *Metal Transactions A*, 1988, **19A**, 2885. Oberursel 1992. p. 83
9 Hagiwara, T., Kobayashi, K. and Sakamoto, T., In *Proceedings of the 4th International Conference on Aluminium-Alloys*, 1986 p. 297
10 Caponetti, E., D'Aguano, E. M. and Triolo, R., *Philosophical Magazine B*, 1991, **63**, 1201
11 Gille, W., *Stereologische Charakterisierung von Mikroteilchensystemen mit der RKWS zu Anwendungen in der Polymeren und Metallphysik*. Doktorarbeit, Halle, 1983

ELSEVIER

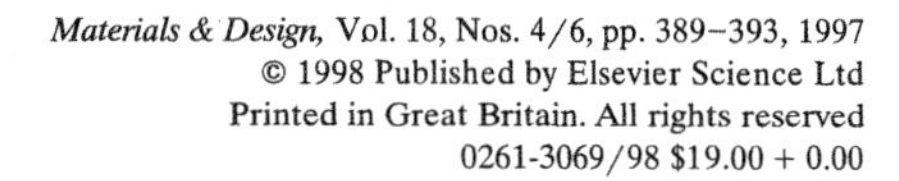

Materials & Design, Vol. 18, Nos. 4/6, pp. 389–393, 1997
© 1998 Published by Elsevier Science Ltd
Printed in Great Britain. All rights reserved
0261-3069/98 $19.00 + 0.00

PII: S0261-3069(97)00081-2

Erosive wear behaviour of aluminium based composites

Q. Fang*, **P. Sidky, M. G. Hocking**

Department of Materials, Imperial College, London SW7 2BP, UK

Received 15 July 1997; accepted 8 August 1997

Al-MMCs reinforced with short fibres or particles of ceramics such as alumina, titanium diboride and silicon carbide result in composites of high specific strength and stiffness, suitable for advanced engineering applications such as in the aerospace and automotive industries. This paper studies the erosion wear behaviour of Al-based composites reinforced with alumina-fibre and in-situ TiB_2 particles using a water/SiC particles slurry jet. From the results of our experiment, the erosion resistance of reinforced Al-MMCs depends on that of the Al-alloy and the reinforcing ceramics, as well as on the bonding strength between the matrix and ceramic fibres or particles. Some design strategies to enhance the erosion resistance of Al-MMCs reinforced with short fibres and particles have been discussed. © 1998 Published by Elsevier Science Ltd. All rights reserved.

Keywords: aluminium based composites; erosive wear behaviour; ceramic fibres and particles; erosion mechanisms

Introduction

In recent years, metal matrix composites (MMCs) have found ever increasing applications as structural materials in various components and engineering systems. The discontinuously reinforced Al-alloy-MMCs with short fibres or particles of ceramics such as alumina, titanium boride and silicon carbide result in composites of high specific strength and stiffness, suitable for advanced engineering applications, such as in the aerospace and automotive industries[1–5]. Using various composite material technologies, hard ceramic particles, whiskers and fibres can be added into a ductile lightweight Al-alloy matrix. It is generally accepted that reinforced Al-MMCs exhibit higher tensile strengths and elastic modulus than unreinforced Al-alloys, but have lower ductility[3,4,6–9]. The axial tensile strength and other mechanical properties of the Al-MMCs are very sensitive to particle size, interfacial bonding, Al-alloy strength as well as processing parameters. The bonding strength between the matrix and ceramic fibres or particles is a major problem during the fabrication and service of MMCs. Recently, a reaction process has received new attention as a better way of producing in-situ ceramic particle in the fabrication of Al-MMCs[22–24]. This self reinforced process appears to have advantages of smaller particle size, uniform distribution and stronger bonding strength between ceramic particles and Al-matrix[24].

Slurry erosion by the impact of liquid-carrying solid particles reduces the life of mechanical components used in many industrial applications, and is one of the main sources of failure of impellers, valves, turbine blades, and pipes used in power plants and oilfield mechanical equipment[10–12]. However, there are few published data on the resistance of Al-MMCs to solid particle erosion. The general laws of erosion for multi-component systems such as fibre reinforced MMCs are not well understood, and an optimum design strategy for erosion-resistant microstructures is not yet available[4]. The study of the erosion of materials has become an important topic in the field of engineering materials. In fact, there has been quite active research on both erosion testing and erosion mechanisms of metals, alloys, brittle materials during the last 30 years[13]. Some recent reviews describe many aspects of the erosion behaviour and erosion mechanisms of materials[14–18]. The present work therefore examines and describes the erosive wear behaviour of two Al-alloy composites reinforced with alumina-fibre and titanium diboride respectively. The erosion response, hardness change on eroded surface and erosion micromechanisms of those composites are investigated and also compared with that of pure-aluminium and unreinforced Al-alloy. Some design strategies to enhance the erosion resistance of Al-MMCs reinforced with short fibres and particles are discussed.

Experimental details

Materials

Two ceramic reinforced Al-composite materials were used in this study:

1. Alumina-fibre reinforced Al-alloy composite; A K10Mg85 Al-alloy (Aluminium + 10% Magnesium)

* Correspondence to Q. Fang

containing 16 vol.% Al_2O_3-fibre (3 μm in diameter and 200 μm in length). The composite was fabricated by gas pressurised liquid metal infiltration of 16% Saffil alumina chopped fibre preform produced by ICI, UK. Prior to the infiltration, the preform was preheated to 450°C, and the Al-10% Mg melt was superheated to 850°C. After pouring the melt onto the preform, 3 MPa gas pressure was introduced into the die cavity to conduct unidirectional infiltration. Post-infiltration, the gas pressure was maintained until the complete solidification of the infiltrated preform. The infiltration facility set-up and procedure have been detailed elsewhere[1,2].

2. In-situ titanium diboride reinforced Al-based composite; The Al-4wt%Cu alloy was first melted at 850°C, then two types of compounds, namely, K_2TiF_6 and KBF_4 were added into the molten Al alloy in atomic ratio of Ti:2B by stirring. The chemical reactions between the two fluorides and the molten Al-alloy form in-situ titanium diboride particle reinforced Al-based composite. Containment of the reaction products such as slag leads to porosity. The average density of the titanium diboride reinforced Al-4Cu-10TiB_2 composite (10 wt% TiB_2 with a mean size of about 1 μm) cast was found to be 2.81, i.e. 90% of the theoretical density.

The Al-/alumina MMC specimens were cut to the dimension of 25 × 25 × 5 mm, while those of titanium boride reinforced Al-TiB_2 composite were cut in to discs of diameter of 20 mm with thickness of 3 mm. The Al-/alumina MMC specimens were cut in two directions, namely, in the direction of parallel and vertical to the fibres for the erosion tests. For comparison, two Al-alloy specimens, namely, Al-10%Mg (Imperial College, UK) and Al-HS15 (British standard, Al-4Cu-0.8Mg-1Si-1Mn, Clerkenwell Ltd., Southampton, UK) were also used in the erosion testing. All the materials tested as shown in *Table 1*.

Experimental set-up and procedure

The room temperature erosion test facility used in the present investigation is illustrated schematically in *Figure 1*. The test rig consists of a stainless steel T tube above an erodent reservoir in the slurry tank. By circulating the water through a loop arrangement, the T tube acts as a suction pump and the SiC-particles are drawn up, resulting in a slurry jet impinging onto the specimen. Both the ejector and the exit nozzle (3.7 mm bore) are made of stainless steel. The system gives a SiC-particle concentration of 5.6 wt.%, a jet velocity of 7.3 m s^{-1} and flow rate of feed water of 4.6 l min^{-1}. The impact angle and distance between the target and the ejector nozzle as well as their relative position, can be easily adjusted. Currently a 15-mm distance is used. The construction details can be found elsewhere[19-21]. The conditions under which the erosion test were carried out are listed in *Table 2*.

To characterise the morphology and hardness changes of the as-received and eroded surfaces and to understand the mode of material removal, the eroded samples were observed under the scanning electron microscope and tested using a Vickers hardness machine.

Results and discussion

Erosion rate

Slurry erosion experiments were conducted on Al-alumina composite specimen with an erosion plane parallel (Al-MMC-P) and vertical (Al-MMC-V) to the fibres. The erosion-angle (E-α) curves for two Al-alumina composite specimens (see *Figure 2*) show typical semi-brittle property, i.e. that the erosion rate increases with impact angle initially, reaching the maximum in impinging angle range of 30–60°, and then decreasing very slowly as erosion-angle α increases further[20,21].

The unreinforced Al-alloy showed a peak at about 30°, and the erosion rate at 90° was about 70% of the peak value. The alumina fibre reinforced composite with different fibre-directions showed different erosion rate against the impact angle, though they have similar maximum erosion at 45°. The Al-MMC-V showed much better erosion resistance than that of unreinforced Al-alloy in the impinging angle range of 0–60°, while Al-MMC-P composite showed intermediate performance at low impact angles (0–35°), but it became worse as the impact angle increased. At normal incidence the erosion resistance of Al-MMC was even poorer than that of the unreinforced aluminium matrix. These results indicate that the erosion resistance of Al-MMC materials is better when the erosion plane is vertical to the fibres rather than parallel with the fibres, and the erosion resistance of the Al-MMC at higher impact angles is not satisfactory, being worse than that of the matrix materials. There may be two main reasons for this erosion behaviour: (1) the discontinuous nature of the reinforcement; (2) the poorer erosion

Table 1 Materials tested

Materials	Composition	$\beta = D_m/D_o^*$	Vickers Hardness
Al-TiB_2 MMC	Al-4Cu-10TiB_2	0.90	43.2 ± 0.5
Al-HE15	Al-4%Cu-1Si-0.5Mg-1Mn	~1	63.1 ± 0.5
Al-10Mg alloy	Al-10Mg	0.96	91 ± 1
Al-MMC-P	Al-10Mg-16 Vol% Al_2O_3	0.94	143 ± 2
Al-MMC-V	Al-10Mg-16 Vol% Al_2O_3	0.94	139 ± 2

D_m: density measured.
D_o: theoretical density.

Table 2 Erosion test conditions

Experimental Parameters	
Erodent	Silicon carbide
Erodent size (μm)	600–850
Erodent shape	Angular
Erodent concentration (Wt %)	5.6
Flow liquid	Water
Jet velocity (m s^{-1})	7.3
Impact angles	15–90°
Distance between nozzle and sample (mm)	15
Test temperature	Room temperature

resistance of alumina itself, compared with SiC, ZrO_2 and sialon[20].

Figure 2 also shows the erosion-impinging angle (E-α) curve for the in-situ titanium diboride reinforced Al-based composite and Al-HS15 (Al-4Cu-0.8Mg-1Si-1Mn). The latter is an alloy with higher hardness and tensile strength. Both demonstrated similar ductile property with peaks at about 17°. It is noted that the maximum erosion rate of Al-TiB_2 MMC reaches about 0.62 mm^3 kg^{-1} at the peak value, which is even bigger than that of Al-HS15 alloy and it is believed that the bigger porosity and slag in the MMCs could be responsible for the poor erosion resistance at low impinging angles. However, the erosion rate of Al-TiB_2 MMC decreases dramatically down to 0.32 mm^3 kg^{-1}, nearly 50% of the peak value at normal impact angle. This implies that the bonding strength between TiB_2 particles and Al-matrix, which forms in situ from the chemical reactions at high temperature, is stronger than that of Al-alumina fibre composites. Microstructure observations (*Figure 3*(*c*)) indicate debonding between alumina fibres and Al-matrix, fibre pull-out and fibre fracture become easier with increasing impact angles.

Microstructures

The microstructures of Al-10%Mg alloy and alumina-fibre reinforced Al-MMC-V, which underwent the erosion test of 1 h at impact angle of 45°, are illustrated in *Figure 3*. The main mechanism of material removal for Al-alloy was the ploughing and cutting (*Figure* 3(*a*)–(*b*)), while that of material removal for alumina fibre reinforced Al-based MMC specimen was both ploughing and fracture. *Figure 3*(*c*) shows the SEM of Al-MMC-P at 45° impact angle for 1 h. The alumina fibres peeling off, cracking of fibres and Al-matrix, and ploughing and cutting of the Al-alloys is evident on the eroded surface of Al-MMC-P, while cracking of fibres and ploughing and cutting of the Al-alloys can mainly be observed on Al-MMC-V (*Figure 3*(*a*)). The microstructures of the eroded Al-MMC specimens also suggests that the bonding strength between alumina fibre and Al-matrix is not strong enough to prevent the alumina fibres from pulling out by SiC-particle erosion of the surrounding matrix. The alumina fibres peeling off in Al-MMC-P may be a main reason of the higher erosion rate, compared with Al-MMC-V.

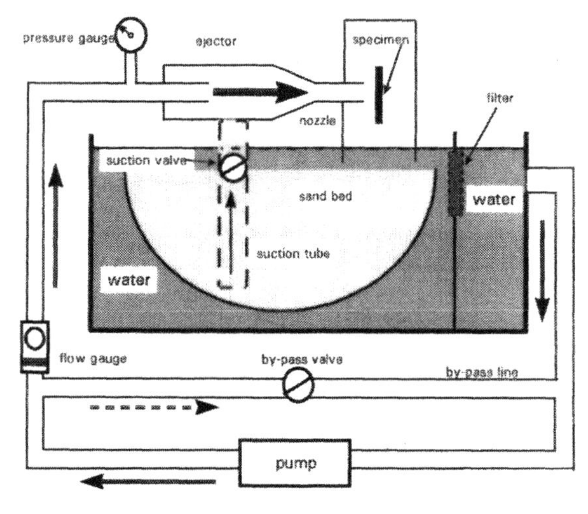

Figure 1 Schematic diagram of the sand/water slurry erosion rig

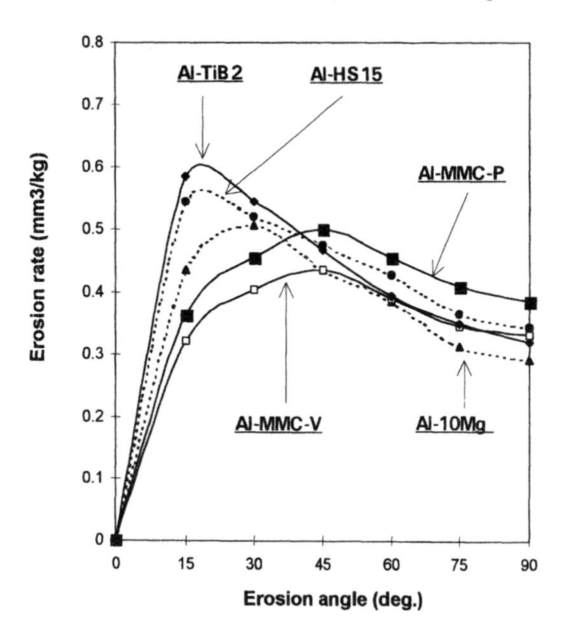

Figure 2 Erosion rate (E-α) for the different materials

Figure 4 show SEM of Al-TiB_2 and Al-HS15 at an impact angle of 45° for 1 h. The morphologies of both specimens show a similar ductile erosive behaviour; scars of cutting and ploughing, and lips from the end of cut and plough exist. The erosion occurs initially by scratching of the surface, then material removal takes place by means of cutting and ploughing. Like ductile materials, such as most metals and alloys, Al-TiB_2 and Al-HS15 show a W-shaped crater at normal impact angle as they suffer more attack by abrasion at lower angles, leaving the central stagnation point (at which the real impact angle is 90°) less attacked than its surroundings. While Al-alumina composite materials in the erosion process show a U-shaped crater, rather than a W-shaped one, indicating a generally semi-brittle nature, although microscopically, both brittle and ductile mechanisms are involved.

Hardness test

Hardness, which represents material resistance to plastic flow, is a critical materials property that strongly controls erosion. A tendency for erosion rate of metals

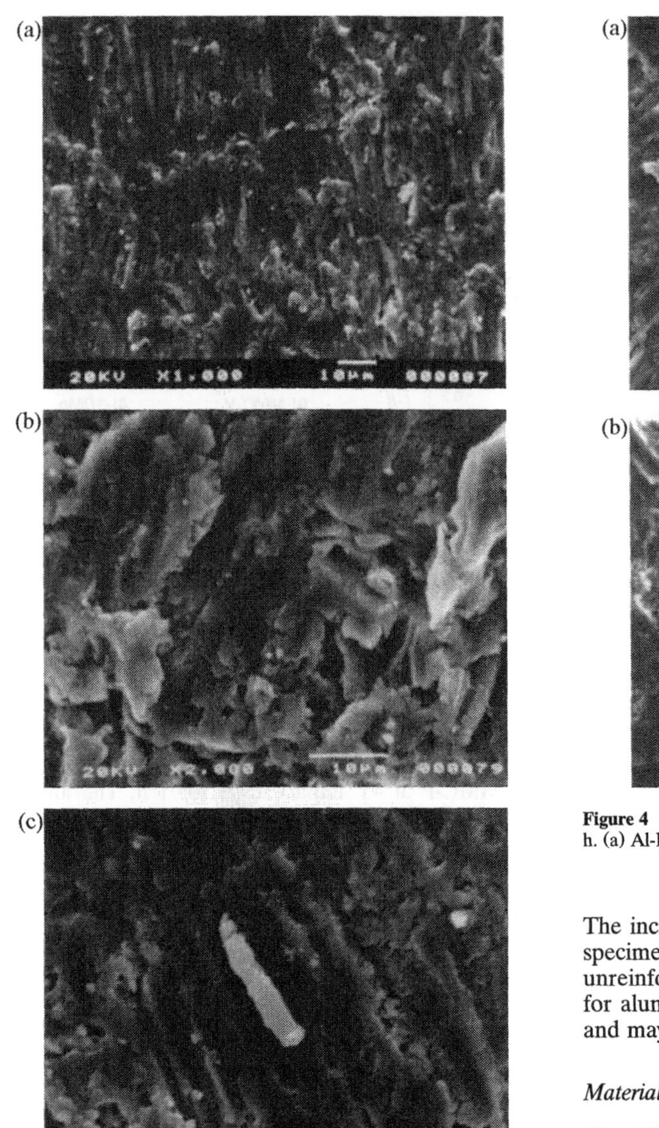

Figure 3 SEM of eroded specimens of Al-10%Mg alloy (a) and alumina-fibre reinforced Al-MMC (b, c) at impact angle of 45° for 1 h. (a) Al-10Mg-alloy, (b) Al-MMC-V, (c) Al-MMC-P

or alloys (ductile materials) to be reduced when the target material hardness is increased has been well established[14]. In the present investigation, the Vickers hardness of the as-received and the eroded specimens were tested. *Table 3* gives the Vickers Hardness of the as-received and eroded specimens of five materials tested. Hardness depends not only on the nature of the material, such as composition, grain size, microstructure, but also on the fabrication process. The in-situ formed Al-composite has the lowest Vickers Hardness owing to some reaction contamination and the higher porosity. *Table 3* also shows that the Vickers hardness of Al-MMC-P and Al-MMC-V increases due to addition of alumina fibres from 91 to 143 and 139, respectively. (The Vickers hardness of alumina is about 1200).

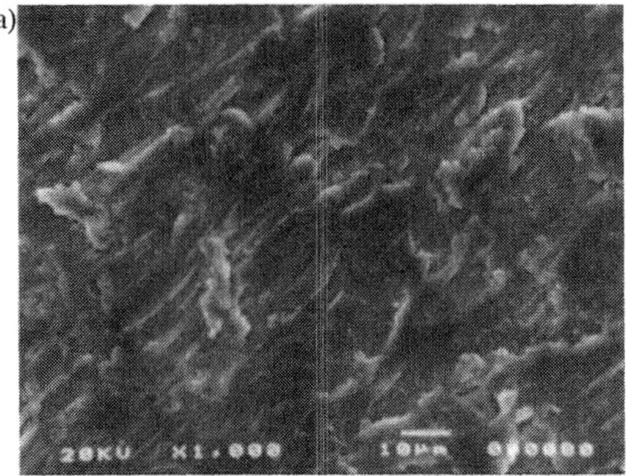

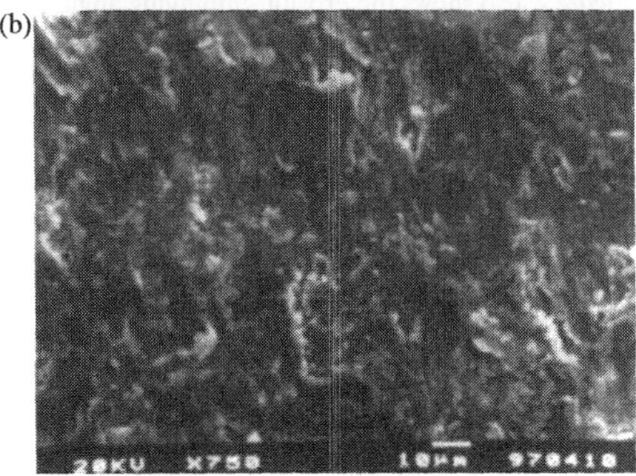

Figure 4 SEM of Al-TiB_2 and Al-HS15 at impact angle of 45° for 1 h. (a) Al-HS15, (b) Al-TiB_2 composites

The increase of Vickers hardness for the three eroded specimens were between 12 and 15%. Comparison with unreinforced Al-10Mg alloy, the increase in hardness for alumina-fibre reinforced Al-MMCs is not dramatic and may be due to the increase in their brittleness.

Materials removal mechanisms

The SEM observation of the eroded surfaces for various specimens used in this work revealed different materials removal mechanisms. In monolithic ductile alloys or ductile composites such as Al-HS15 and Al-TiB_2, which have low hardness but where the bonding between the ceramic particles and Al-matrix is strong, erosion occurs principally by scratching of the surface. Material removal takes place more easily at low impact angle owing to ploughing and cutting by SiC particle flow. At higher impingement angles, however, impact of SiC particle flow on the specimen surface may result in a harder surface and less efficient ploughing and cutting, and thus less material removal (*Table 3*).

The alumina fibre reinforced Al-based MMC specimen eroded in a different way at normal impact angles and at oblique angles. At normal impact the alumina fibres were fragmented into short pieces before being removed. The material was detached in fragmented pieces which contained both aluminium matrix and broken fibres. This semi-brittle manner of fracture is due to the discontinuous nature of the material. At oblique impact angles, the fibres were less likely to break, and the loss of material was mainly due to the

Table 3 Vickers hardness of the as-received and the eroded specimens

Materials	As-received	Eroded (90°)	Increase (%)
Al-TiB_2 MMC	43.2 ± 0.5	49.2 ± 0.5	13.9
Al-HE15	63.1 ± 0.5	69.3 ± 0.5	10
Al-10%Mg alloy	91 ± 1	106 ± 1	15
Al-MMC-P	143 ± 2	161 ± 2	12
Al-MMC-V	139 ± 2	157 ± 2	13

removal of the matrix in a ductile manner and the detachment of fibres when they became exposed (*Figure 3(c)*). But with increase of the impact angle, the breaking and pulling out of alumina fibres dominates the materials removal mechanism owing to the weaker bonding strength between fibres and Al-matrix and the poorer erosion resistance of the alumina ceramic itself[20]. Once a fibre is removed, thereby forming a channel on the surface of the matrix; this may result in more efficient removal of the matrix. The main mechanism of material removal for the alumina fibre reinforced Al-based MMCs were the fracture and detachment of the alumina fibres, and the ploughing and cutting of the Al-based matrix.

Design strategies

The discontinuously reinforced Al-alloy-MMCs with short fibres and particles of ceramic offer a possibility of high erosion-resistant materials through appropriate design of multicomponent systems, including microstructures of composites, interfacial properties between matrix and additives. The role played by the reinforcement in composites varies depending on the properties, size and sort of additives used, microstructure of the composite and test conditions. From the results of our experiments, some design strategies for fabrication of a high erosion resistance Al-composite can be suggested:

- Use high specific strength Al-alloys and use highly erosion resistant ceramic fibres which possess high specific strength and stiffness
- Enhance the bond strength between reinforced fibres or particles and Al-matrix such as by means of in-situ reaction process could play a critical role in increasing erosion resistance
- Design the fibre reinforced composite components, such that the erosion plane is vertical to the general fibre orientation could enhance the erosion resistance of Al-MMC materials

Conclusions

In conclusion, the erosion wear behaviour of the alumina fibre reinforced Al-MMCs and in-situ titanium boride reinforced Al-based composite by a slurry erosion rig has been demonstrated. The surface characteristics and damage micromechanisms have been described and discussed. Further work is under way to obtain a better fundamental understanding of erosion wear on different MMC materials with an attempt to find a general law for multicomponent systems such as fibre reinforced MMCs to improve their erosion resistance.

Acknowledgements

The authors would like to thank Dr S.Y. Long of Imperial College and Dr L.Lu of the National University, Singapore for providing Al-MMC specimens and valuable discussion.

References

1 Long, S., Zhang, Z. and Flower, H.M., *Acta Metall.*, 1995, **43**, 3489–3498
2 Long, S., and Flower, H.M. Interfacial structures of SiC fibre-Mg containing aluminium fabricated under systematically varied conditions. In *Interfacial Phenomena of Composites*, Eindhoven, Holland, 1995
3 Thomas, M.P. and King, J.E., *Materials Science and Technology*, 1993, **9**, 742–753
4 Morrison, C.T., Routbort, J.L., Scattergood, R.O. and Warren, R., *Wear*, 1993, **160**, 345–350
5 Humphreys, F.J., Miller, W.S. and Djazeb, M.R., *Materials Science and Technology* 1990, **6**, 1157–1166
6 Kamat, S.V., Hirth, J.P. and Mehrabian, R., *Acta Metall.*, 1989, **37**, 2395
7 Papazian, J.M. and Adler, P.N., *Metall. Trans.*, 1990, **21A**, 401
8 Lewandowski, J.J. and Liu, C., *Processing and properties for powder metallurgy composites*, ed. P. Kumar et al. PA, TMS, Warrendale, 1988, pp. 117
9 Organista, A., Chevet, F., Thery, S., Heritier, J. and Levaillant, C., In Proc. *On The materials revolution through the 90s: powders, metal matrix composites, magnetics, Oxon*, British Nuclear Fuels Metals Technology centre, 1989, 2, p. 36
10 Qureshi, J., Characterization of coating process for steam turbine blades, *Journal of Vac. Sci. Technol.*, 1986, **A4** (6), 2638–2647
11 Yee, K.K., *Int. Metall. Rev.*, 1978, **226**, 19–41
12 Cooper, D., Davis, F.A. and Wood, R.J.K., Selection of wear-resistant materials for the petrochemical industry. *J. Phys. D: Appl. Phys.* 1992, **25**, A195–A204
13 Finnie, I., Some reflections on the past and future of erosion. *Wear*, 1995, **186/187**, 1–10
14 Ruff, A.W. and Wiederhorn, S.M., Erosion by solid particle impact. In *Treatise on Materials Science and Technology*, ed. C.M. Preece, Academic, New York, 1979, **16**, pp. 69–126
15 Levy, A.V. (ed.), *Corrosion-Erosion-Wear of Materials at Elevated Temperatures*. NACE, Houston, 1987
16 Adler, W.F. (ed.), *Erosion: Prevention and Useful Applications*. ASTM, Philadelphia, PA, STP 1979, pp. 664
17 Field, J.E. and Dear, J.P. (eds.), *Erosion by Liquid and Solid Impact*. Cavendish Lab., Cambridge, 1987
18 Ritter, J.E. (ed.), Erosion of ceramic materials. *Trans Tech Publ.*, Switzerland, 1992
19 Zu, B.L., Huchings, I.M. and Burstein, G.T., *Wear*, 1990, **140**, 331–344
20 Xu, H., Fang, Q., Hocking, M.G. and Sidky, P.S., Erosion of ceramic materials by a sand/water slurry jet. *Wear*, submitted, 1996
21 Fang, Q., Sidky, P.S. and Hocking, M.G., Effect of corrosion on erosion wear behaviour of ceramic materials. *Corrosion Science*, 1997, **39** (3), 511–527
22 Wood, J.V., Davies, P. and Kellie, J.L.F., *Materials Science and Technology*, 1993, **9**, 833
23 Gotman, I. and Koczak, M.J., *Materials Science and Engineering-A*, 1994, **A187**, 189
24 Qin, S., Lu, L., Lai, M.O. and Chen, F.L., *Third Int. Conference on Composites Engineering*. ICCE/3, New Orleans, USA, 1996, pp. 693

Materials & Design, Vol. 18, Nos. 4/6, pp. 395–399, 1997

0261-3069/98 $19.00 + 0.00

PII: S0261-3069(97)00082-4

Synthesis and characterization of the Me–HTSC composite

Zvonimir D. Stanković[a], Lidija T. Mančić*[,b]

[a]*University of Belgrade, Technical faculty in Bor, 19210 Bor, Yugoslavia*
[b]*Institute of Technical Sciences of the Serbian Academy of Sciences and Arts, Knez Mihailova 35/IV, Belgrade, Yugoslavia*

Received 11 July 1997; accepted 21 July 1997

This study shows that homogeneous composite compacts with improved superconducting and mechanical properties can be successfully prepared from pure Me–HTSC powders (Me=Ag, Cu, Al; HTSC=$YBA_2Cu_3O_{7-x}$). Composites were produced under the following conditions: the mass fraction of the metal phase was in the range 2.5–40 wt%, the compacting pressure was 30 kN/cm^2, the sintering temperature was from 270 to 950°C and the cooling rate was 50°C/h and 220°C/min. Microstructure and X-ray diffraction analysis have been done, alongside the T_c value determination.

Keywords: homogeneous; composite; Me–HTSC powders

Introduction

Since the discovery of high temperature superconducting oxides (HTSC), tremendous efforts have been made in the synthesis, processing and characterization of these materials in an effort to move toward applications. Both good superconducting and mechanical properties are desirable, for most practical applications in areas where commercialization of products are still several years off, e.g. large high-field magnets and power transmission lines in power generation, transportation and storage, experimental levitation trains, superconducting generators and motors in transportation, cellular base stations and microwave filters in microelectronics[1–3]. Previous studies have observed that the $YBa_2Cu_3O_{7-x}$ compounds are generally very brittle with unacceptably low strength, but also, that some problems connected with these characteristics can be solved by producing metal — high temperature superconducting ceramic composites[1,2]. It is known that the inclusion of a ductile phase in a ceramic matrix can improve both the strength and the fracture toughness of the material. Unfortunately, $YBa_2Cu_3O_{7-x}$ compounds react with most materials and only a few of them do not degrade the superconducting properties. Therefore, numerous studies have been performed and silver and copper were found to be the most desirable metal phases in the ceramic matrix[3–6]. It has also been shown that the presence of Ag and Cu inclusions improves the HTSCs mechanical and superconducting properties and reduces the amount of microcracking in sintered specimens. The observed mechanical improvements in the composite materials are believed to be due to the increasing toughness and relieving of the anisotropic thermal stress in $YBa_2Cu_3O_{7-x}$ grains, while the increase in the critical current density value (J_c) indicates that metal inclusions promote the joining between HTSC grains[7]. To establish the processing of composite materials, a relationship between the amount of metal phase and both superconducting and mechanical properties must be determined. Microstructural control is also essential for developing practical superconducting materials. In view of these problems, our effort has been directed on the synthesis and characterization of Me–$YBa_2Cu_3O_{7-x}$ composite samples with different amounts of the metal phase. Silver, copper and aluminium (for the first time), were used as the metal phase in the ceramic matrix. Aluminium is widely used in electronic devices and it has been reported that the substitution of up to 20 atom percent of cations with aluminium in Y–Ba–Cu–O and Bi–Sr–Ca–Cu–O systems seems to have no significant effect on the T_c value[8]. The influence of the metal content in the ceramic matrix on critical temperature (T_c) and microstructure of composites was determined by measuring T_c values by X-ray and metallographic analysis.

*Correspondence to L. T. Mančić

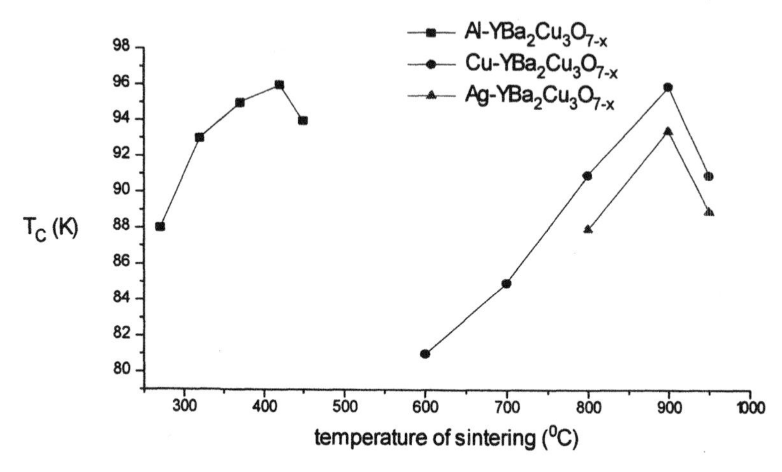

Figure 1 Critical transition temperature vs. sintering temperature for Me–HTSC composites

Experimental

$YBa_2Cu_3O_{7-x}$ powder was prepared by a solid rate reaction using powders of Y_2O_3, $BaCO_3$ and CuO. The powders were calcined at 900°C for 4 h in air and were then ground and recalcined at 950°C under the same conditions. The cooling rate after thermal treatment was 50°C/h. Composites of metal–$YBa_2Cu_3O_{7-x}$ were made by mixing the metal and ceramic phases in different weight ratios (the metal phase content ranged from 2.5 to 40 wt%). Obtained homogenized mixtures were passed at 300 MPa into pellets 10 mm in diameter and then sintered for 4 h at optimal temperatures in air, followed by slow cooling to room temperature. Optimal thermal treatments for metal–HTSC composites were determined from sintering experiments with 20% Me–HTSC specimens sintering using different temperature regimes. The maximal temperature values were 900, 900 and 420°C, for Ag-HTSC, Cu-HTSC and Al–HTSC composites, respectively. The examined specimens were cooled slowly to room temperature (cooling rate was 50°C/h) in the first case and rapidly (220°C/min) by directly dipping the specimens into liquid nitrogen, in the second case.

Measurements of T_c values were made using an indirect method based on monitoring and measuring the time of Maissner's effect[9]. The composition of the samples was investigated by X-ray analysis. Optical microscopy was used to determine the microstructure of samples, previously prepared by ordinary polishing methods for this kind of investigation[10].

Results and discussion

The enhancement of the critical transition temperature, T_c, is strongly affected by the sintering temperature of the composite. From *Figure 1*, optimal values of the sintering temperature for Me–$YBa_2Cu_3O_{7-x}$ composites were found to be 900°C, for Ag and Cu composites and 420°C for Al composite materials. It is obvious that the addition of a metal phase reduced the sintering temperature of the $YBa_2Cu_3O_{7-x}$ phase. X-ray analysis of sintering specimens indicate that an increase of the sintering temperature up to optimal values affected changes of the peak intensity which is related to the $YBa_2Cu_3O_{7-x}$ phase, in accordance with changes of T_c values. Namely, the intensity of the (110), (104) and (123) diffraction lines are maximum values for samples sintered at optimal temperatures in all investigated systems.

Figure 2 shows an optical micrograph of the 20% Al–$YBa_2Cu_3O_{7-x}$ composite sintered at different temperature regimes (300 × magnification). The sample sintered at the optimal temperature of 420°C (*Figure 2d*), is characterized by twinning of $YBa_2Cu_3O_{7-x}$ grains, while a higher temperature of sintering leads to partial melting of the sample (*Figure 2e*). The sample prepared at the optimal temperature followed by rapid cooling (220°C/min), has randomly distributed Ag and superconducting grains, both smaller in diameter (*Figure 2f*). The total volume fraction of the superconducting phase compared to other phases present in this sample is reduced. As a consequence, the measured T_c value for this sample was 88 K. This is another confirmation that slow cooling is necessary for improvement of superconducting properties[10].

Measurements of T_c as a function of the metal content in the composites are shown in *Figure 3*. The addition of metal powders to $YBa_2Cu_3O_{7-x}$ composites up to 20 wt% increased the T_c value. The maximum T_c (95 K) for Ag composites is obtained with 5 and 10 wt% of a metal phase, while these values are slightly higher for Cu and Al composites and is 96 K for a metal content of 10 and 20 wt%. The obtained increase of the T_c value in Ag and Cu composites is in agreement with earlier published data[6]. XRD patterns of samples with the 20 wt% of the Me phase (*Figure 4*) have diffraction lines for the orthorhombic 1–2–3 phase and also, peak of (200) plane which is related to pure metals in all investigated samples, confirming that Ag, Cu and Al were not built into perovskite ceramic cells.

From the micrographs, presented in *Figure 5* (300 × magnification), $YBa_2Cu_3O_{7-x}$ grains are visible in a

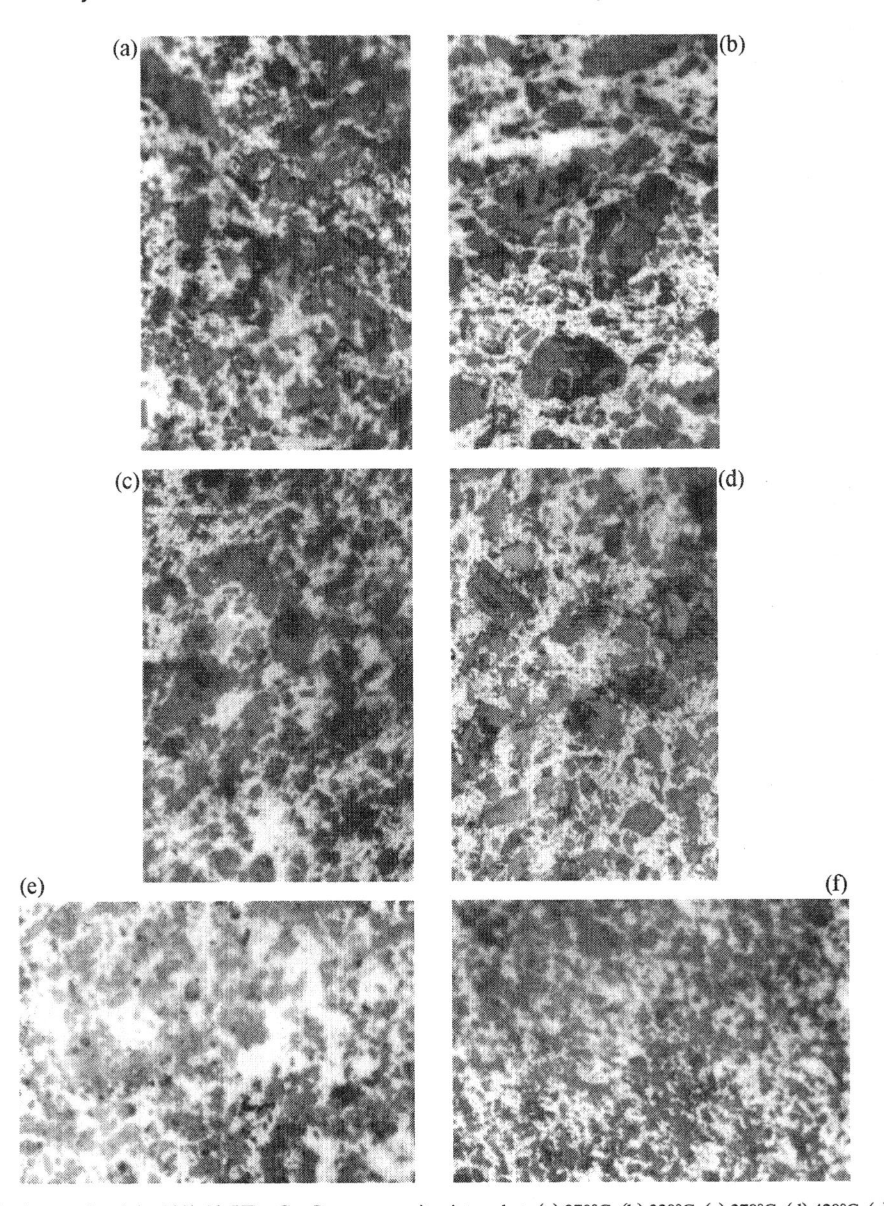

Figure 2 Optical micrographs of the 20% Al–$YBa_2Cu_3O_{7-x}$ composite sintered at: (a) 270°C; (b) 320°C; (c) 370°C; (d) 420°C; (e) 450°C; and (f) 420°C, followed by rapid cooling

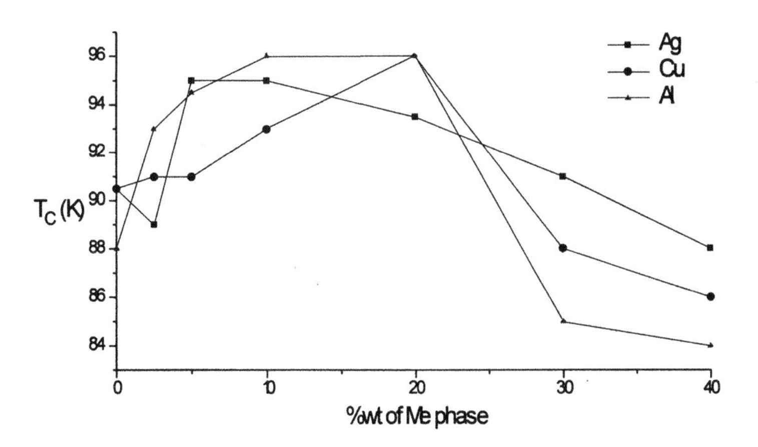

Figure 3 T_c vs. metal content in the Me–HTSC composites

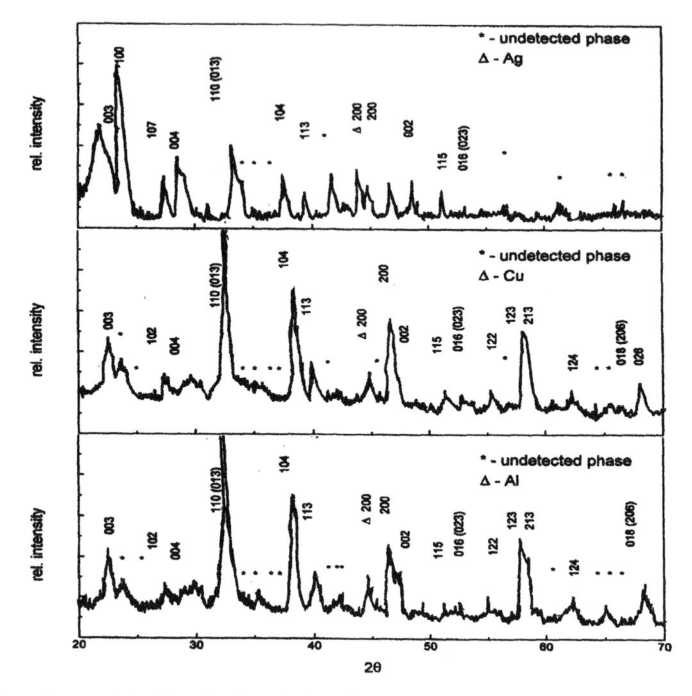

Figure 4 XRD patterns of samples containing 20 wt% of metal phase (marked diffraction lines related to $YBa_2Cu_3O_{7-x}$ compounds)

plate-like form. The grain sizes covered a wide range and the tendency of grain growth with the increase of the Me content is present. The absence of microcracking is assumed to be a consequence of metal inclusions into the ceramic matrix. The sharper edged grains and joining between the grains consists of the superconducting phase in the samples containing 20 wt% of Cu, are in agreement with the highest T_c value measured for this system. Inspection of a cross-section of the composite with 30 wt% of Al (*Figure 5j*), indicates that $YBa_2Cu_3O_{7-x}$ grains are mostly isolated by bigger metal grains, causing a decrease in T_c. From the results obtained, it can be concluded that so long as the metal content in composites is such that superconducting grains can form a continuous network, the influence of metal addition on the T_c value is positive. The increase in T_c is believed to be due to the increase of area contact between $YBa_2Cu_3O_{7-x}$ grains[4].

Conclusions

Ag, Cu and Al powders were used in developing sintered Me– $YBa_2Cu_3O_{7-x}$ composites, because applications require further improvement in mechanical and physical properties of bulk superconducting elements. The optimum weight ratios of the metal powders for $YBa_2Cu_3O_{7-x}$ were found to be 20% for Cu and Al composites and 10% for Ag composites. It was shown that addition of a Me phase reduced the sintering temperature of the composites. No change in cell parameters was determined in all investigated samples. The enhancement of grain size and form in samples were obvious with the change of Me content. So long as the metal content in composites was up to 20 wt%, the $YBa_2Cu_3O_{7-x}$ grains formed a continuous network and the T_c value increased. A higher content of the metal phase degraded the superconducting properties of the composite. The absence of microcracking suggests that Me inclusions play an important role in relieving residual stress in $YBa_2Cu_3O_{7-x}$ grains.

Acknowledgements

This research was financially supported by the Republic of Serbia Science Foundation.

References

1 Singh, J. P., Leu, H. J., Poeppel, R. B., Van Voorhees, E., Goudey, G. T., Winslay, K. and Shi, D. *Journal of Applied Physics*, 1989, **66** (7), 3145
2 Yamamoto, T., Terni, G. and Ueyama, T. *Japanese Journal of Applied Physics*, 1993, **32**, 4496
3 Vipulandan, C. and Salib, S. *Journal of Materials Science*, 1995, **30**, 763
4 Kang, W. J., Handa, S., Yoshi, I. and Nagata, A, 1992, **31**, 3312
5 Tang, F., Yuan, Q. and Yuan, Y. *Journal of Materials Science*, 1995, **30**, 374
6 Salama, K., Ravi-Chander, K., Selvamanickam, V., Lee, D. F., Reddy, P. K. and Rele, S. V. *Journal of Metals*, 1988, **8**, 6
7 Matsumoto, Y., Homba, J. and Yamaguchi, Y., 40th International Society of Electrochemistry Meeting, Extended Abstracts, Vol I, Kyoto, 1989, p. 102

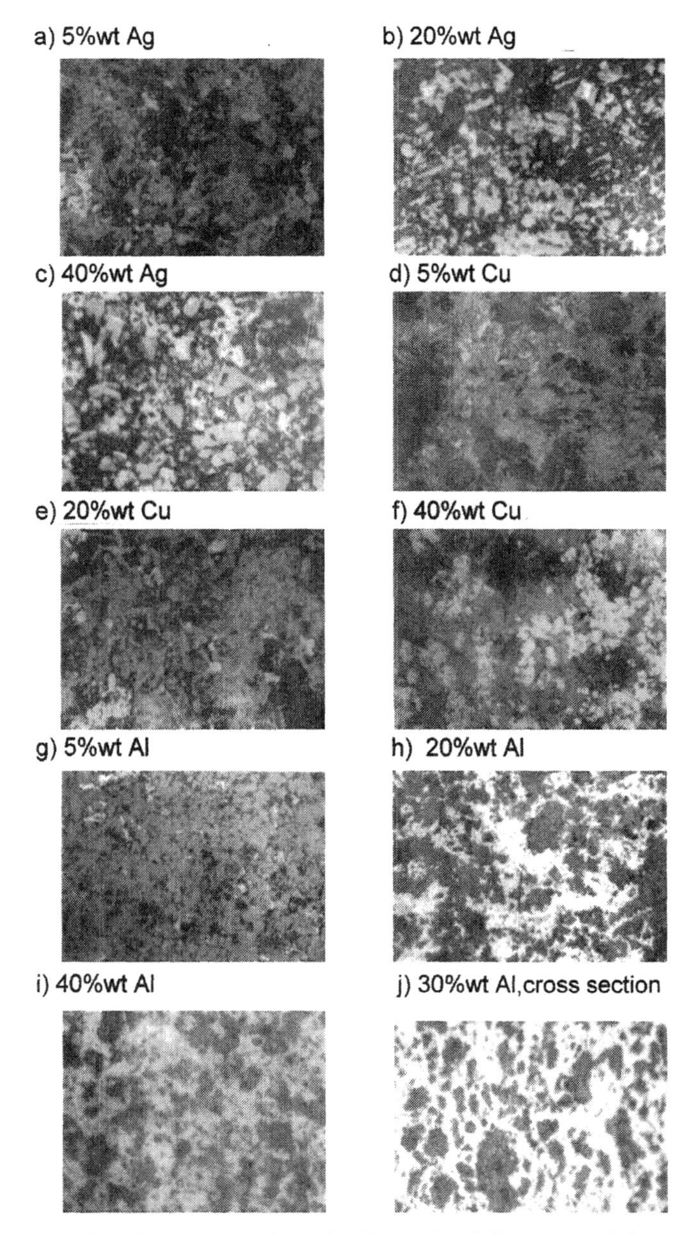

Figure 5 Optical micrographs of Me–$YBa_2Cu_3O_{7-x}$ composites with 5, 20 and 40 wt% (from a to i and h) cross-section of samples with Al — 30 wt%

8 Lin, W. T., Kao, H. P., Fang, Y. K. and Pan, F. M. *Journal of Electrochemical Society*, 1991, **138** (12), 3686
9 Stanković, Z., Mančić, L., Stević, Z., Rajčić-Vujasinović, M., Gusković, D. and Zivković, D., XXIV October Mining and Metallurgy Conference, Extended Abstracts. Bor, 1992, p. 598
10 Gopalakrishanan, I. K., Yakhmi, J. V., Vaidya, M. A. and Iyer, R. M. *Applied Physics Letters*, 1987, **51**, 1367

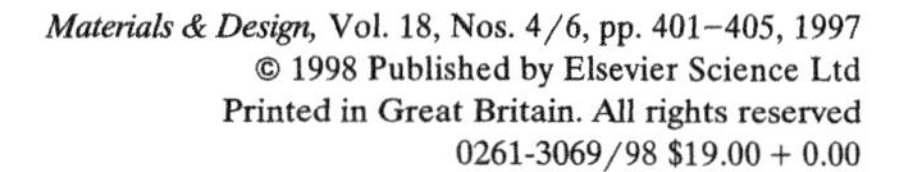

Materials & Design, Vol. 18, Nos. 4/6, pp. 401–405, 1997
© 1998 Published by Elsevier Science Ltd
Printed in Great Britain. All rights reserved
0261-3069/98 $19.00 + 0.00

PII: S0261-3069(97)00090-3

Technological aspects of particle-reinforced composites production

Miroslaw Cholewa[a,*] Józef Gawroński[a], Zenon Ignaszak[b]

[a]*Foundry Division, Silesian Technical University, 7 Towarowa Street, 44-100 Gliwice, Poland*
[b]*Foundry Division, Poznan University of Technology, 5 Piotrowo Street, 61-138 Poznan, Poland*

Received 30 August 1997; accepted 5 September 1997

Technological aspects of particle-reinforced composite production together with their geometrical and thermal analysis have been presented in this article. An estimation of metallurgical properties of aluminium alloy matrix composite has been revealed. This study concerns the influence of technological manufacturing parameters on the crystallization, solidification and exploitation features of MMC (metal matrix composite) with aluminium alloy matrix, reinforced with C-graphite, SiC and Al_2O_3 particles, in the amount up to 5% by weight. The composites here investigated are designed as a material for mechanical parts, which should be characterized by considerable wear resistance, low or high friction coefficient, as well as erosion and corrosive wear resistance. © 1998 Published by Elsevier Science Ltd. All rights reserved.

Keywords: aluminium composite; particle reinforcement; erosive–corrosive wear

Introduction

Cast composites with aluminium alloy matrix, reinforced with C-graphite, SiC and Al_2O_3 particles have different rheological features, which influence the production process parameters: temperature and mixing time, amount, particle size and kind of reinforcement. This research characterizes the cast composites made with inorganic surface-acting substances and with electromagnetic fields.

Technological aspects of composite casts

The production of composites in the liquid state is based on the activation of surface phenomena directly on the reinforcement surface, without intervention in Cu or Fe matrix alloy, particularly in Al matrix alloy. The mechanical mixing that was applied can be aided by magnetic field. Gravity method of casting may also be used with lost pattern moulding. The tendency to segregation can be limited by movement of the metallic mixture in the mould during solidification, using a magnetic field. Segregation occurs particularly for large reinforcing particles. The manufacturing parameters are mainly: reinforcement quantity, up to 12%; particle size, up to 120 μm; mixing temperature range, 580–1700°C; magnetic field induction, up to 10 mT; amount of surface-acting substance, 0.05–0.7‰ by weight[1–3].

* Correspondence to Dr M. Cholewa. Tel.: + 48 32 316031; fax: + 48 32 1703697

Aluminium alloy matrix characteristic

In the worldwide foundry industry there is an increasing interest in Al and its alloys as autonomous alloys and also as composite matrix. Modern metallurgical-refining measures and new casting technologies resulted in better useful properties of this constructional material and its products. Porosity decreases the useful properties of Al alloys casts and comes from two basic sources (hydrogen desorption and solidification shrinkage). Microporosity may originate from the following sources:

1. differentiated solubility of hydrogen in liquid and in solid phase;
2. impeded liquid metal feeding; and
3. partial destruction and dilatation of the mould.

An additional factor that influences porosity levels is Al alloys inoculation. Inoculation also influences the nucleation process and the solidification morphology and hence changes the feeding conditions.The author carried out work predicting shrinkage defects[4].

For cast-iron alloys, quality may by estimated by the presence of porosity and thus may be anticipated by gradient criterion application. All alloys should be treated in a different way. Their gaseous state (content of hydrogen) before casting ought to be taken into account. Otherwise, the described prognosticating of porosity in Al alloys and in MMC will be burdened with error[5].

Geometrical analysis of SiC, Al_2O_3 reinforcing particles

The analysis was carried out for medium-level particle sizes: 115, 97, 81 μm. Measurements and experimental data recording were performed:

- maximum surface of particle projection on plane;
- characteristic particle dimensions: length, L; width, D; and
- particle projection circumference on measurement plane.

Diagrams of particle length-to-width ratio, L/D, are shown in *Figure 1*. The curves were made depending on particle grain size: fractions of 120, 150 and 180 correspond to the following grain values 115, 97, 81 μm.

In the *Table 1* measurement results and comparative calculations are given. As a result of the analysis presented in *Table 1*:

1. relation between particle circumference and its projection field surface is typical for regular-size particles;
2. relation between external surface of a particle and its volume indicates that the observed particle form can be considered rather of cubic than spherical shape;
3. Al_2O_3 particles have a better contact surface with matrix than SiC particles (preferred conditions for heterogenous and catalytic nucleation); and
4. SiC particles are much more regular and more differentiated in the particular grain size, in comparison to Al_2O_3 (control possibility of technological and exploitation properties is higher).

Theoretical analysis of solidification process in micro-areas of composites

Chosen thermophysical properties of matrix and reinforcement for thermal calculation were defined as follows:

- c_w, specific heat [kJ/(kg K)]: matrix, 0.96; Al_2O_3, 1.07; SiC, 1.03;
- λ_w, thermal conductivity [W/(m K)]: matrix, 167; Al_2O_3, 3.7; SiC, 16.5; and
- b_w, thermal effusivity [W $s^{1/2}/(m^2$ K)]: matrix, 2.07 $\times 10^4$; Al_2O_3, 3.93×10^3; SiC, 7.26×10^3; and thermal conductivity for mould material, $\lambda =$ 0.13–0.17 W/(m K).

Average cooling rate of composite casts for the above conditions: $R = 0.31°/s$.

An estimation of MMC solidification process was performed with Simtec simulating code (*Figure 2*). A single SiC particle surrounded by metallic matrix with ideal mutual thermal contact was analyzed. Introducing

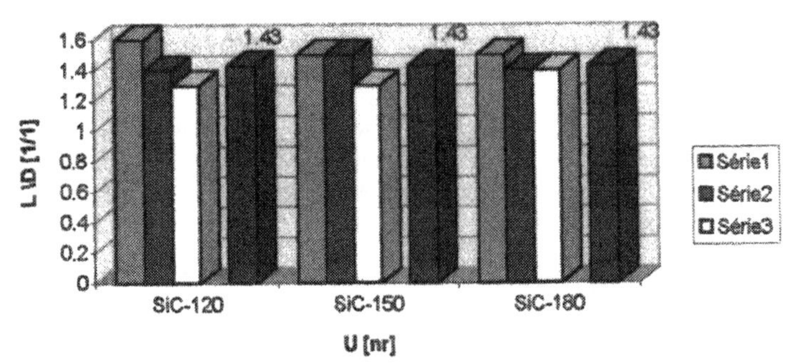

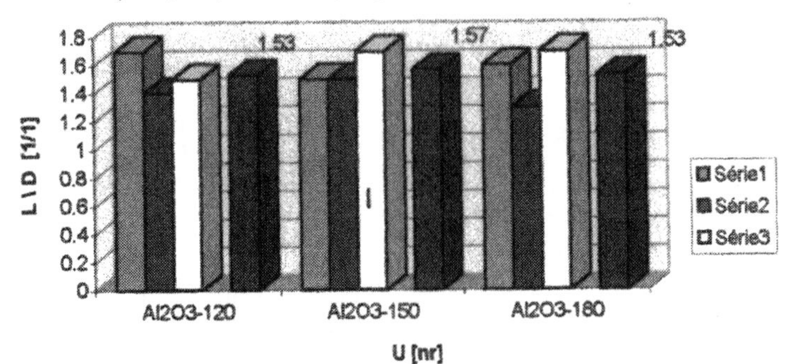

Figure 1 Length-to-width ratio (L/D) of reinforcing particles: (a) SiC; (b) Al_2O_3

Table 1 List of reinforced particle projection field to its circumference ratio

Kind of reinforcement	P, field (μm^2)[a]	O, circumference (μm)[a]	i = P/O (μm)[a]	$i^{sp} = 0.25(P)^{0.5}$ (μm)[b]	i^{so} = O/16 (μm)[b]	$i^{kp} = 0.5(P/\pi)^{0.5}$ (μm)[b]	$i^{ko} = 0.25(O/\pi)$ (μm)[b]
SiC-120	20 700	570	36	35.97	35.63	40.58	45.36
SiC-150	9900	420	24	24.88	26.25	28.07	33.42
SiC-180	5100	280	18	17.85	17.50	20.15	22.28
Al_2O_3-120	11 000	430	26	26.22	26.88	29.58	34.22
Al_2O_3-150	7700	360	21	21.94	22.50	24.75	28.65
Al_2O_3-180	7000	340	20	20.92	21.25	23.60	27.06

[a] Measured.
[b] Calculated.
Note: i, area-to-circumference ratio from measurement; i^{sp}, i^{so}, computational ratio, based on the assumption that characteristic dimension of particle corresponds to dimension of cubic edge (calculated properly with a section and circumference); i^{kp}, i^{ko}, computational ratio, based on the assumption that characteristic dimension of particle corresponds to dimension of sphere's diameter (calculated properly with a section and circumference) [e.g. $i^{kp} = P/O = D/4$; $D = 2\,(P/\pi)^{0.5} = 2\,(20\,700/\pi)^{0.5} \Rightarrow i^{kp} = 0.5\,(20700//\pi)^{0.5} = 40.58\ \mu m$].

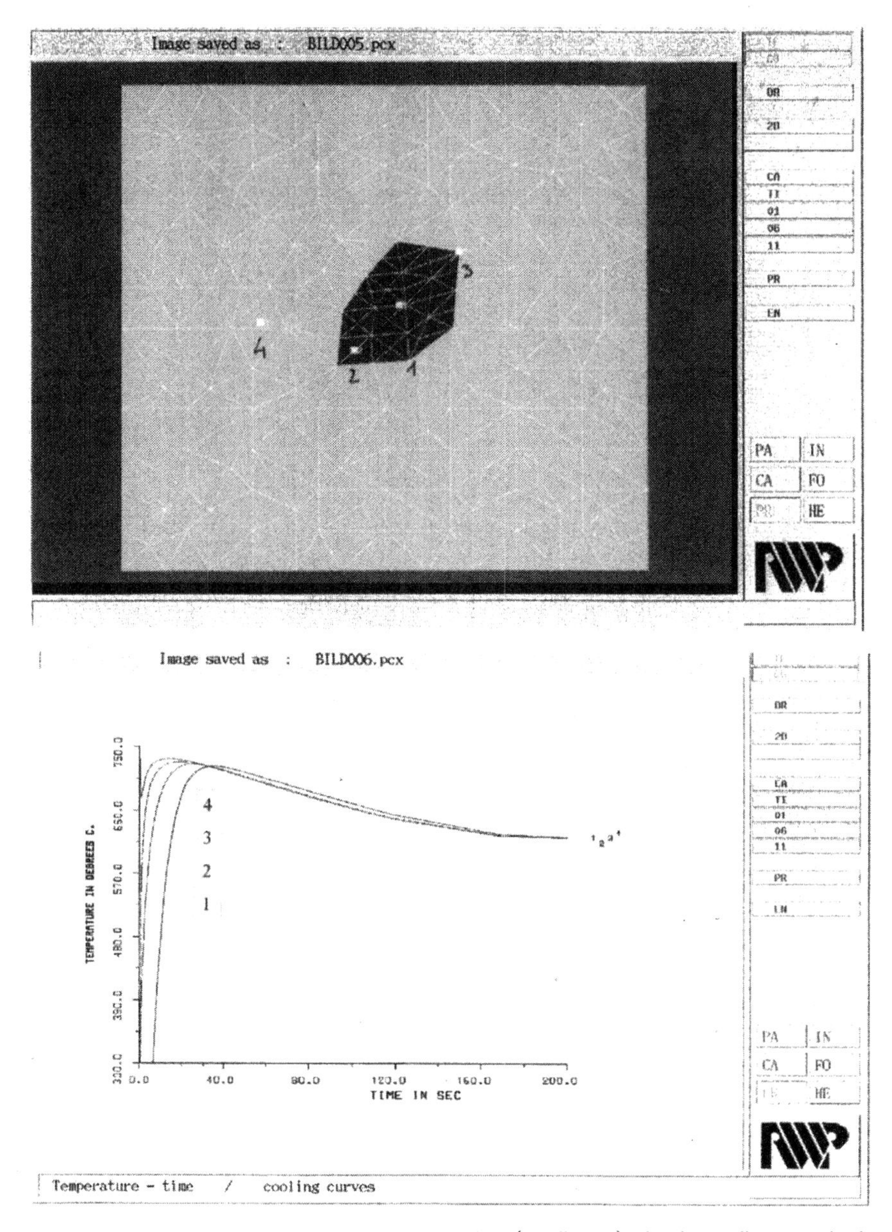

Figure 2 Position of examined points of SiC particle: 1, 2, 3 and, in matrix, 4 (top diagram) related to cooling curves in chosen places (bottom graph)

the particle geometry definition in the code, the above-mentioned results of geometrical analysis were applied. To analyze the transfer mechanism, the micro-area was thermally insulated by the enveloping mould material (95% of quartz sand). As a result of calculations, the particle surface reaches the temperature of high surface activity (650°C) after just 5 s of contact with metallic matrix. This accumulating particle warming effect was confirmed in the model by the highest particle temperature in the interval of time 30–180 s.

From the simulation it appears that:

1. reinforcing SiC and Al_2O_3 particles are local thermal centers (micro hot spots) in the matrix of the composite; in the surroundings of the particle, the matrix crystallization, solidification and cooling are retarded;
2. heat exchange is slow, typical for ceramic matrix mould, the composite matrix structure near the particle reveals a tendency to crystallization in a state of coarse eutectics;
3. in the case of non-inoculated matrix of eutectic alloy, primary Si precipitation in the near-neighbourhood of reinforcing particles is expected; and
4. knowing that the MMC reinforced by SiC and Al_2O_3 have comparable properties, the SiC MMC is advantageous in respect of casting requirements, the thermal characteristics and repeatability of geometric features in the analyzed values range being decisive.

a.

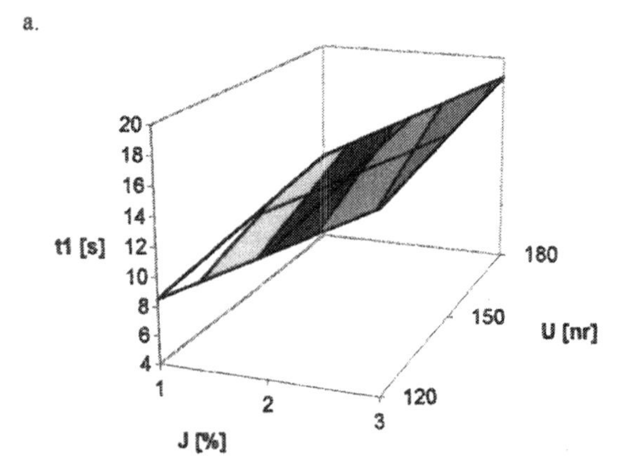

b.

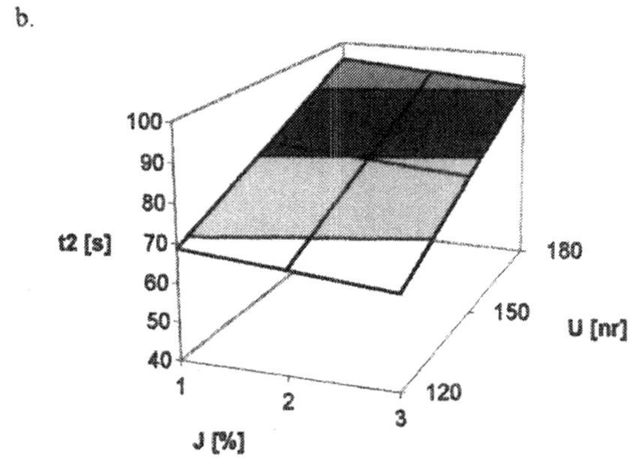

Figure 3 Dependance between crystallization times and particle size value of Al_2O_3, for different proportion of reinforcement: (a) time of thermal effect of phase α crystallization; (b) time of eutectic finish crystallization ($\alpha + Si$)

Crystallization process analysis

The amount of the solid phase, solidified from the liquid in a time unit, is a function of crystallization velocity and nucleation surface value. The larger is this surface, the lower is the respective crystallization velocity and so is superfusion. The substance activating the reinforcing particle surface increases the number of crystallization nuclei[1]. The describing activation has a catalytic character. The mechanism of the relationship has both the physical and chemical character. The physical one, where the point is the formation of a solid solution with matrix metal oxides. The chemical character leads to maximization of adhesion energy in the border zone of the matrix-reinforcement interface, forming in its area more profitable thermodynamical potentials of reaction.

In this investigation the thermal and derivative analysis was used. The results contain the temperature variation curves $T = f(t)$ and their derivatives $dT/dt = f'(t)$. The results are also calculated from Hartley's experiment plan, the co-ordinates of the early stage and final crystallization points. The analysis mentioned concerns the composite with Al_2O_3, with eutectic composition of industrial Al–Si alloy matrix. In *Figure 3* examples of crystallization time formula depending on quantity and particle size of reinforcement — SiC particles are shown.

From some observations we can notice that:

1. the composites start and finish their crystallization process at a temperature higher than that of matrix alloy without reinforcing particles; the smaller the particles, the higher is the crystallization temperature of the material composite;
2. crystallization time depends on thermal properties of reinforcement, mainly on thermal conductivity; the lower the thermal conductivity, the shorter is the period of crystallization and it happens earlier; and
3. addition of reinforcing particles, changing the crystallization process, results in higher composite tendency to volumic solidification morphology (endogenic).

Exploitation properties — examples

Al–Mg–Mn + SiC composite — resistant to erosive-corrosive wear (ECW)

An investigation of composite sample use was carried out by admission of extremal conditions of water salinity (120 kg of salt/1 m^3 of water); water included also loose abrasive particles of quartz sand (main fraction of grains: 0.63/0.40/0.32 mm and homogeneity factor 70%). The erosive–corrosive environment was a deposit consisting of 67% of sand by weight, with acidity at pH 6 and the average temperature 26°C. Medium ECW concerns the composite including 5% by weight of SiC of particle size 71 μm.

On the grounds of the study the following remarks can be made:

1. the dominant parameter responsible for the use of ECW is erosion;
2. an increase (approx. 30%) of wear resistance in relation to Al–Si alloys traditionally applied in pump carcass construction was obtained; and
3. further increase of resistance to use can be obtained, e.g. by introduction of a greater proportion of reinforcing particles whose shape is approximately spherical. The wear is in proportion to the amount of reinforcement.

Al–Si + C_{gr} composite — resistant to abrasive wear

A comparison of abrasive wear and composite friction coefficient with materials traditionally applied in piston rings and cylindrical sleeves was performed (e.g. cast iron Gh 190 B, Italian standard). As a reinforcement, synthetic graphite of particle size 63 μm in the amount 3.34% was used. The remaining parameters correspond to the conditions of cold starting of a combustion engine with restricted lubrication:

- typical oil batch, 122 mg;
- pressure, 1 MPa;
- to-and-fro motion with linear average velocity, 1 m/s; race-way, 0.15 m; and
- friction path, 180 km.

Table 2 Comparison of tribological properties of cast iron and Al–Si + C_{gr} composite

Tested material	Wear [g]	Friction coefficient μ[1/1]	Oil quantity q [mg]
Cast iron (Gh 190 B)	0.02001	0.18	—
Cast iron (Gh 190 B)	0.03314	0.18	122
Al–Si + C_{gr}	0.00321	0.02	44
Al–Si + C_{gr}	0.00011	0.01	122

The results were drawn up in *Table 2*. From the conducted studies it follows that:

1. a composite under the conditions of technically dry friction is not subject to seizing, contrary to the material of the matrix;
2. the amount of wear for composites is lower by approximately two orders; with the amount of oil decreased three times it is still lower by one order than for cast iron; and
3. the friction factor is lower by one order in comparison to that of cast iron.

Conclusions

This article describes some practical and technological aspects concerning MMC production useful during design analysis of their application in chosen technical domains. The detailed conclusions on the micro and macro phenomena in MMC structure are given above. On the examples it was proved that the method from micro to macro modelling gives good results providing advantageous exploitation features of MMC.

References

1 Gawroński, J. and Cholewa, M., Surface phenomena at graphite-aluminium boundary in composites resistant to wearing out. *TBM Archive*, 1992, **1**

2 Cholewa, M., Gawroński, J. and Szajnar, J., Aluminium-SiC particles composites. In *The Technology of Shape Composites Production, Proceedings of the IV Conf. CADCOMP'94, 4–6 July 1994.* Wessex Institute of Technology, University of Portsmouth, Southampton, 1994, pp. 321–328

3 Cholewa, M., Gawroński, J., Szajnar, J., Composite bodies of the mining pumps. *KMiS (Solidifiction of Metals and Alloys)*, 1996, **24**

4 Ignaszak, Z., Comparative study of casting feeding criteria. *KMiS (Solidifiction of metals and alloys)*, 1996, **24**

5 Ignaszak, Z., Wołujewicz, P., Possibility analysis of shrinkage porosity identification in the Al alloy casts. *TBM Archive* 1994, **13**

PART II

Symposium E
on
Material Aspects for Electric Vehicles including Batteries and Fuel Cells

MATERIAL ASPECTS FOR ELECTRIC VEHICLES INCLUDING BATTERIES AND FUEL CELLS

PROCEEDINGS OF THE ICAM/E-MRS 1997 SPRING MEETING
SYMPOSIUM E ON MATERIAL ASPECTS FOR ELECTRIC VEHICLES
INCLUDING BATTERIES AND FUEL CELLS
STRASBOURG, FRANCE, JUNE 16-20, 1997

T. HARTKOPF
Technische Hochschule, Darmstadt, Germany

A. MORETTI
RENAULT, Research Department, Trappes, France

J.G. WURM
Waterloo, Belgium

M. WAKIHARA
Tokyo Institute of Technology, Tokyo, Japan

Sponsors

This conference was held under the auspices of:

The Council of Europe
The Commission of European Communities

It is our pleasure to acknowledge with gratitude the financial assistance provided by:

Banque Populaire	(France)
Elsevier Science	(The Netherlands)
Office du Tourisme, Strasbourg	(France)

ELSEVIER

Journal of Power Sources 72, Number 1, 30 March 1998

Contents

ELSEVIER

Journal of Power Sources 72 (1998) 1–8

Electrochemical performance and chemical properties of oxidic cathode materials for 4 V rechargeable Li-ion batteries

A. Ott [a], P. Endres [a], V. Klein [a], B. Fuchs [a], A. Jäger [a], H.A. Mayer [a], S. Kemmler-Sack [a,*], H.-W. Praas [b], K. Brandt [b], G. Filoti [c], V. Kunczer [c], M. Rosenberg [d]

[a] *Institut für Anorganische Chemie, Auf der Morgenstelle 18, 72076 Tübingen, Germany*
[b] *VARTA Batterie AG, Gundelhardtstr. 72, 65779 Kelkheim, Germany*
[c] *Institute of Physics and Technology of Materials, 76900 Bucharest, Romania*
[d] *Ruhr-Universität Bochum NB03/32, 44780 Bochum, Germany*

Received 28 July 1997; accepted 14 October 1997

Abstract

The systems $Li_{1+x}Mn_{2-x}O_{4-\delta}$ and $Li_{1-x}Ni_{1+x}O_2$ were studied as oxidic cathode materials for 4 V rechargeable Li-ion cells. From a comparison of the results it follows that the spinel system can be subdivided into three regions: in region I ($0 \leq x < 0.05$) the electrochemical performance is unsatisfying due to a large volume reduction of about 7.5%, inducing stress, multiphase materials and capacity fading. Region II ($0.05 \lesssim x \lesssim 0.2$) contains the best cathode material with a composition near $x = 0.1$. The spinel framework is stable against Li extraction. In region III ($0.2 \leq x \leq 1/3$) the capacity is too low for an application in the 4 V region. In the system $Li_{1-x}Ni_{1+x}O_2$ the highest capacity is observed for $LiNiO_2$ with $x = 0$. Experimental difficulties during material synthesis were overcome by suppression of the decomposition of $LiNiO_2$ into $Li_{1-x}Ni_{1+x}O_2$, Li_2O and O_2.

Keywords: Li-ion cells; Oxidic materials; Rechargeable batteries; 4V material

1. Introduction

Oxidic materials which can be used as cathodes in 4 V rechargeable Li cells are the layered oxides $LiCoO_2$ and $LiNiO_2$ of α-$NaFeO_2$ type (space group $R\bar{3}m$ [1]) and the cubic spinel $LiMn_2O_4$ ($Fd\bar{3}m$ [2–6]). $LiNiO_2$ is the Li rich endmember of the system $Li_{1-x}Ni_{1+x}O_2$ and difficult to prepare [7,8]. Solid solutions of $LiNi_{1-x}Co_xO_2$ have also been investigated as electrodes in such cells [9]. For $x = 0.5$ the compound is easier to obtain than $LiNiO_2$ but has almost the same voltage profile. The voltage of $Li/LiNiO_2$ cells is about 0.25 V lower than $Li/LiCoO_2$ cells making the former less prone to electrolyte oxidation problem to high temperatures [10]. In addition, Ni is less expensive than Co and is more abundant, so $LiNiO_2$ and its Ni rich solid solutions may be more useful than $LiCoO_2$. From the point of view of starting material, price and toxicity, the $LiMn_2O_4$ spinel has a considerable advantage. However, its rechargeable capacity and cyclability in the 4 V region are inferior to that of the layered oxides. Several attempts to improve the electrochemical performance have been reported. The most effective candidate is low valent cation doping to 16d sites (denoted in square brackets), $Li[M_xM_{2-x}]O_4$ [11]. Monovalent Li doping seems to be the best since a small fraction of dopant is effective (due to the large charge difference relative to Mn) and lattice disturbances are minimized. A second advantage of Li doping is the prevention of foreign ions from occupying tetrahedral 8a sites which will hamper the migration of Li between the 8a and 16c sites, resulting in a decrease of rechargeable capacity. Investigations of the recently synthesized Li-rich spinels of type $Li[Li_xMn_{2-x}]O_4$ [4,12] reveal a slight oxygen deficiency: $Li_{1+x}Mn_{2-x}O_{4-\delta}$ [5,6,13]. Furthermore, the rechargeable capacity is strongly dependent on the Li/Mn ratio [4,6,13].

To get a better insight into the electrochemical performance the aim of the present investigation was to study the influence of the Li content on the electrochemical, chemical and magnetic properties of several members of the spinel phase $Li_{1+x}Mn_{2-x}O_{4-\delta}$ and the ordered rocksalt α-$NaFeO_2$ type of the Li-rich phase $Li_{1-x}Ni_{1+x}O_2$.

* Corresponding author. Fax: +49-0-7071-296918.

0378-7753/98/$19.00
PII S0378-7753(97)02772-9

2. Experimental

For the synthesis of the series $Li_{1+x}Mn_{2-x}O_{4-\delta}$ the starting materials β-MnO_2 (Selectipur, Merck) and $LiOH \cdot H_2O$ (Alfa Ventron) were thoroughly mixed in an agate mortar and heated in corundum boats (Degussit Al 23) in air between 400 and 850°C for 4 h and 16 h with an intermittent regrinding and X-ray analysis (XRD, Philips powder diffractometer, Ni-filtered CuKα radiation, Au standard). All materials were quenched to room temperature. The correlation between the reaction temperature and the parameter x as well as the experimental conditions for Li extraction with Br_2/CH_3CN and H_2SO_4/H_2O are indicated in Ref. [13].

For the synthesis of Li-rich members of the system $Li_{1-x}Ni_{1+x}O_2$ the thoroughly ground starting materials Li_2O_2 (Alfa Ventron corrected for a content of 5% Li_2CO_3) and NiO (Johnson Matthey) were pelletized. Two pellets were put one on top of the other in an alumina container and fired in flowing oxygen for 48 h at 700°C with three intermittent regrindings. The upper pellet was employed for the further investigations.

Susceptibility and Mössbauer spectroscopic measurement were performed according to Ref. [14]. NMR spectra were recorded as in Ref. [13].

3. Results and discussion

3.1. System $Li_{1+x}Mn_{2-x}O_{4-\delta}$

3.1.1. XRD and electrochemical performance

In the series $Li_{1+x}Mn_{2-x}O_{4-\delta}$ a complete series of solid solutions is existing for $0 \leq x \leq 0.33$. All materials crystallize in the cubic spinel structure of space group $Fd\bar{3}m$. The oxygen deficiency increases with increasing x.

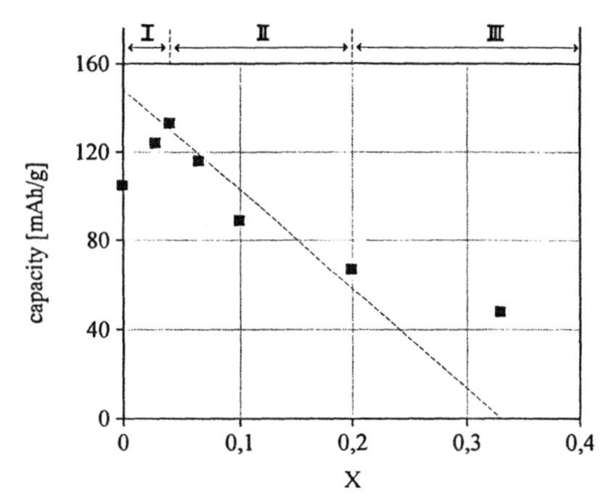

Fig. 1. The first charge capacity (3.3–4.3 V) of the spinel phase $Li_{1+x}Mn_{2-x}O_{4-\delta}$. The broken line is the theoretical capacity for $\delta = 0$.

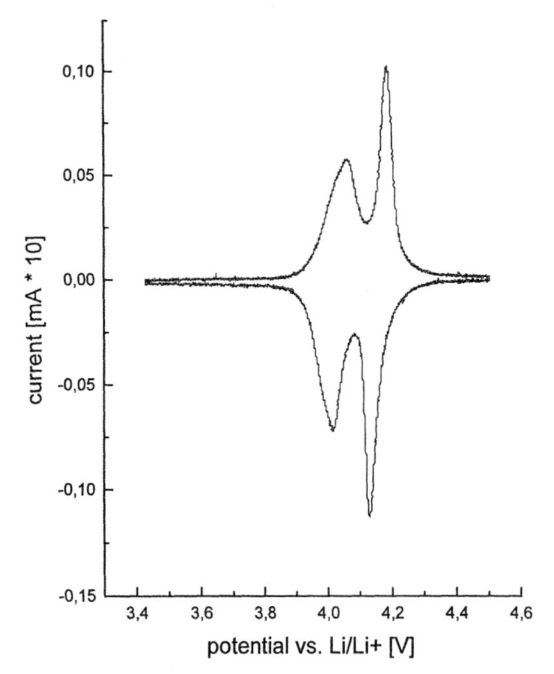

Fig. 2. Cyclic voltammogram of $LiMn_2O_4$.

The lattice constants vary linearly between 8.247 Å ($x = 0$, $\delta = 0$) and 8.157 Å ($x = 0.33$, $\delta = 0.13$ [13]).

The electrochemical Li extraction from the system $Li_{1+x}Mn_{2-x}O_{4-\delta}$ is accompanied by an oxidation of Mn^{3+} to Mn^{4+}. The theoretical capacity for the case of $\delta = 0$ is indicated in Fig. 1 together with values of the first charge capacity (3.3–4.3 V). The electrochemical performance can be roughly divided in three regions: in region I ($x \lesssim 0.05$) the experimental capacity is inferior to the theoretical value, in region II ($0.05 \lesssim x \lesssim 0.2$) both values are in fairly good agreement whereas in region III ($0.2 \lesssim x \leq 0.33$) the electrochemical performance is better than expected for the case of $\delta = 0$.

Li/Mn-spinels are three-dimensional Li-ion conductors. Li^+ extraction from $LiMn_2O_4$ is accompanied by a high volume reduction of 7.5% between the starting material and the fully extracted spinel. This large volume decrease is considered as one reason for the pronounced fading of the rechargeable capacity [6,15,16]. Li^+ ions are inserted and extracted into the spinel phase by a two-step process; illustrated by two pairs of redox peaks in the cyclic voltammogram of Fig. 2. From the crystallographic point of view this behaviour is explained by the splitting of the original Li positions in 8a with the coordinates 0, 0, 0; 1/4, 1/4, 1/4, (space group $Fd\bar{3}m$) in two sets of sites: 4a (0, 0, 0) and 4c (1/4, 1/4, 1/4); space group $F23$. Firstly, Li is removed from half of the tetrahedral positions with strong Li–Li interaction (site 4a), affording less energy (oxidation potential at about 4.0 V vs. Li/Li^+) than for the second set (4c) with the higher potential of about 4.15 V. Li extraction from 4c affords more energy due to the stronger attractive forces as a result of the increasing content of higher valent Mn^{4+}.

By application of Br_2/CH_3CN with a redox potential of Br_2/Br^- in CH_3CN vs. Li^+/Li of 4.1 V [17] as extraction medium for the system $Li_{1+x}Mn_{2-x}O_{4-\delta}$ we were able to remove only the first set of Li ions with a potential of about 4.0 V, whereas the deeper positions remain unaffected. Accordingly, the decrease of the unit cell volume is relatively small and situated between ≈ 3% for $x = 0.0$ (lattice constant of the extraction product: $a = 8.165$ Å, main phase) and ≈ 1% for $x = 0.33$ (8.157 Å [13]). The XRD diagrams of the extraction residues in Fig. 3 indicate that Li-rich spinel with $0.1 \leq x \leq 0.33$ yield single phase material with the typical diffraction pattern of a cubic spinel. However, for $x = 0$ and 0.05 multiphase material is formed, easily visible from the appearance of several satellite peaks in the XRD. By acidic Li extraction with e.g., H_2SO_4/H_2O both Li positions are accessible. However, even deep acidic Li extraction results uniformly in multiphase samples [6,13]. It is very likely that the tendency of Li ordering is responsible for this effect. Due to the fact that certain Li:□ ratios on the tetrahedral position will be more suitable for cationic ordering than others several ordered phases of different stability are simultaneously produced.

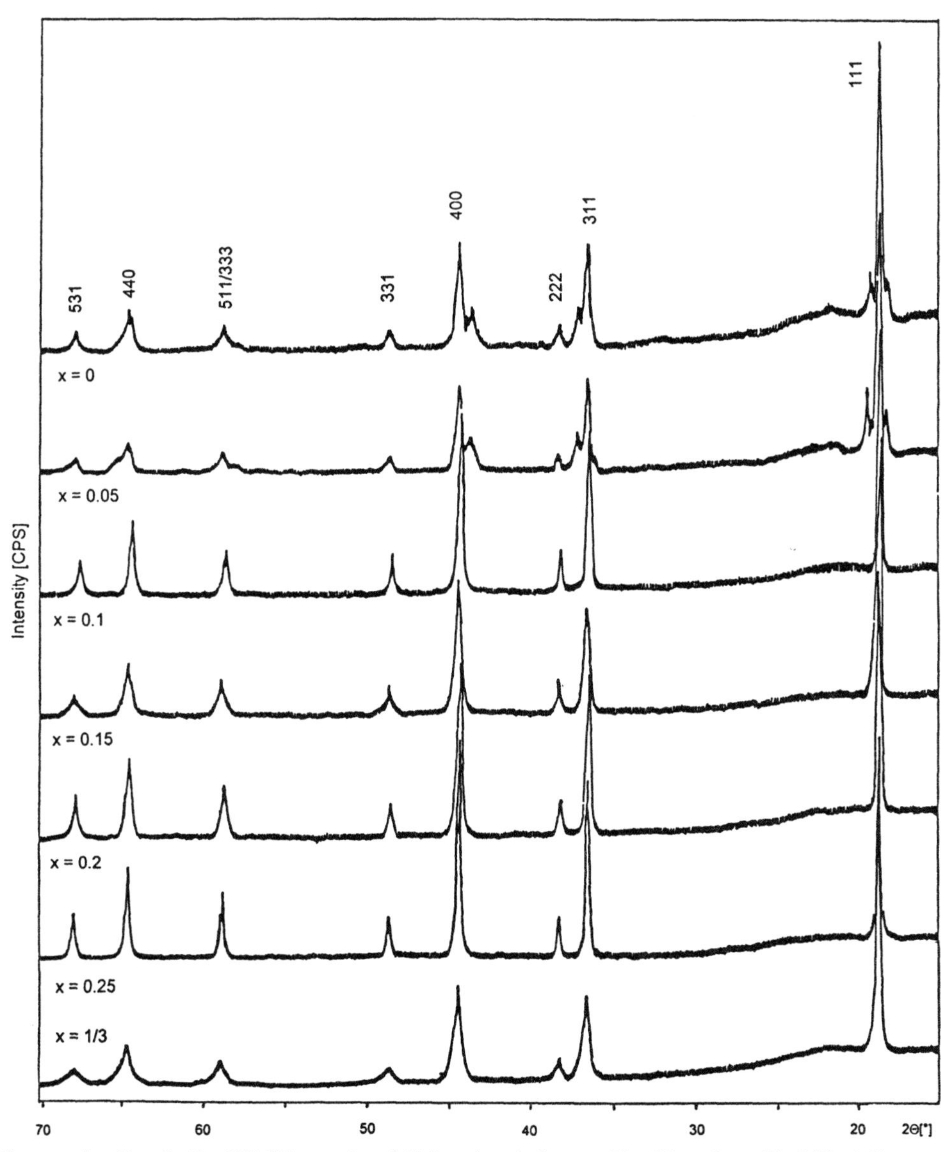

Fig. 3. XRD diagrams of residues for Br_2/CH_3CN extraction of Li from the spinel system $Li_{1+x}Mn_{2-x}O_{4-\delta}$. The Miller indices are given for each Bragg peak with reference to the setting of the cubic spinel cell ($Fd\bar{3}m$). For clarity the spectra are shifted along the y axis.

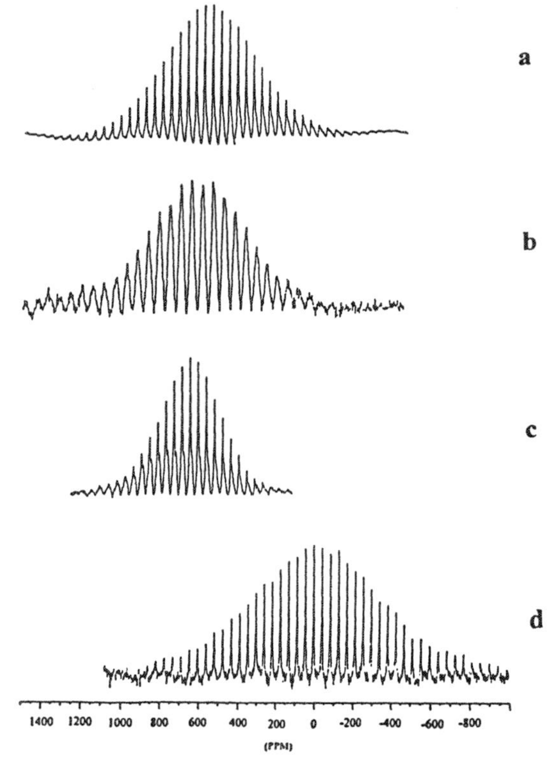

Fig. 4. {^{7}Li} MAS-NMR spectra of (a) $LiMn_2O_4$ ($x = 0$), (b) $Li_{1.2}Mn_{1.8}O_{3.93}$ ($x = 0.2$), (c) $Li_{0.6}Mn_2O_4$ (Br_2/CH_3CN extraction) and (d) $LiCrMnO_4$.

3.1.2. {^{7}Li} MAS-NMR spectra

The ^{7}Li NMR spectra of some members of the system $Li_{1+x}Mn_{2-x}O_{4-\delta}$ and its extraction products are shown in Fig. 4 together with the spectrum of $LiCrMnO_4$. For all compounds the NMR spectrum exhibits a single line spaced by spinning side bands. The Knight shift for the system $Li_{1+x}Mn_{2-x}O_{4-\delta}$ are uniformly situated around 550 ppm (cf. the spectra for $x = 0$ and 0.2 in Fig. 4a,b). Moreover, Li extraction does not influence the position of the ^{7}Li NMR signal of the remaining Li; situated for e.g., $Li_{0.6}Mn_2O_4$ and other extraction products invariably in the 550 ppm region (Fig. 4c). The quantity of the Knight shift of $LiMn_2O_4$ is in agreement with the data from the literature [18,19]. The calculated electron density of ≈ 0.9 (electron density of Li in the compound/electron density of Li in atomic state) is higher than for Li metal (0.44 [20]). This means that an average of 90% of 2s electrons is located at the lithium nucleus and Li exists in an almost atomic state.

For comparison we synthesized the well known spinel $LiCrMnO_4$ with the fixed cationic valences Cr^{3+} and Mn^{4+} [21,22]. This compound does not show any Knight shift and lithium exists in completely ionic state with an apparent valency of $+1.0$. Interestingly, this spinel is not accessible for a Li extraction with Br_2/CH_3CN.

3.1.3. Magnetic susceptibility and Mössbauer spectroscopy

The temperature dependence of the magnetic susceptibility of spinels of the system $Li_{1+x}Mn_{2-x}O_{4-\delta}$ indicate

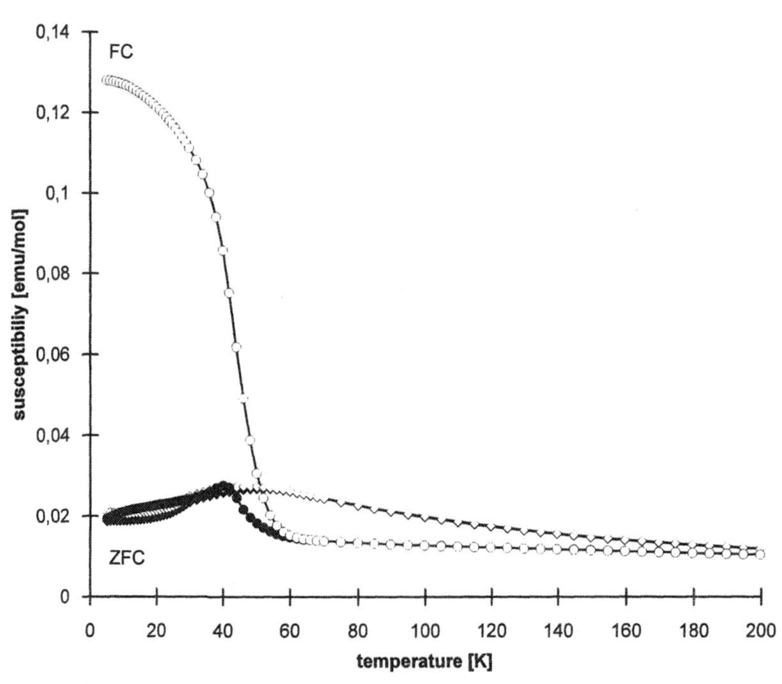

Fig. 5. χ vs. T for ^{57}Fe doped (○) $LiMn_2O_4$ and (◇) $Li_{0.15}Mn_2O_4$; FC: field cooled, ZFC: zero field cooled; $B = 0.1$ T.

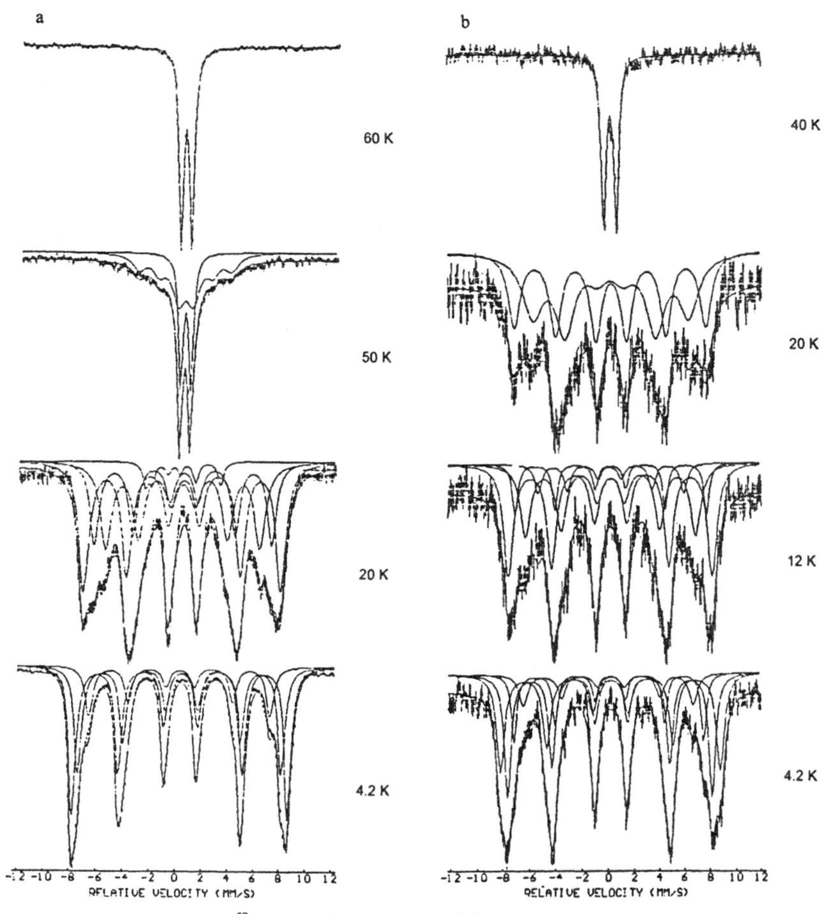

Fig. 6. Mössbauer spectra of ^{57}Fe doped (a) $LiMn_2O_4$ and (b) $Li_{0.15}Mn_2O_4$ taken at several temperatures.

the validity of the Curie–Weiss law at higher temperatures. Below a freezing temperature T_F a spin glass behaviour develops. T_F decreases with increasing x from around 55 K ($x = 0$) to about 10 K ($x = 0.33$ [5]). In the delithiated material the reduction of the Mn^{3+} content reduces the

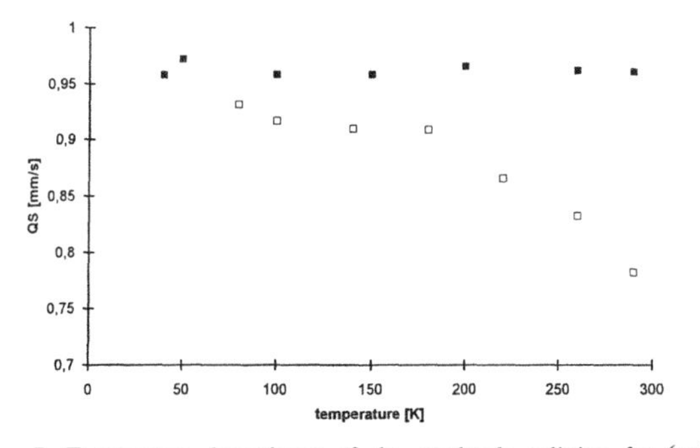

Fig. 7. Temperature dependence of the quadruple splitting for (□) $LiMn_2O_4$ and (■) $Li_{0.15}Mn_2O_4$.

total magnetical susceptibility and induces a practically antiferromagnetic behaviour below about 50 K. Fig. 5 indicates the χ vs. T traces for $LiMn_2O_4$ and $Li_{0.15}Mn_2O_4$.

From the Mössbauer spectra of ^{57}Fe doped $LiMn_2O_4$ and $Li_{0.15}Mn_2O_4$ in Fig. 6 it follows the existence of a low temperature state with frozen and possibly ordered spins in a kind of antiferromagnetic configuration. A good fit of the 4.2 K spectra needed either a distribution of hyperfine fields or four different Lorenzian sextets with hyperfine fields between 52 and 43 T. The hyperfine fields decrease with increasing temperature but for some Fe spins a relaxation occurs which gives rise to a strong intensive Mössbauer doublet in the central part well below the critical temperature.

A very interesting finding is a Mössbauer spectroscopic evidence for the Jahn–Teller effect in $LiMn_2O_4$. Its content of 50% of Mn^{3+} in octahedral positions is very close to the critical concentration for the occurrence of the Jahn–Teller distortion. As was shown previously [23] the

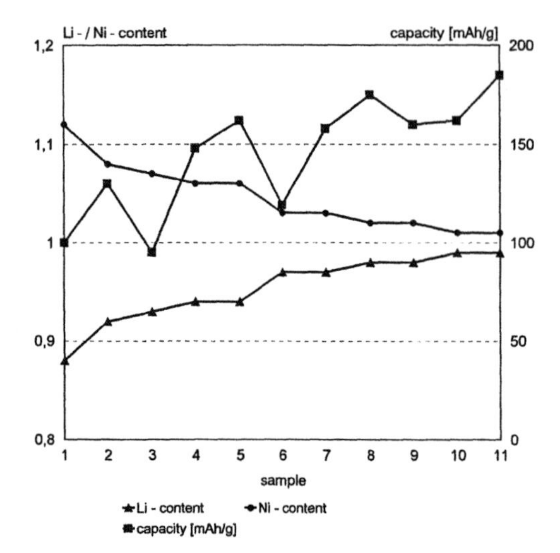

Fig. 8. Rechargeable capacity vs. Li/Ni content for several materials of the system $Li_{1-x}Ni_{1+x}O_2$.

quadrupole splitting (QS) of Fe ions in the Mn surroundings of the spinels $CdMn_2O_4$ and $ZnMn_2O_4$ is a measure for the local Jahn–Teller distortion around the Fe ion. As one can see from Fig. 7 the QS of $LiMn_2O_4$ starts to decrease rapidly above 200 K due to a transition from the Jahn–Teller distorted to a non-distorted cubic structure. In contrast the QS of $Li_{0.15}Mn_2O_4$, where one has a very low Mn^{3+} content of 15%, is practically independent of temperature, indicating the absence of any Jahn–Teller effect.

Table 1
Average oxidation state of Ni (Ox(Ni)), Li/Ni ratio and resoluting composition for some Li-rich materials of the system $Li_{1-x}Ni_{1+x}O_2$

Ox(Ni)	Li/Ni ratio	Composition
+2.86	0.88	$Li_{0.94}Ni_{1.06}O_2$
+2.95	0.94	$Li_{0.97}Ni_{1.03}O_2$
+3.00	1.00	$Li_{1.00}Ni_{1.00}O_2$

Our results are in good agreement with the data of Yamada and Tanaka [24], who found evidence for a Jahn–Teller distortion for $LiMn_2O_4$ around 280 K from thermal analysis and XRD. The observed tremendous increase up to 60% between 280 and 260 K with about 65% of the distorted tetragonal phase at 220 K, what we also see in the ^{57}Fe QS.

3.2. System $Li_{1-x}Ni_{1+x}O_2$

3.2.1. XRD and electrochemical performance

Several Li-rich members of the system $Li_{1-x}Ni_{1+x}O_2$ with $0.0 \leq x \lesssim 0.1$ were studied. They can be prepared as single phase materials if the decomposition into Li_2O, oxygen and a Ni-rich phase is suppressed. For $LiNiO_2$ the following relation is valid:

$$2LiNi^{3+}O_2 \rightarrow Li_{1-x}Ni^{3+}_{1-x}Ni^{2+}_{2x}O_2 + xLi_2O + x/2O_2$$

All prepared Li-rich materials crystallize in the α-$NaFeO_2$ type of space group $R\bar{3}m$. In this Li-rich region the lattice constants are very similar and not indicative for

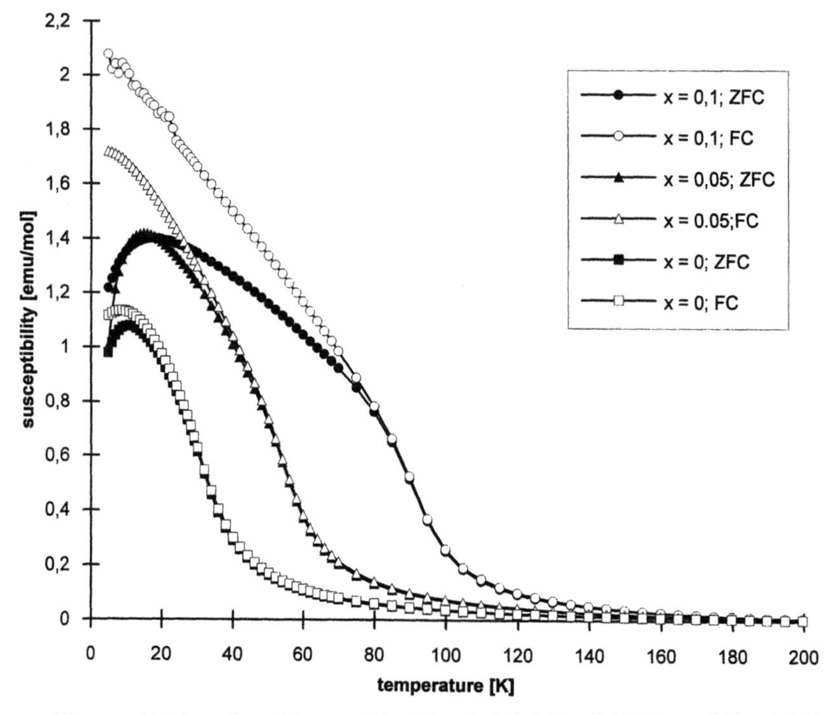

Fig. 9. χ vs. T for some Li-rich members of the system $Li_{1-x}Ni_{1+x}O_2$; FC: field cooled, ZFC: zero field cooled; $B = 0.1$ T.

the individual composition (e.g., $x = 0.08$: $a = 2.885$; $c = 14.21$ Å; $x = 0.01$: $a = 2.884$, $c = 14.21$ Å (hexagonal setting)). However, the electrochemical performance strongly depends on the Li/Ni ratio since an excess of Ni enters into the Li layers with the result of a blockage of the two-dimensional Li migration paths. In agreement with several observations of other authors (e.g., [25,26]) the rechargeable capacity increases with increasing Li/Ni ratio (Fig. 8). The best values are observed for x around 0.0.

Due to the fact that the $x = 0.0$ compound $LiNiO_2$ contains a maximal Ni^{3+} content the determination of the average oxidation state of Ni by redox titration is a very reliable proof for the composition of the synthesized material. Table 1 gives some values together with the Li/Ni ratio (from AAS analysis) and the resulting composition.

3.2.2. Magnetic susceptibility

In the ideal rhombohedral structure of $LiNiO_2$ ($R\bar{3}m$, Li in 3a, Ni in 3b, O in 6c) the NiO_6 octahedra share edges to form a triangular Ni lattice. Nonmagnetic Li layers alternate with magnetic Ni layers, making the interlayer distance (4.73 Å) much longer than the intralayer distance (2.88 Å). So magnetic correlations between Ni ions are considered as two dimensional [27] and a frustrated antiferromagnetic $S = 1/2$ triangular lattice [28] or spin glass [29] may be anticipated. For $x > 0$ the excess of Ni occupies the Li positions 3a; thus creating a three dimensional connection between the two dimensional Ni layers (in 3b). Accordingly, the magnetic properties are strongly influenced by the Ni amount in site 3a. Moreover, the two-dimensionality of the magnetic interaction may be destroyed by a certain degree of disorder between Li in 3a and Ni in 3b. Again a three-dimensional magnetic correlation is introduced.

From the temperature dependence of the magnetic susceptibility of some Li-rich members of the system $Li_{1-x}Ni_{1+x}O_2$ in Fig. 9 it follows that the extend of the magnetic interactions increase with increasing x. An analysis of the magnetic properties reveal strong ferromagnetic correlations and spin freezing at low temperatures, reminding many characteristic features of spin-glass behaviour. The freezing temperature increases with increasing Ni content in 3a from 30 K for $x \approx 0.0$ via 54 K ($x = 0.05$) to 90 K for $x = 0.10$.

4. Conclusions

The electrochemical performance in the 4 V region of the spinel system $Li_{1+x}Mn_{2-x}O_{4-\delta}$ and of the ordered rocksalt phase $Li_{1-x}Ni_{1+x}O_2$ is strongly influenced by the parameter x. According to our studies of various chemical and physical properties the spinel system $Li_{1+x}Mn_{2-x}O_{4-\delta}$ can be subdivided into three regions:

In region I ($0 \leq x < 0.05$) the spinel framework is not stable against Li extraction. The electrochemical performance is unsatisfying due to the large volume reduction of about 7.5% between the starting material and the fully extracted spinel, including stress and capacity fading. The extraction products are always multiphase. Region II ($0.05 \lesssim x \lesssim 0.2$) contains the best cathode materials. The optimal composition is situated near $x = 0.1$. The spinel framework is stable against Li extraction. At different extraction levels always single phase materials are present. The volume drop decreases to about 5.5% ($x = 0.1$) and 4.5% ($x = 0.2$). All materials are slightly oxygen deficient. In region III ($0.2 \lesssim x \leq 0.33$) the host matrix is again stable against Li extraction. Due to an oxygen deficiency even spinels with $x = 0.33$ conserve a minor ability for Li extraction. However, the observed capacity is too low for an application in the 4 V region.

In the system $Li_{1-x}Ni_{1+x}O_2$ of α-$NaFeO_2$ type the highest capacity is observed for the ideal composition $LiNiO_2$ ($x = 0$). Due to the absence of Ni in the Li sheets the Li ions are not hampered from their two-dimensional migration. Experimental difficulties during material synthesis were overcome by suppressing the decomposition reaction 2 $LiNiO_2 \rightarrow Li_{1-x}Ni_{1+x}O_2 + Li_2O + x/2O_2$.

Acknowledgements

This work was supported by the Bundesministerium für Bildung, Wissenschaft, Forschung und Technologie as well as by the Verband der Chemischen Industrie.

References

[1] K. Mizushima, P.C. Jones, P.J. Wiseman, J.B. Goodenough, Mater. Res. Bull. 15 (1980) 783.
[2] M.M. Thackeray, W.I.F. David, P.G. Bruce, J.B. Goodenough, Mater. Res. Bull. 18 (1983) 461.
[3] M.M. Thackeray, P. Johnson, L.A. de Piciotto, P. Bruce, J.B. Goodenough, Mater. Res. Bull. 19 (1984) 179.
[4] R.J. Gummov, A. de Kock, M.M. Thackeray, J. Solid State Chem. 69 (1994) 59.
[5] P. Endres, B. Fuchs, S. Kemmler-Sack, K. Brandt, G. Faust-Becker, H.-W. Praas, J. Solid State Ionics 89 (1996) 221, and Refs. cited therein.
[6] S. Kemmler-Sack, B. Fuchs, P. Endres, H.-W. Praas, K. Brandt, GdCH-Monographie, in: F. Beck (Ed.), Vol. 3, Elektrochemie der Elektronenleiter, 1996, p. 342.
[7] R. Kanno, H. Kubo, Y. Kawamoto, T. Kaniyama, F. Izumi, Y. Takeda, M. Takano, J. Solid State Chem. 110 (1994) 219.
[8] A. Hirano, R. Kanno, Y. Kawamoto, Y. Takeda, K. Yamaura, M. Takano, K. Okyama, M. Ohashi, Y. Yamaguchi, Solid State Ionics 78 (1995) 123.
[9] T. Ohzuku, H. Komori, K. Sawai, T. Hirai, Chem. Express 5 (1990) 733.
[10] S.A. Campbell, C. Bowes, R.S. McMillan, J. Electroanal. Chem. 284 (1990) 195.
[11] A. Yamada, J. Solid State Chem. 122 (1996) 160.
[12] B. Fuchs, Diplomarbeit, Universität of Tübingen, 1990.
[13] P. Endres, A. Ott, S. Kemmler-Sack, A. Jäger, H.A. Mayer, H.-W. Praas, K. Brandt, J. Power Sources 69 (1997) 149.

[14] B. Seling, C. Schinzer, A. Ehmann, S. Kemmler-Sack, G. Filoti, M. Rosenberg, J. Linhart, W. Reimers, Physica C251 (1995) 238.
[15] T. Ohzuku, M. Kitagawa, T. Hirai, J. Electrochem. Soc. 137 (1990) 769.
[16] Y. Xia, M. Yoshio, J. Electrochem. Soc. 143 (1996) 825.
[17] A.R. Wizansky, P.E. Rauch, F.J. Disalvo, J. Solid State Chem. 81 (1989) 203.
[18] N. Kumagai, T. Fujiwara, K. Tanno, T. Horiba, J. Electrochem. Soc. 143 (1996) 1007, and Refs. cited therein.
[19] Y. Kanazaki, A. Tamiguchi, M. Abe, J. Electrochem. Soc. 138 (1991) 333.
[20] C. Ryter, Phys. Rev. Lett. 5 (1960) 10.
[21] A. Ott, Diplomarbeit, University of Tübingen, 1996.
[22] G. Blasse, J. Inorg. Nucl. Chem. 25 (1963) 743.
[23] G. Filoti, A. Geiberg, V. Gomolea, M. Rosenberg, Int. J. Magn. 2 (1972) 65.
[24] A. Yamada, M. Tanaka, Mater. Res. Bull. 30 (1995) 715.
[25] H. Arai, S. Okada, H. Ohtsuka, M. Ichimura, J. Yamaki, Solid State Ionics 80 (1995) 261.
[26] V. Klein, Dissertation, Universität of Tübingen, 1995.
[27] K. Yamaura, M. Takano, A. Hirano, R. Kanno, J. Solid State Chem. 127 (1996) 109, and Refs. therein.
[28] K. Hirakawa, H. Kadwaki, K. Ubrekoshi, J. Phys. Soc. Jpn. 54 (1985) 3526.
[29] M. Rosenberg, P. Stelmaszyk, V. Klein, S. Kemmler-Sack, G. Filoti, J. Appl. Phys. 75 (1994) 6813.

ELSEVIER

Journal of Power Sources 72 (1998) 9–13

Application of plasma spray deposited coatings for seawater activated batteries

S. Tamulevičius *, R. Dargis

Kaunas University of Technology, Physics Department, Studentų 50, Kaunas 3031, Lithuania

Received 28 July 1997; revised 15 November 1997

Abstract

Seawater activated batteries based on Mg and Ni/Al electrodes were constructed and investigated at different electrolyte temperatures. The Ni/Al coatings which were applied as the cathodes for seawater activated batteries were produced by plasma spray deposition. Voltage–time ($U = E - IR(t)$) dependence was measured for the galvanic pair Mg–Ni/Al, where I was constant current, E the electromotive force of the galvanic pair and $R(t)$ the variable resistance. It was found that $U(t)$ inclination depends on the anode corrosion rate, and the mass of the anode is the only parameter that restricts the life time of the seawater activated cell. The current density of this cell was found to be a linear function of the temperature of the seawater. Output power density dependence on the spacing between electrodes and number of cells was investigated for cells with different electrode area. A maximum output power density of 3×10^4 W/m^3 was obtained for these cells.

Keywords: Seawater batteries; Plasma spray; Ni/Al coatings

1. Introduction

Batteries which use seawater as the electrolyte are attractive power sources as they have moderate cost, excellent safety characteristics, infinite shelf life under dry conditions, and they are environmentally friendly [1]. Seawater activated batteries can power remote sea buoys and emergency signals, or underwater sensors and navigation aids.

This type of galvanic cell consists of an anode (negative terminal) and an inert cathode where hydrogen evolves [2]. Seawater batteries have an advantage over alkaline batteries because they do not need to be contained in water- and pressure-resistant vessels when they are used for subsea applications. Such cells need a continuous supply of sea water to remove the products of the cell reaction. The cathode functions as a hydrogen electrode. The magnesium anode dissolves in the seawater to form magnesium hydroxide that diffuses away. The electrochemical reactions taking place in the cell can be described in the following way:

Anodic reaction: $Mg + 2OH^- = Mg(OH)_2 + 2e^-$

Cathodic reaction: $2H_2O + 2e^- = 2OH^- + H_2$

The corrosion rate of the magnesium alloy anode is usually proportional to the operating current density (current per unit area of electrode) [3]. Therefore, for a given power output the life is determined by the mass of the anode. The output voltage depends on the difference between the electrochemical potentials of the materials used to produce the anode and the cathode. This potential can be slightly influenced by the alloying of the anode material. The maximum current output that is available for the cell depends on the electronic structure and surface area of the cathode. A large area of a cathode is required to avoid polarization of the cell.

In this work, we present preliminary experimental results concerning Ni/Al coatings applied as a cathode for seawater activated batteries. These coatings were produced by a plasma spray technique which permits application of sufficiently thick, well bounded coatings of virtually any material that melts without decomposing (including most metals, ceramics and some polymers). The properties of the coating are highly dependent on the temperature of the

* Corresponding author.

0378-7753/98/$19.00
PII S0378-7753(97)02808-5

powder on impact, as well as on its velocity. These parameters are also greatly influenced by the output level of the plasma gun [4].

The production of an active and nanoporous nickel surface by preparing of precursor alloys composed of nickel and a nonnoble metal with mole fraction of nickel lower than 0.5 ($NiAl_3$, Ni_2Al_3, NiSi, $NiSi_2$, Ni_5Zn_{21}) and subsequent leaching has been well known for long time. Raney nickel is prepared in this way, which is widely used as an efficient cathodic electrocatalyst [5].

2. Experimental

The plasma gun used in our experiments was the SG-100 from Miller Thermal. This plasma gun operates at power levels of 100 kW maximum. An output power level (that defines the plasma temperature) of $P = 17.7$ kW was used. The powders were fed to the plasma torch by a Miller Thermal Model 1270 Computerized Powder Hopper. The powder manufactured at Miller Thermal. (Al(AI-1020) and Ni (AI-1023)), was applied for Ni–Al coatings. The diameter of particles was ~ 40 μm. Ni (98%) and Al (2%) powders were mechanically mixed. The coatings were deposited at low-vacuum (~ 1 Pa) to avoid incorporation of air into the flame and oxidation of the powder [6,7], which can sufficiently change the electrochemical properties of the coatings. Mixture of the plasma gases $Ar + N_2$, and corresponding power output were used to achieve complete melting of the powder (in our experiment the flux of argon gas was 24×10^{-3} m^3/s, and nitrogen 2.5×10^{-3} m^3/s). More details of the deposition of the Ni/Al coatings can be found elsewhere [6]. The morphology of the plasma spray coatings was investigated by Scanning Electron Microscope (SEM). The electrochemical properties of the coatings were examined using a galvanic couple magnesium-deposited coating. Electromotive force and short circuit current were measured using NaCl solution (3.0%) as electrolyte (synthetic seawater) in a 200-l vessel where the electrolyte was mechanically mixed to achieve a circulation which was kept constant through all the experiments. Cathode deterioration by calcium carbonate, that takes place in the real seawater, was avoided in the case of NaCl solution. The coatings were chemically treated (etched in the KOH) to enlarge their porosity—to produce extended surface.

The seawater activated test cell consisted of an anode made from a magnesium alloy produced at Marc Metals (Mg doped with Al $< 9\%$, Zn $< 6\%$, Zr $< 1\%$). The thickness of the anode was 1 mm. The thickness of the cathode was 40 μm, (Fig. 1).

Typical electrochemical testing including the output power dependence on operating time, power density dependence on the space between electrodes, and number of cells, dependence of the current density on temperature of electrolyte were evaluated.

3. Results and discussion

Fig. 2 shows SEM photographs of the Ni/Al coatings produced. It is seen that the coatings consist of many droplets that have been torn off the solidified particles and have solidified on the surface (Fig. 2a). The developed

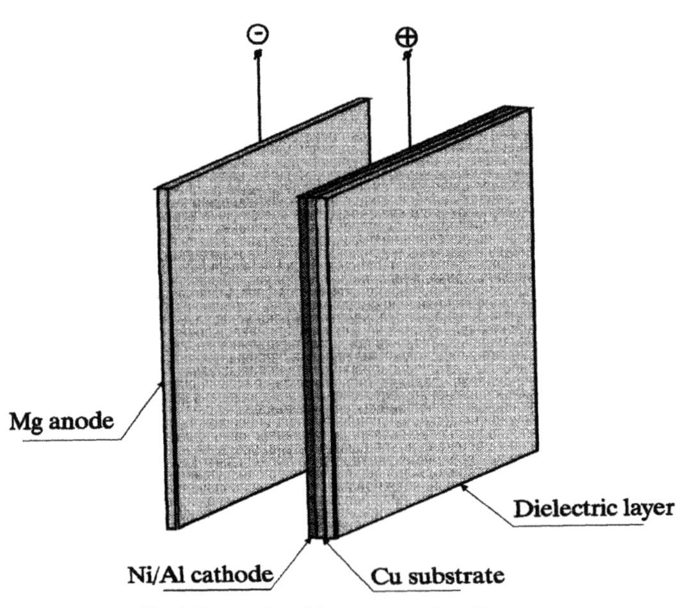

Fig. 1. Construction of the seawater activated battery.

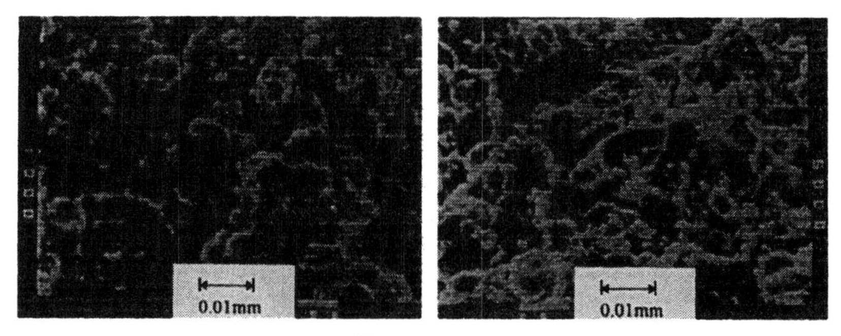

Fig. 2. The SEM photographs of plasma spray deposited coatings: (a) before chemical treatment in KOH solution; (b) after chemical treatment in KOH solution.

surface and large porosity can be seen. The porosity is mainly due to solidification shrinkage.

The electrochemical properties of the coatings were measured using a galvanic couple which consisted of a magnesium anode and the produced Ni/Al coating. Short circuit current density of the galvanic pair was $j = 13.5$ mA/cm^2.

After that coatings were chemically treated in KOH. As one can see in Fig. 2b, the Ni/Al coating is smoother and its porosity is larger than it was before chemical treatment. The short circuit current density of the galvanic couple which consisted of chemically treated Ni/Al coating as a cathode was $j = 45.7$ mA/cm^2.

One of the possible reasons for the increase of short circuit current density could be the enlargement of the surface area of the cathode due to enlargement of porosity by the chemical treatment. During this process Al reacts with KOH.

The composition of the coatings was investigated by Auger Electron Spectroscopy (Fig. 3). Only the Al peak was detected during the investigation of the surface of the Ni/Al coating (Fig. 3a). From the position of the peaks with energies lower than the energies of pure Al, it can be concluded that Al and nonstoichiometric Al oxides dominate the surface of the coating. The Ni peak could be detected after etching by an ion beam (Ar^+, 3 keV, 5×10^{16} cm^2) (Fig. 3b). It is seen that energies of the Ni peaks are lower than for pure Ni which indicates the existence of a mixture of Ni and its compounds with Al and O. So it is evident that during plasma spray deposition,

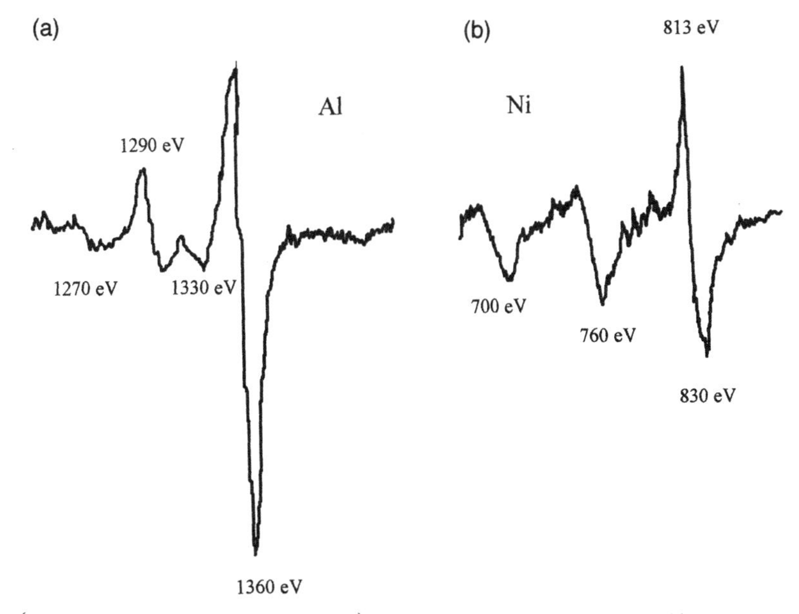

Fig. 3. The Auger spectra (typical energies and peak-to-peak intensities) of the plasma spray deposited coatings: (a) Al peaks of the plasma spray deposited coatings before etching by ion beam (as deposited); (b) Ni peaks of the plasma spray deposited coatings after etching by ion beam (Ar^+, 3 keV, 5×10^{16} cm^{-2}).

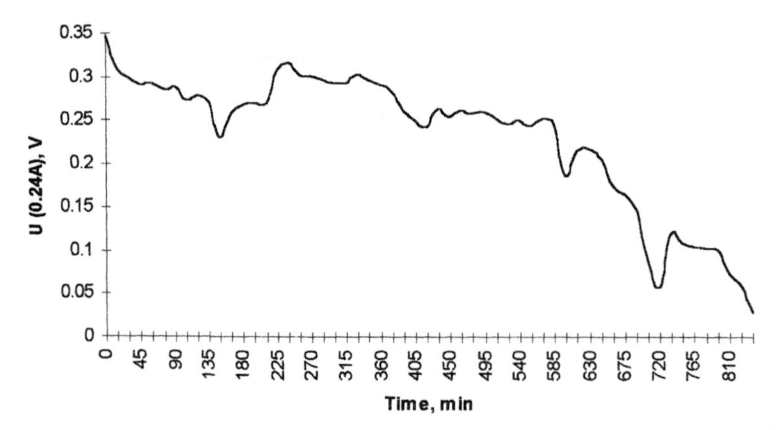

Fig. 4. The voltage–time dependence of galvanic pair Mg–Ni/Al, measured for the constant current $I = 0.24$ A.

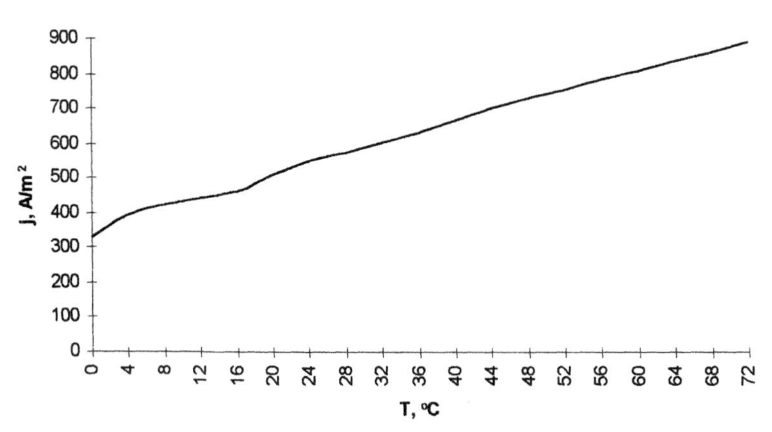

Fig. 5. The dependence of the short connection current density of the seawater battery on temperature of electrolyte.

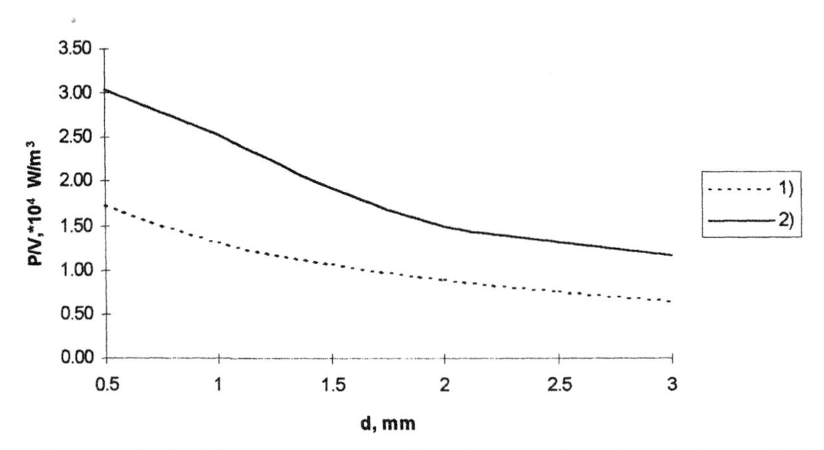

Fig. 6. The dependence of the power density of the seawater activated cells on the space between the electrodes for cells with different area of the electrodes: (1) area of the electrodes, 4.20×10^{-3} m^2; (2) area of the electrodes, 1.75×10^{-3} m^2.

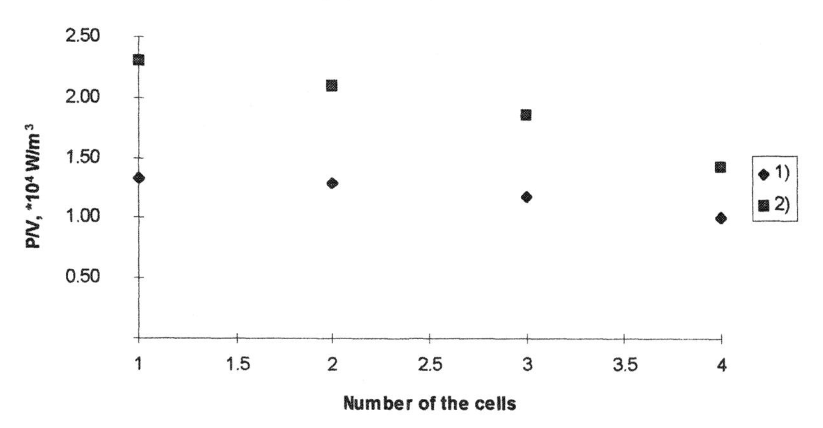

Fig. 7. The dependence of the power density of the seawater activated batteries on the number of cells for batteries with different area of the electrodes: (1) area of the electrodes, 420×10^{-3} m^2; (2) area of the electrodes, 1.75×10^{-3} m^2.

the diffusion of Al to the surface of the coating occurs. The mechanism of this diffusion has yet to be understood completely and needs to be investigated more deeply.

For further examination of the electrochemical properties of the seawater cell, chemically treated Ni/Al coatings were applied as a cathode. Fig. 4 shows the voltage–time ($U_{(t)} = E - Ir(t)$) dependence for the Mg–Ni/Al pair measured for the constant current I, where E is the electromotive force of the galvanic couple (equal to 0.7 V), $r(t)$ is the variable resistance, which was varied electronically to keep the value of I constant. Since the NaCl solution was used as an electrolyte, the deterioration of the cathode by calcium carbonate, which takes place in real seawater [1], was avoided. Masking of the electrodes by magnesium hydroxide, which is formed during the discharge, had no great effect because of its high porosity and poor adhesion to the electrode surface. Probably the $U(t)$ slope mainly depends on the anode erosion rate and it can be changed by applying different surface areas of the cells, i.e., in our experiments the mass of the anode is the only factor that restricts the life time of the seawater cell.

Short circuit current density of this cell was found to be a linear function of the temperature of seawater (Fig. 5).

The power density dependence on the space between electrodes and number of cells was investigated for the cells with different area of the electrodes, (Figs. 6 and 7, respectively). Fig. 6 illustrates that current density or output power for the Mg–Ni/Al pair is limited by the resistance of the electrolyte, i.e., salinity. Temperature will be the parameter that influences available output for the fixed surface area of electrodes. As seen from Fig. 6, for the cell which has the smaller area, power density decreases faster. To increase the output voltage of the galvanic cell connection of many cells in series can be used [6]. With the increased number of cells, the power density decreases because of leakage currents, and this decrease can be 20% or more. In our case, we have found that this connection is efficient enough for smaller areas of electrodes and for a small number of cells. Connection in series of three and even four cells can be used.

In summary, we have demonstrated that plasma spray technology can be used effectively to produce (Ni/Al) cathodes for seawater activated cells. Plasma gun output power is found to be a key technological parameter that defines electrochemical properties (short circuit current, available power density) of the Mg–Ni/Al cell. It is shown that output power for this type of batteries is defined by the erosion rate of the anode and the maximum output power density of 3×10^4 W/m^3 was obtained for these cells.

References

[1] O. Hasvold, Proceedings of the 34th Intern. Power Sources Symp., (1990) 413.
[2] C.L. Opitz, U.S. Pat. 3,401,063 (1968).
[3] E. Buzzelli, J. Jackovitz, J. Lauer, Sea Technol. 32 (1991) 66.
[4] H. Herman, MRS 13 (1988) 60.
[5] S. Rausch, H. Wendt, J. Electrochem. Soc. 143 (1996) 2852.
[6] S. Tamulevičius, R. Dargis, S. Smetona, K. Šlapikas Medžiagotyra, Mater. Sci. 2 (1996) 23.
[7] W. Cai, H. Liu, A. Sickinger, E. Muehlberger, D. Bailey, E.J. Lavernia, J. Thermal Spray Technol. 3 (1994) 135.

ELSEVIER

Journal of Power Sources 72 (1998) 14–19

Interfacial electric properties of beta″-alumina and electrode by AC impedance

Q. Fang [a,*], J-Y. Zhang [b]

[a] *Department of Materials, Imperial College, London SW7 2BP, UK*

[b] *Department of Electronics and Electrical Engineering, University College London, Torrington Place, London WC1E 7JE, UK*

Received 28 July 1997; accepted 14 October 1997

Abstract

In order to enhance cell power density and to study the interfacial electric property between beta″-alumina and an electrode, test cells of Na(l)/beta″-alumina/M, where M = TiN or TiB_2 or Na–Sn or Na–Pb molten alloys as electrode materials, were set up and run within the temperature range of 400°–800°C. The performance of the test cells and the interfacial electric properties were investigated by measuring current–voltage characteristics and AC impedance. The maximum power density of 0.18 W cm^{-2} for TiN and 0.24 W cm^{-2} for TiB_2 could be achieved with a large electrode-area of 30 cm^2 at 800°C. A simplified model and equivalent circuit were given, based on the impedance data. The effect of microstructure of the porous electrode and roughness of the beta″-tube on the cell electric performance and impedance has been studied and discussed. The electron-transport through the porous electrode to the interface of the electrode and the beta″-tube surface is the control step for the electrode reaction, $Na^+ + e \rightarrow Na$, rather than the mass-transport step, for a cell of Na(l)/beta″-alumina/porous thin film electrode. The AC impedance data demonstrated that wetting of the beta″-alumina electrolyte plays an important roll in reducing the cell resistance for the molten Na–Sn or Na–Pb electrode, and the molten alloy electrodes have a smaller cell-resistance, 0.3–0.35 Ω cm^2, at 700°C after 10–20 h. The comparison with sputtered thin, porous film electrodes, showed that the microstructure and thickness of electrode, and the interfacial resistance between electrode and the surface of the beta″-alumina is crucial to enhance cell power density.

Keywords: Beta″-alumina; AC impedance; Interfacial electric properties

1. Introduction

Beta″-alumina has been extensively studied in the last three decades due to its potential applications such as electrochemical sensors, and particularly, as a high energy battery—sodium sulphur battery for electric vehicles [1,2]. Another application for heat–electricity converter is the Alkali Metal Thermoelectric Converter (AMTEC). AMTEC is a system, in which beta″-alumina acts as a solid state electrolyte and sodium is used as working medium, for direct conversion of heat into electricity [3–6]. Its main advantages are high efficiency, high power density, modular design, low maintenance and low manufacturing cost. An AMTEC cell with an efficiency of 19% [7,8], lifetimes to 10,000 h [4,5], and cell power destiny to 1 W cm^{-2} [9] has been reported.

The AMTEC essentially consists of two compartments (Fig. 1): a high-pressure compartment, where sodium is heated by suitable heat source to 500 to 1000°C and a low-pressure compartment, where sodium vapour is condensed at a temperature of about 300°C, corresponding to a saturated vapour pressure of about 10^{-5} bar. In principle, electricity is generated in the AMTEC based on sodium ions transport from the high-pressure to the low-pressure compartment through beta″-alumina electrolyte, while electrons flow through the load, delivering suitable electric power. The driving force for this process is the pressure difference between the two compartments. Separation and recombination of ions and electrons take place at electrodes on both sides of the solid state electrolyte.

The electrodes play a very important role in electrochemical process of the AMTEC. The anode is normally liquid sodium metal itself, and the cathode is often prepared using materials with high melt point, good electric conductivity and high electrochemical corrosion resistance. In order to improve the stability of electric power output

* Corresponding author. Tel.: +44-0171-594-6805; fax: +44-0171-584-3194; e-mail: q.fang@ic.ac.uk.

0378-7753/98/$19.00 © 1998 Elsevier Science S.A. All rights reserved.
PII S0378-7753(97)02773-0

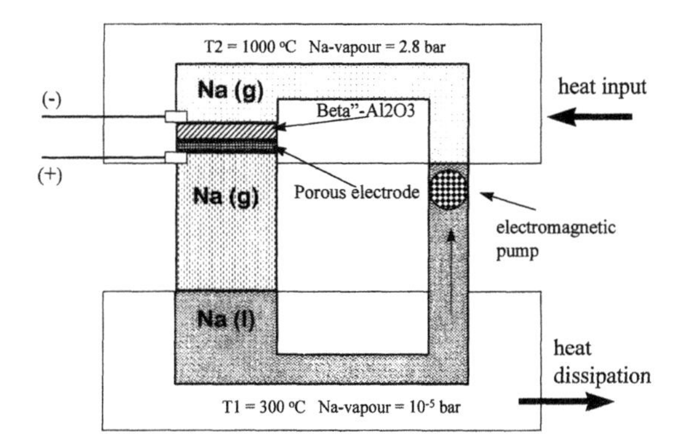

Fig. 1. Schematic diagram of the principle of the AMTEC.

and electrochemistry, the AMTEC study with TiN, TiB_2 thin films as electrodes has been carried out [10–12]. The results from sputtered Mo, TiN or TiB_2 thin-film electrodes indicated that the power density of a AMTEC cell depends not only on thickness, porosity and electric conductivity of the thin film used, but also on the contact between the electrode and the electrolyte, and the connection of the electrode and the electricity collector [13,14]. The contact-interface resistance has a strong influence on the electric power density for the AMTEC cell.

Use of the polycrystalline sodium beta″-alumina ceramic as a solid state electrolyte for the sodium sulphur battery or AMTEC, requires consideration of not only the effects of the interior of the grains and grain boundaries, but also the effects of the electrode and interface between the electrode and the ceramic materials. This work is an attempt to determine the electrical properties of a sodium/beta″-alumina/electrode test cell and the contributions of the electrolyte, electrode and interface between the electrode and the beta″-alumina ceramic by impedance spectroscopy and current–voltage curves. Previous work using molten alloy Na–Sn and Na–Pb as electrode materials in order to study the contact- and interface-impedance of an AMTEC cell has been reported [15]. The aims of this work are mainly to investigate the impedance data and electric properties of different electrode materials, to assess the performance of AMTEC-cells with different electrode materials, to determine the influence of the contact- and interface-impedance between the electrode and the electrolyte and of the electrode itself and to look for a new route to enhance AMTEC-cell power density.

2. Experimental details

2.1. Beta″-alumina ceramic electrolyte tubes

The beta″-alumina tube of 25-mm diameter and 200-mm length, with a 1.3-mm wall thickness were obtained from BAT, ABB, Germany.

2.2. Electrode materials

TiN and TiB_2 were chosen as electrode materials for investigating a sodium gas cell. The TiN and TiB_2 electrode thin films were deposited onto the beta″-alumina tube using a magnetic sputtering technique with a TiN and a TiB_2 target, respectively. The sputtering power was 500 W at an Ar pressure of 8×10^{-3} mbar. The two sputtered electrode thin films show a similar columnar microstructure, which is important for sodium flow through the cathode. The thickness was varied from 0.2 to 10 μm for TiN and from 0.5 to 4 μm for TiB_2, and the area on to which the electrode materials were sputtered was 30 cm^2.

The sodium alloys were prepared by the following two methods: (1) ex situ using pure sodium and tin or lead in a nitrogen atmosphere and; (2) in situ by a coulometric transfer of sodium through the cell into pure molten Sn or Pb metal.

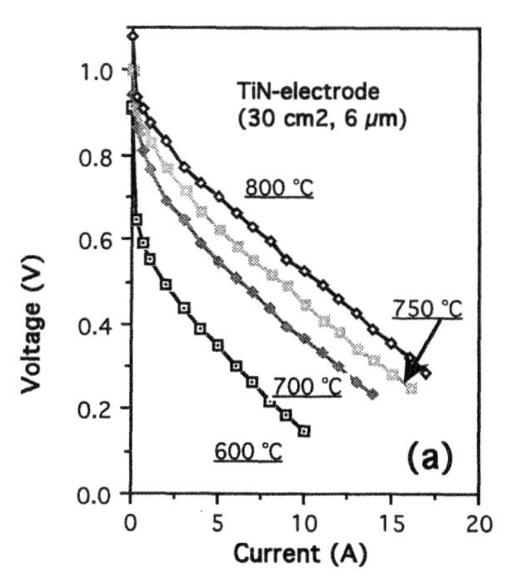

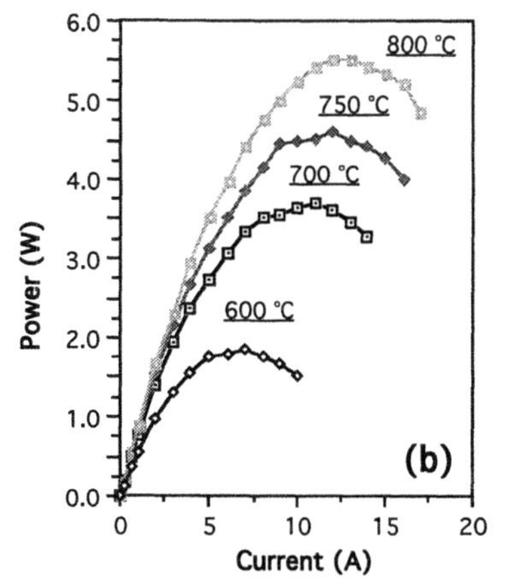

Fig. 2. Current–voltage and current–power curves of TiN electrode at different temperatures.

2.3. Test cells

The tubes with different electrodes were tested in special cell arrangements. The details of the test cells can be found elsewhere [10–15]. Current–voltage curves were measured under steady state condition in a temperature range from 500–800°C. The cell impedance was determined by a commercial impedance spectrometer with potentiostatic control (Zahner Electronic IM 5d, Germany).

3. Results and discussion

3.1. Na(l) / betd″-Al_2O_3 / porous thin film electrode

The current–voltage relationship and the corresponding current–power curves at different temperatures for the TiN electrode are shown in Fig. 2. It can be seen that a maximum power density of 0.18 W cm^{-2} could be achieved with a large electrode-area of 30 cm^2 at 800°C.

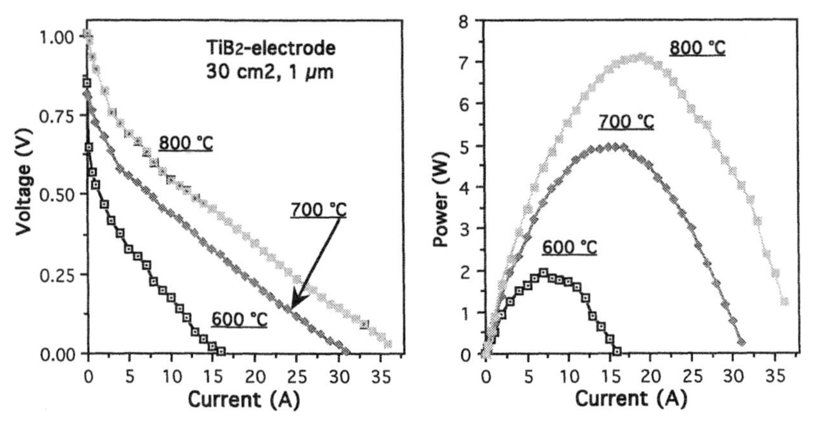

Fig. 3. Current–voltage and current–power curves of TiB_2 electrode at different temperatures.

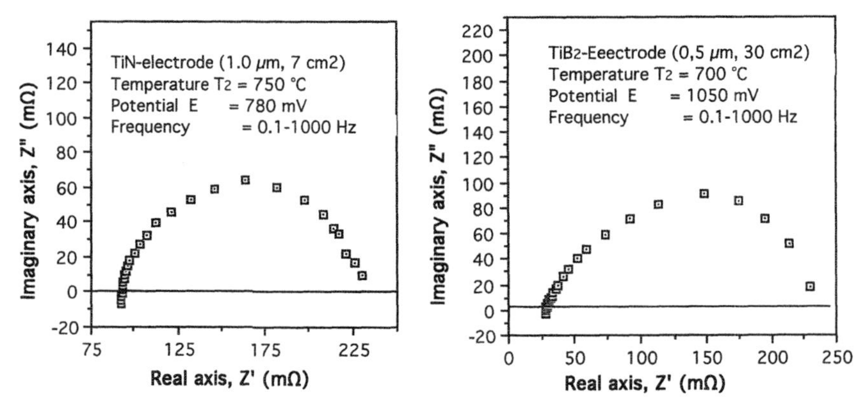

Fig. 4. Typical complex plane plots of impedance data of TiN and TiB_2 electrodes.

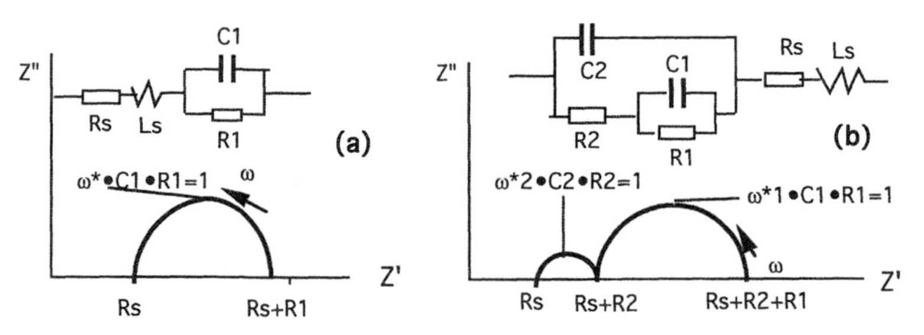

Fig. 5. Equivalent circuit for AMTEC Na–gas cell: (a) four parameters; (b) six parameters.

Fig. 3 shows the current–voltage and the corresponding current–power curves for the TiB_2 electrode at different temperatures. Maximum current density and maximum power density were 1.3 A cm^{-2} and 0.24 W cm^{-2}, respectively.

The impedance measurements were carried out for test cells with different electrode materials. Fig. 4 shows typical complex plane plots of impedance data for TiB_2 and TiN electrodes at open circuit. These complex plane plots have similar features in that the impedance spectrum consists of a flatted half-circle, which can be considered as an RC-loop with a resistance and an inductance (Fig. 5a). This series inductance, which results from the electrode as well as the wrapped Mo-wire used as a current collector, is independent of the measuring potential. According to the impedance measurements of the test cell, an equivalent circuit for the Na–gas cell is shown in Fig. 5b, where L_s is a series inductance, R_s is a series resistance. Sodium ionic charge is considered to pass easily through the grain and the grain boundaries in beta″-alumina, the resistance and dielectric properties of the interface between beta″-alumina and electrode are described by C_2, R_2. The R_1C_1-loop represents the impedance contribution from a porous electrode. As a simplified model of this work, it is convenient to ignore the contribution of the crystal grain interior and the interface between electrode and current collector, but in practice, the contact between electrode and current collector plays an important role to enhance the cell performance [13,14].

Values of interfacial capacitance and both kinetic and mass-transport parameters were obtained from the impedance data of two thin, porous TiN and TiB_2 electrodes and of Na–M melted alloy electrodes over large temperature ranges. Using the simplified model of the equivalent circuit above, impedance data was fit to Bode or complex plane plots. Typical impedance data of TiN and TiB_2 are shown in Fig. 6. Data for TiB_2 are shown in Table 1. The results reveal that the frequency and potential dependence are useful in evaluating detailed models of the

Table 1
Impedance data of TiB_2-electrode (3.9 cm^2; 1 μm; T_2 = 750°C)

Potential (V)	1.043	0.843	0.740	0.710
Voltage (V)	1.043	0.800	0.600	0.500
Current (A cm^{-2})	0.0003	0.0411	0.1328	0.2002
R_s (Ω cm^2)	0.600	0.588	0.582	0.577
L_s (μH cm^2)	1.4	1.4	1.4	1.4
R_1 (Ω cm^2)	1.920	0.504	0.208	0.158
C_1 (mF cm^{-2})	29.00	27.59	19.51	15.82
R_2 (Ω cm^2)	12.195	1.533	0.486	0.337
C_2 (mF cm^{-2})	65.31	82.59	90.31	87.59

properties of the electrode/beta″-ceramics interface and of the porous electrode itself.

The magnetic sputter parameters such as sputtering power, Ar pressure and the morphology of the beta″-alumina surface have a strong influence on the microstructure of the porous electrode [11]. The effect of microstructure of the electrode on the electric performance can be investigated by AC impedance measurements and current–voltage curves. Table 2 shows the measured data for TiB_2 thin electrodes, which were prepared in different Ar-pressures from 5×10^{-3} mbar to 1.6×10^{-2} mbar, the thickness of the TiB_2 electrode films were 0.5 μm. The resistances R_1, R_2, and the capacitance decrease with the decreasing Ar-pressure. Suppose the resistance and the capacitance here are directly stated to the microstructures (grain size, porosity and interface area) of the electrodes and thus, to the interface of the electrode and the beta″alumina-tube surface, then the fact that the poorer cell electric performance comes with an electrode sputtered at a higher sputtering pressure, in which the electrode thin film with higher porosity is deposited, shows that the electron-transport through the porous electrode to the interface of the electrode and the beta″alumina-tube surface is the control step for the electrode reaction, $Na^+ + e \rightarrow Na$, rather than mass-transport step. It has been reported that the maximum thickness of TiB_2 and TiN thin film for

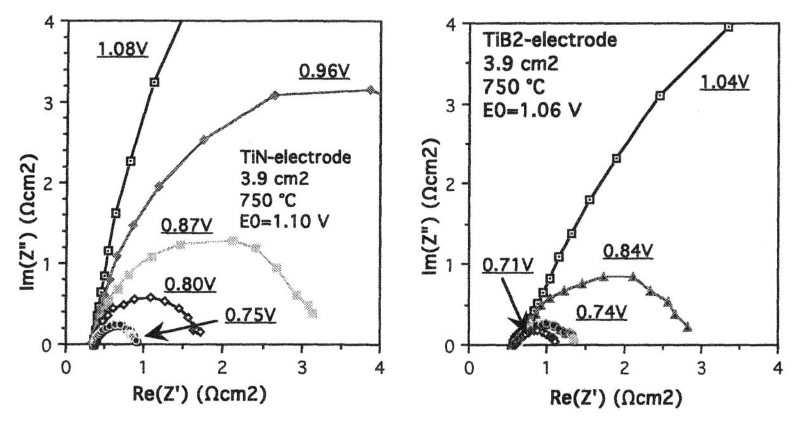

Fig. 6. Complex plane plots of impedance data of TiN and TiB_2 electrodes using different measuring voltages.

Table 2
Impedance data of the test cell with TiB_2-electrode sputtered under different Ar-pressures

Cell number	B-9	B-6	B-8
Ar-pressure (mbar)	1.6×10^{-2}	8×10^{-3}	5×10^{-3}
R_s (Ω cm^2)	0.93	0.87	0.89
L_s (μH cm^2)	3.09	3.29	3.04
R_1 (Ω cm^2)	2.12	1.51	0.84
C_1 (mF cm^{-2})	26.00	11.42	6.11
R_2 (Ω cm^2)	5.85	4.17	2.65
C_2 (mF cm^{-2})	79.50	27.6	22.86
Maximum power density (W cm^{-2})	0.13	0.17	0.18

maximum power output is up to 3.0 μm and 10 μm, respectively [12].

The effect of roughness of beta″alumina-tube on the cell electric performance is shown in Table 3. 'As-received' beta″-alumina (B6) tube has a roughness of about 15 μm. One cell test was also carried out with a TiB_2 thin film (0.5 μm) on a polished beta″alumina-tube. The roughness of the polished beta″alumina-tube (B11) was about 2–3 μm. The result shows that the maximum power output of the polished beta″alumina-tube cell is smaller than that of normal one. Though the polished beta″alumina-tube can reduce the surface resistance (from 36 to 30 Ω cm^{-1}), the decreased interface contact area between electrode and electrolyte leads to bigger resistances R_1, R_2, and smaller capacitances C_1 and C_2, which result in a smaller power output from the test cell.

3.2. *Na(l) / betd″-alumina / Na–M (M = Sn, Pb) cells*

It is worth noting that the wetting of the solid state electrolyte by molten metals (Sn or Pb) or molten alloys is of highest importance. The relatively high impedance of the cell which is measured immediately after set-up of the cell shows that in the beginning the beta″-alumina tube is only poorly wetted. A better wetting of the electrolyte tube and a lower contact-resistance between electrolyte and molten electrode would be obtained by longer heating or by flowing current for a longer time of cell operation. Table 4 shows the wetting effect, in which the cell-impedance measurements were carried out at 700°C after the new cell was set up.

Table 3
Impedance data of the test cell of TiB_2-electrode with different roughness

Number	B-11	B-6
Ar-pressure (mbar) (mbar)	8×10^{-3}	8×10^{-3}
R_s (Ω cm^2)	0.90	0.87
L_s (μH cm^2)	3.02	3.29
R_1 (Ω cm^2)	1.53	1.51
C_1 (mF cm^{-2})	10.67	11.42
R_2 (Ω cm^2)	5.87	4.17
C_2 (mF cm^{-2})	21.14	27.16
$R_{surface}$ (Ω cm^{-1})	30	36
Maximum power density (W cm^{-2})	0.153	0.17

Table 4
Contact-resistance of the cell with molten Na–Sn and Na–Pb electrode by 700°C

Time (min)	5	30	60	120	180	240	300
R_c of Sn (Ω cm^2)	1.10	0.80	0.65	0.58	0.50	0.48	0.48
R_c of Pb (Ω cm^2)	1.20	1.00	0.85	0.76	0.68	0.65	0.64

The electrode process with Na–Me molten electrode is relatively simple, compared with the electrode process in a porous, thin film electrode, which includes interfacial transfer of Na^+ ions, electrochemical reaction and diffusion of Na–gas in the holes of the electrode. The impedance spectrum of a test cell with a molten Na–M electrode is also simple. Since the interface capacitance is very small, the impedance is only an ohmic resistance. This resistance is the sum of the ohmic resistance of the electrolyte and the interfacial charge transfer resistance. The resistance (impedance) depends on the temperature and the current density, but is almost independent of the Na concentration in the Na–M alloy. The cell resistance for various molten alloy electrodes are definitely different at lower temperature but they become practically the same in all cases by 700°C (resistance range 0.3–0.35 Ω cm^2).

The maximum power density and cell-resistance for an AMTEC cell with an Na–M electrode are strongly dependent on the Na-concentration in the Na–M electrode. The maximum power density for 0.5 mol% Na and 700°C for Na–Pb electrode reaches 0.30 W cm^{-2}, while for Na–Sn electrode the maximum power density for 0.5 mol% Na and 700°C is 0.21 W cm^{-2}. The maximum power densities decrease with the increasing Na-concentration. These maximum power densities for Na–Me alloy electrodes are comparable with, or better than, that for the sputtered Mo and TiB_2 electrodes, in which it is just 0.13 W cm^{-2} and 0.17 W cm^{-2} at 700°C, respectively. The fact, that AMTEC with a molten alloy electrode has good power density, is mainly due the lower interface resistance. Cell resistances of 0.75, 0.66 and 0.51 Ω cm^2 for TiN, Mo and TiB_2 sputtered porous electrodes at 800°C have been reported [11,12].

4. Conclusions

Test cells Na(l)/beta″-alumina/M, where M = TiN or TiB_2 with Na–Sn or Na–Pb, as electrode materials, were set up and run with temperature range of 400°–800°C. The performance of the test cell and the interfacial electric properties were investigated by measuring current–voltage characteristics and AC impedance. The maximum power density of 0.18 W cm^{-2} for TiN and 0.24 W cm^{-2} for TiB_2 could be achieved with a large electrode-area of 30 cm^2 at 800°C. A simplified model and equivalent circuit was given, based on the impedance data. The effect of microstructure of the porous electrode, roughness of beta″alumina-tube on the cell electric performance and impedance has been studied and discussed. Electron-transport through the porous electrode to the interface of the electrode and the beta″alumina-tube surface is the control step for the electrode reaction, $Na^{+} + e^{-} \rightarrow Na$, rather than mass-transport step for a cell of Na(l)/beta″-alumina/porous thin film electrode. The AC impedance data demonstrated that wetting of the beta″-alumina electrolyte plays an important roll in reducing the cell resistance for molten Na–Sn or Na–Pb electrodes, and the molten alloy electrodes have a smaller cell-resistance, 0.3–0.35 Ω cm^2 at 700°C after 10–20 h. The comparison with the sputtered thin, porous film electrodes, showed that the microstructure and thickness of the electrode, and the interfacial resistance between the electrode and the surface of beta″-alumina is crucial to enhance cell power density.

Acknowledgements

The authors would like to thank Professor H. Wendt of Tech. Uni. Darmstadt, Drs R. Knodler and F. Harbach of the Research Centre Heidelberg, ABB for their help and valuable discussion.

References

[1] N. Weber, J.T. Kummer, Advances in Energy Conversion Engineering ASME Conference, Floride (1967) p. 913.

[2] D.A.J. Rand, J. Power Sour. 4 (1979) 101.

[3] T.K. Hunt, N. Weber, T. Cole, Research on the Sodium Heat Engine, Proc. 13th Intersoc. Energy Convers. Eng. Conf., 1978, p. 2011.

[4] T. Cole, Thermoelectric energy conversion with solid electrolytes, Science 221 (4614) (1983) 915.

[5] J.T. Kummer, N. Weber, US patent 3,458,356 (1969).

[6] N. Weber, Energy Conver. 14 (1974) 1–8.

[7] C.P. Bankston, T. Cole, S.K. Khanna, A.P. Thakoor, Alkali metal thermoelectric conversion (AMTEC) for space nuclear power systems, in: M.S. El-Gentz, M.D. Hoover (Eds.), Space Nuclear Power Systems 1984, Orbit Book, Malabar, FL, 1985, p. 393.

[8] T.K. Hunt, N. Weber, T. Cole, High efficiency thermoelectric conversion with beta″-alumina electrolytes, the sodium heat engine, in: J.B. Bates, G.C. Farrington (Eds.), Fast Ionic Transport in Solids, North-Holland, Amsterdam, 1981, p. 263.

[9] R.M. Williams, B. Jeffries-Nakamura, M.L. Underwood, B.L. Wheeler, M.E. Loveland, S.J. Kikkert, J.L. Lamb, T. Cole, J.T. Kummer, C.P. Bankston, High power density performance of WPt and WRh electrodes in the alkali metal thermoelectric converter, J. Electrochem. Soc. 136 (1989) 893.

[10] H-P. Bossmann, Q. Fang, R. Knodler, F. Harbach, Test Cells for the Development of the AMTEC 26th IECEC, Vol. 5, Boston, Aug. 1991, pp. 481–486.

[11] Q. Fang, Untersuchungen zur Material und Betriebstechnik des AMTEC, PhD Thesis, Technische Hochschule Darmstadt, 1993.

[12] Q. Fang, Reinhard Knödler, Porous TiB_2-electrodes for the alkali metal thermo–electric converter (AMTEC), J. Mater. Sci. 27 (1992) 6725.

[13] R. Knödler, K. Reiβ, B. Westhoven, J. Mater. Sci. Lett. 11 (1992) 343.

[14] R. Knödler, A. Kranzmann, H-P. Boβmann, F. Harbach, Performance of single cells of an alkali metal thermoelectric converter (AMTEC), J. Electrochem. Soc. 139 (1992) 3030.

[15] Q. Fang, H. Wendt, Performance and thermodynamic properties of Na–Sn and Na–Pb molten alloy electrode for AMTEC, J. Appl. Electrochem. 26 (1996) 43–52.

ELSEVIER

Journal of Power Sources 72 (1998) 20–21

Bipolar plate materials development using Fe-based alloys for solid polymer fuel cells

R. Hornung [a,*], G. Kappelt [b]

[a] *Siemens, Corporate Technology, Paul-Gossen-Str. 100, 91052 Erlangen, Germany*
[b] *Institute for Material Sciences, Department of Corrosion and Surface Techniques, Martensstr. 7, 91058 Erlangen, Germany*

Received 28 July 1997; accepted 14 October 1997

Abstract

Due to its high efficiency and the relatively low working temperature of 80°C, the solid polymer fuel cell (SPFC) is mainly intended for transport applications. In this work the authors present results on bipolar plate materials development using economical Fe-based alloys. The construction materials are exposed to very different potentials by electrochemical contact with the electrodes. Great demands are made on the bipolar plates with regard to the corrosion behaviour and contact resistance. Electrochemical investigations regarding corrosion behaviour showed that, in principle, Fe-based alloys can be employed. Future work will concentrate on reduction of the contact resistance between the construction material and the current collector ensuring a high efficiency of SPFC.

Keywords: Solid polymer fuel cell; Bipolar plate material; Fe-based alloy

1. Introduction

The development of alternative electricity generator technologies, such as fuel cells, has been stimulated by the discussion about the reduction of pollutant and CO_2 emissions. Due to its high efficiency and the relatively low working temperature of 80°C, the solid polymer fuel cell (SPFC) is mainly intended for transport applications. Within the design of the SPFC [1,2], one can distinguish between the functional materials (membrane, Pt-catalyst, and current collector) and the construction materials for the bipolar plates (Fig. 1). At the moment, an Au-plated Ni-based alloy (NiB) is used for the bipolar plates. The Au-coating is applied in order to reduce the contact resistance between the construction material and the current collector.

The general objective of this investigation is to examine the suitability of economical corrosion resistant Fe-based alloys (FeBs) for construction of the bipolar plates, avoiding cost-intensive surface coating. The problem to be solved is that on the surface of corrosion-resistant FeBs a passivating oxide-layer (mainly Cr_2O_3) is generated that does protect from corrosion, but does cause a high contact resistance.

The construction materials are exposed to very different potentials (0–1000 mV/RHE) by the electrochemical contact with the electrodes. Great demands are made on the bipolar plates with regard to the corrosion behaviour and the minimisation of the contact resistance between the

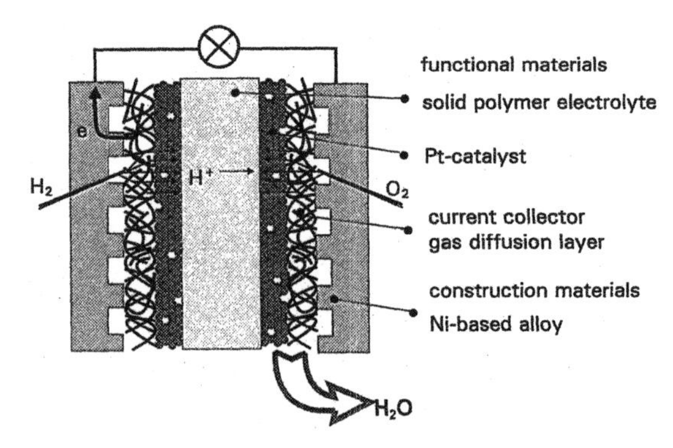

Fig. 1. Design of solid polymer fuel cell.

* Corresponding author. Siemens AG, ZT EN 1, P.O. Box 3220, D-91050 Erlangen. Tel.: +49-91317-32912; fax: +49-91317-31747; e-mail: regina.hornung@erls.siemens.de.

0378-7753/98/$19.00 © 1998 Elsevier Science S.A. All rights reserved.
PII S0378-7753(97)02774-2

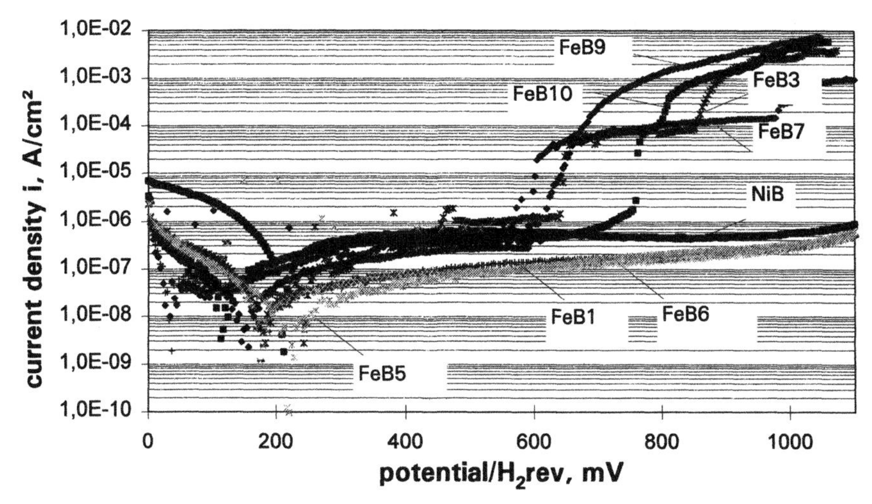

Fig. 2. Polarization curve of Fe-based alloys in HCl (0.1 mol/l).

construction material and the current collector. An increased contact resistance influences the cell performance negatively by decreasing the efficiency of the fuel cell.

2. Experimental work and results

The selection criteria for FeBs were mainly the Cr, Mo and N content in accordance with the pitting resistance equivalent (PRE = %Cr + 3.3% Mo + 30% N), combined with a relatively high Ni content. The selected FeBs were electrochemically investigated in different electrolytes. According to the analysis of the product water of a running SPFC test station, a SPFC model electrolyte was defined containing F^-, Cl^-, NO_3^-, SO_4^{2-} ions (ppm). The polarization curves indicated that most of the FeBs exhibited a characteristic comparable to that of the Ni-based alloy. The measurements in HCl (0.1 mol/l) showed that some FeBs are susceptible to pitting corrosion, because at a potential of 600 mV/RHE the current densities increase gradually by 3–4 orders of magnitude (Fig. 2). The resistant FeBs stand out due to a higher pitting corrosion resistance PRE $\geq$ 25. Long-term tests (load condition 0.3 A/cm^2, 350 h, H_2/O_2 pressure 2 bar, membrane Nafion 117) were performed in 3 cm^2 SPFC cells. Regarding aging, these experiments showed that some FeBs performed comparably with the NiB.

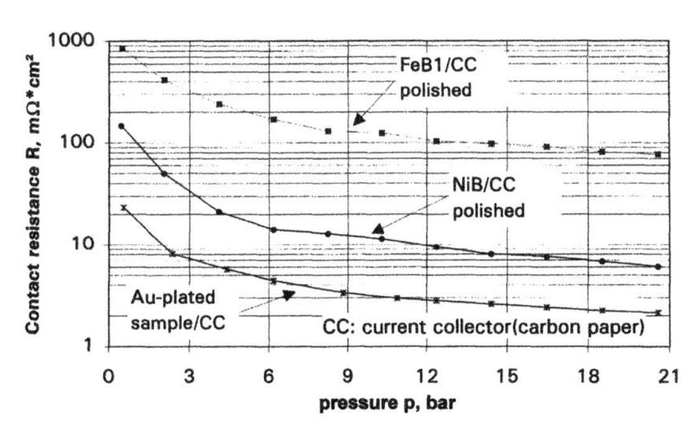

Fig. 3. Pressure dependence of the contact resistance.

The pressure dependence of the contact resistance NiB/current collector is shown in comparison with that of FeB/current collector and Au-coated sample/current collector in Fig. 3. From here it is to be seen that, at present, only the contact resistance of the coated sample is low enough.

In principle, production of low-cost SPFCs could be accomplished using FeBs bipolar plates. Future work will concentrate on economical methods for decreasing the contact resistance FeB/current collector down to an acceptably low value.

References

[1] R.v. Helmolt, R. Hornung, M. Waidhas, Presented at the 12th meeting of Paul Scherrer Institute 'Elektrochemische Energiespeicherung', CH-Villigen, November 1996.

[2] M. Waidhas, A. Datz, U. Gebhardt, R.v. Helmolt, R. Hornung, G. Luft, To be published in the Proceedings of Joint ISE/ECS Meeting, Paris, 31 August–5 September 1997.

Journal of Power Sources 72 (1998) 22–26

Synthesis and characterization of $LiMn_2O_4$ for use in Li-ion batteries

D.I. Siapkas [a], C.L. Mitsas [a,*], I. Samaras [a], T.T. Zorba [a], G. Moumouzias [a], D. Terzidis [a], E. Hatzikraniotis [a], S. Kokkou [a], A. Voulgaropoulos [b], K.M. Paraskevopoulos [a]

[a] *Department of Physics, Aristotle University of Thessaloniki, Thessaloniki 54006, Greece*
[b] *Department of Chemistry, Aristotle University of Thessaloniki, Thessaloniki 54006, Greece*

Received 28 July 1997; revised 1 December 1997

Abstract

Lithiated spinel manganese dioxide was synthesised from electrochemical MnO_2 and Li_2CO_3 with deficiency or excess lithium ($Li_xMn_2O_4$, $0.8 < x < 1.3$) for use in $Li/Li_xMn_2O_4$ and Li-ion cells. Micron-sized $Li_{1.05}Mn_2O_4$ prepared at 730°C showed high Li utilization, excellent cyclability and good rate capability with an initial discharge capacity of 123 mA h/g and 10% discharge capacity reduction after 20 cycles. Different types of commercial carbonaceous materials were also investigated with respect to their electrochemical performance vs. Li. Unoptimised Li-ion cells, using $Li_{1.05}Mn_2O_4$ prepared at 730°C as the cathode material, EC-DMC-$LiPF_6$ electrolyte and carbon fibres, showed promising performance characteristics.

Keywords: $Li_xMn_2O_4$; Li-ion cells; Synthesis

1. Introduction

Since the introduction by Sony of its rechargeable lithium-ion battery in 1990 [1], much research effort has been directed to the development of lithium-ion systems by several research laboratories such as Bellcore and battery manufacturers worldwide in order to meet the demands of reliable, safe, high-energy density and environmentally friendly portable power sources. It seems that in the late 1990s lithium-ion batteries have become the state of the art in portable consumer electronics and are prime candidates for use in powering electric vehicles [2–4]. This is quite understandable since lithium-ion cells, with cathode material one of the high-voltage oxides $LiNiO_2$, $LiCoO_2$, $LiMn_2O_4$ and anode material some form of carbon, have a specific capacity comparable to that of Ni–Cd rechargable batteries, with an energy density almost 3 times as high. Of the three above-mentioned transition metal oxides, $LiMn_2O_4$ shows exceptional promise due to its low cost, low toxicity and excellent voltage profile characteristics. On the other hand, many types of carbon (graphite natural or synthetic, petroleum coke, mesocarbon, carbon black, etc.) have been proposed and used for anode materials with their performance being greatly dependent on their compatibility with the electrolyte system [5]. Although currently available Li-ion systems use some form of disordered carbon, it seems that the future trend is the use of graphite or graphitised carbon, which delivers specific capacities close to the theoretical value of 370 mA h/g [6]. The disadvantage of the highly ordered graphites is that they are more sensitive to the electrolyte solution than the disordered carbons [5]. In addition, different forms of disordered carbon such as carbon fibres or mesocarbon microbeads are also considered mainly due to the good cyclability that they exhibit [2,7,8].

In this work, which is part of a project in collaboration with Germanos Batteries S.A., we present an overview of our attempt to develop a Li-ion system with $Li_xMn_2O_4$ as the cathode material. To this extent, we report on the optimization of synthesised spinel $Li_xMn_2O_4$ with respect to its lithium utilization, cyclability and rate capability by controlling the synthesis conditions and particle size of the prepared material and by correlating its structural, physical and transport properties with its electrochemical performance. In addition, we present the evaluation of the electrochemical performance of different commercial carbonaceous materials as evidenced by their cycling characteristics and comment on their compatibility with several organic solvent electrolyte systems. Finally, we show preliminary results of unoptimised Li-ion systems under in-

* Corresponding author.

0378-7753/98/$19.00
PII S0378-7753(97)02809-7

vestigation, employing graphite and carbon fibre anode electrodes.

2. Experimental

Lithiated spinel manganese dioxide ($Li_xMn_2O_4$, $0.8 \leq x \leq 1.3$) was prepared by solid-state reaction of commercially available electrochemical MnO_2 (EMD) (Tosoh Hellas), widely used for Zn–MnO_2 batteries, with Li_2CO_3 (Aldrich) at temperatures ranging from 450 to 850°C. Having determined the actual manganese content of the EMD to be 58.8% ± 0.5% w/w by energy dispersive spectroscopy (EDS), the appropriate amount of Li_2CO_3 was determined in order to obtain the required excess or deficiency of lithium in $Li_xMn_2O_4$. The Li content of some of the final products was determined by atomic absorption spectroscopy (AAS) and was within 2–5% of the nominal values.

The electrolytes used were $LiPF_6$ 1 M with ethylene carbonate (EC)-dimethyl carbonate (DMC) (2:1 v/v), $LiPF_6$ 1 M with propylene carbonate (PC)-EC-DMC (1:1:2 v/v) and $LiClO_4$ 1 M with EC-diethoxy ethane (DEE) (1:1 v/v). $LiPF_6$ (Aldrich, 99.99 + %) and $LiClO_4$ (Fluka, > 99%) was used as received. PC (Aldrich, 99%), EC (Fluka, > 99%), DEE (Aldrich, 98%) and DMC (Fluka, > 99%) were dried with molecular sieves 4 Å. PC was also distilled under reduced pressure and the middle fraction was collected. All mixtures were made in a glove box, under argon atmosphere, where moisture content was less than 2 ppm.

Five different commercial carbonaceous materials were investigated: (a) Synthetic graphite 1–2 μm (Aldrich), (b) synthetic graphite < 325 mesh (Johnson Matthey), (c) mesocarbon microbeads MCMB-6-28S (Osaka Gas), (d) carbon fibre AGM-98 and synthetic graphite A-625 < 200 mesh (Asbury Mills).

Cathode electrodes were formed by pressing a mixture of the active material with acetylene black and EPDM binder with a 89:10:1 ratio onto aluminium disk current collectors. Anode electrodes were made by mixing the carbon powders with acetylene black and PVDF binder at a 90:5:5 ratio in cyclopentanone. The resulting slurry was deposited onto a copper disk, and the electrodes were pressed. Electrochemical measurements were conducted galvanostatically using an Arbin battery test system and two electrode cells. Cells were assembled in an argon filled glove box.

3. Results and discussion

3.1. Optimised lithiated MnO_2 as cathode material

The X-ray diffraction patterns of $LiMn_2O_4$ prepared at temperatures below 730°C exhibit extra peaks corresponding to a Mn_2O_3 phase present, the proportion of which decreases with increasing firing temperature. The lattice parameter of the sample prepared at 730°C obtained after fitting was 8.232 Å, which is in agreement with that reported in the literature for cubic $LiMn_2O_4$ [9]. Fig. 1 shows the X-ray diffraction patterns of samples prepared with deficiency or excess lithium ($0.8 \leq x \leq 1.3$) at a firing temperature of 700°C. It can be seen that within the range $1.1 < x < 1.25$, the as-prepared material is single-phase spinel $Li_xMn_2O_4$. At other stoichiometries, phases of Mn_2O_3 and Li_2MnO_3 appear as impurities for $x < 1.1$ and $x > 1.25$ respectively, as identified by the corresponding peaks in the XRD patterns of Fig. 1. The Li_2MnO_3 phase, the formation of which has been proposed to be due to a disproportionation reaction of $Li_xMn_2O_4$ [10], is usually identified by either of the two XRD peaks at $2\theta \sim 37°$ or $\sim 45°$ which are of comparable intensity [10,11]. In Fig. 1, it is seen that the former can clearly be identified as a peak whereas the latter is observed as a sideband of the $LiMn_2O_4$ (004) peak. This peak broadening is probably a consequence of the small particle size and poorer crystallinity as compared to other work [12,10] in which the synthesis temperature ranged from 750–900°C. The preparation of single phase lithiated manganese dioxide with excess Li up to $x = 1.15$ and $x = 1.20$ has been reported previously by Tarascon et al. [12] and Gao and Dahn [10], respectively. In the second of these reports, it is determined that the amount of excess Li incorporated into the cubic spinel phase is synthesis temperature and cooling rate-dependent. In particular, lower synthesis temperature and/or cooling rates favour the preparation of single-phase $Li_xMn_2O_4$.

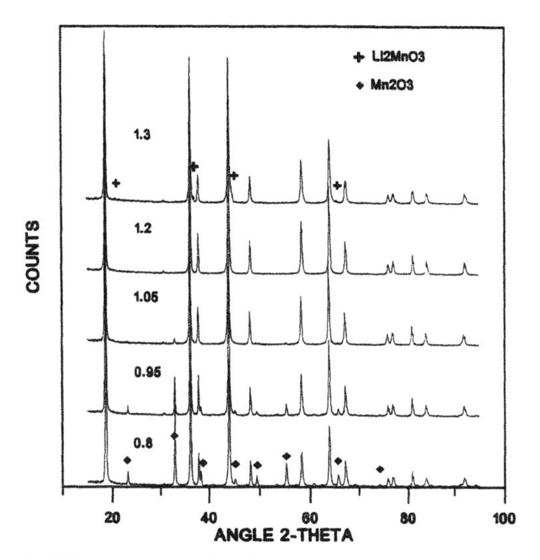

Fig. 1. XRD patterns of $Li_xMn_2O_4$ with various initial Li stoichiometry ($0.8 < x < 1.3$) prepared at 700°C. The peaks corresponding to the Mn_2O_3 and the Li_2MnO_3 phases are marked with ♦ and +, respectively.

FTIR spectroscopy was used in different aspects of the characterization of the as-prepared materials such as the determination of the completion of the solid-state reaction, structural identification of the stoichiometric material ($LiMn_2O_4$) that exhibits 4 bands in the mid- and far infrared spectral regions (620, 500, 350, 250 cm^{-1}) and information on the electrical conductivity from the plasma edge [13]. The dc and ac conductivity of MnO_2 [14] is 1–2 orders of magnitude higher than that of the spinel $LiMn_2O_4$ material, suggestive of hopping conduction via the small polaron in the former [15,16].

Galvanostatic measurements for the $Li/Li_xMn_2O_4$ cells were performed between the limits 3.5–4.3 and 3.5–4.7 V with a charge/discharge rate of C/8. The results presented were obtained for at least four different cells for any of the experiments referred to. The voltage profiles obtained with the 4.7 and 4.3 V upper limits were very similar demonstrating the good oxidative stability of the electrolyte. Fig. 2 shows the first cycle voltage profiles corresponding to cells charged to 4.7 V with cathodes of $LiMn_2O_4$ and $Li_{1.05}Mn_2O_4$, prepared at 730°C, with first discharge capacities of 120 mA h/g and 127 mA h/g respectively corresponding to 0.82 and 0.87 mol of Li per mol of active cathode material. These values are reduced to 115 mA h/g and 123 mA h/g when the cells are cycled between 3.5 and 4.3 V. Furthermore, Fig. 3 shows the cyclability of these two cathode materials. It can be seen that the discharge capacity of the stoichiometric material decreases by about 13% after 20 cycles while that of the $x = 1.05$ material, by about 10%. This is in agreement with previous results showing that addition of excess Li to stoichiometric $LiMn_2O_4$ increases the materials' cyclability [12,17,18].

Spinel $Li_{1.05}Mn_2O_4$ was ground in a vibrating ball mill in order to reduce the particle size and its distribution. Material ground for 9 h showed a tenfold increase in specific surface area with a narrow particle size distribu-

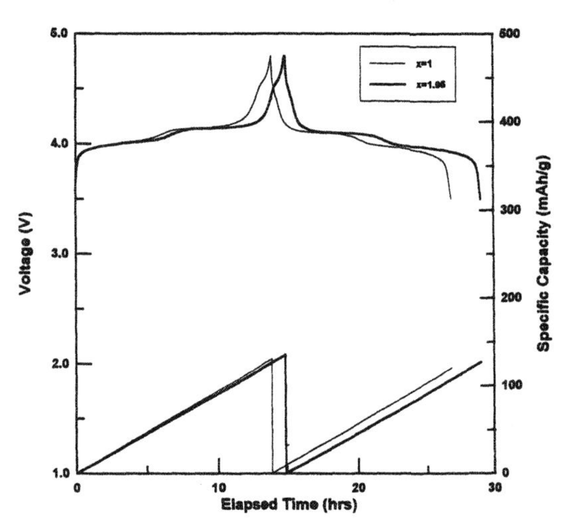

Fig. 2. Typical first cycles between 3.5 and 4.7 V for the $Li/Li_xMn_2O_4$ cells obtained with different Li stoichiometries ($x = 1.0$ and 1.05).

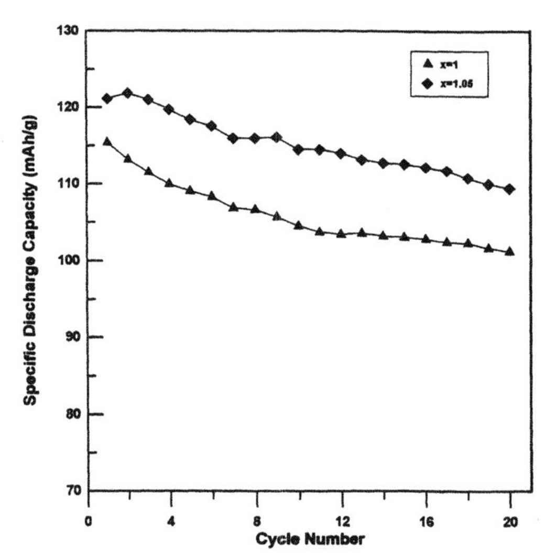

Fig. 3. Specific discharge capacity vs. cycle number for $Li/Li_xMn_2O_4$ cells cycled between 3.5 and 4.3 V at a C/8 rate.

tion [13]. The increase of the discharge current rate from C/8 to 1C decreases the discharge capacity of as-prepared $LiMn_2O_4$ by about 10% (0.78 to 0.70). The same rate increase on the $Li_{1.05}Mn_2O_4$ ground material reduces its discharge capacity by about 5% (0.83 to 0.79). Thus, material prepared with 0.05 Li excess at 730°C with uniform, micron-sized particles shows high Li utilization, excellent cyclability and good rate capability.

3.2. Different carbons as anode materials

Three types of synthetic graphite, mesocarbon microbeads and carbon fibre, all commercially available, were investigated. As confirmed by XRD measurements, the mesocarbon microbead sample was highly graphitised, whereas the carbon fibre was disordered. Electrochemical measurements were performed in galvanostatic mode at a current density of 100 $\mu A/cm^2$ both for charge and discharge. The specific capacity corresponding to the first discharge and the charge:discharge ratio of subsequent cycles of the five different carbon samples examined, is summarised in Table 1. The low ratios of the first charge-to-discharge cycle is related to the initial formation of the 'solid-electrolyte interface' (SEI), as discussed later. In subsequent cycling, the cycling efficiency is significantly enhanced, and as the number of cycles is increased, the efficiency increases approaching unity. At the 20th cycle, the charge:discharge ratio is between 0.95 and 0.98 for all types of carbon.

In Fig. 4, we present the voltage profile corresponding to the initial discharge/charge behaviour of GAL graphite. Initially, the cell voltage drops sharply from the OCV at assembly to about 1.4 V. At this voltage, a small plateau

Table 1
Specific capacity corresponding to the first discharge and charge:discharge ratios for subsequent cycles of the five different samples of commercial carbon investigated

Company	Type	Code	Electrolyte	First specific discharge capacity (mA h/g)	Cycling efficiency				
					1st	2nd	5th	10th	20th
Aldrich	Synthetic Graphite	(GAL)	EC-DMC + $LiPF_6$	398	0.82	0.90	0.95	0.94	0.98
Asbury	Carbon fibres	(CF)	EC-DMC + $LiPF_6$	327	0.59	0.90	0.96	0.98	0.97
Asbury	Synthetic Graphite	(GAS)	EC-DMC + $LiPF_6$	349	0.56	0.84	0.92	0.97	0.97
Osaka	Mesocarbon Microbeads	(MCMB)	EC-DMC + $LiPF_6$	330	0.79	0.80	0.97	0.98	0.98
Osaka	Mesocarbon Microbeads	(MCMB)	PC-EC-DMC + $LiPF_6$	311	0.85	0.87	0.85	0.89	0.95
Johnson Matthey	Carbon Graphite	(GJM)	EC-DMC + $LiPF_6$	380	0.83	0.96	0.82	0.91	0.96

appears, followed by a second one much more pronounced at a voltage of about 0.8 V. These two plateaus are manifested by peaks in the incremental capacity, calculated for the first lithiation (inset of Fig. 4), and are attributed to the formation of the 'SEI'. It is an established fact [6] that the reduction of EC occurs at 0.8 V, suggesting that the latter peak is due to this reaction. By the fact that the former peak has approximately 30 times less capacity than this, it can be assumed that it is due to a secondary passivating reaction most probably to the reduction of the DMC component of the electrolyte system [19]. Both these peaks do not appear in the subsequent charge/discharge cycles and correspond to the irreversible capacity of the first lithiation. The lower voltage region (0.3–0.02 V) corresponds to the intercalation of carbon and is also manifested by a series of peaks in the incremental capacity curve. These peaks are attributed to staging phenomena in the graphitic structure [20] and appear similar both at the first lithiation as well as at the subsequent charge/discharge cycles, but their position shows a dependence on the type of graphitic material.

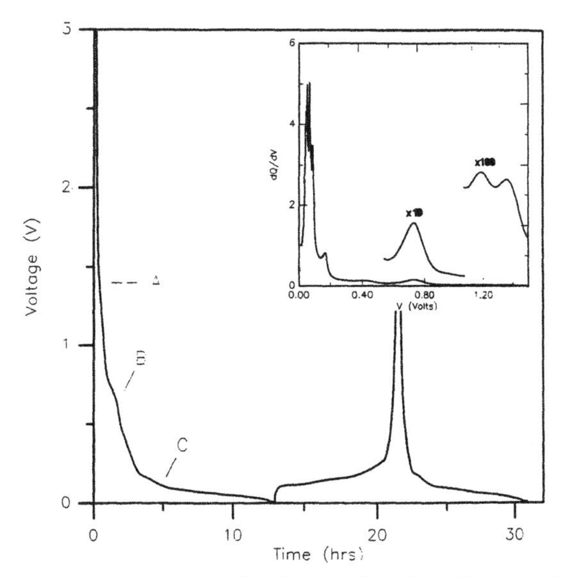

Fig. 4. The first lithiation followed by the first charge/discharge of synthetic graphite (GAL). In the inset is shown the incremental capacity for the first lithiation.

3.3. Li-ion cells

Two of the carbonaceous materials investigated, namely the carbon fibres (CF) and synthetic graphite (GAL), were used in laboratory Li-ion cells incorporating the optimised cathode material and EC-DMC-$LiPF_6$ as electrolyte. Carbon fibres were chosen, since recently it has been pointed out that microfibres as anode materials are expected to increase Li-ion cell efficiency [2].

These preliminary results based on our $C/Li_xMn_2O_4$ system are presented in Fig. 5a,b. The positive and negative electrode masses of these cells were estimated by taking the values that give equal capacities for the first discharge of Li_xC_6 vs. Li to 20 mV and the first charge of $Li_{1.05}Mn_2O_4$ vs. Li to 4.3 V giving mass ratios of 3.3 and 2.7 for graphite and carbon fibres, respectively. The current density used was 0.1 mA/cm^2 and the cells were cycled between 2.3 and 4.3 V exhibiting an average voltage of 3.9 V. The comparison of the performance of the two unoptimised cells clearly shows improved characteristics of the carbon fibre containing cell relative to that having a graphite anode. The latter exhibits specific capacity values significantly lower than expected [21] with very poor stability manifested by a 40% capacity decrease after 20 cycles. Furthermore, the capacity shows a continuously decreasing trend with cycling. Contrary to this, the former cells show a much lower capacity fading after 20 cycles (~ 20%), with approximately half of this reduction occurring between the first two cycles. The reversible specific capacity shows a tendency to stabilise at about 50 mA h/g after 25 cycles which is about the same as the capacity of

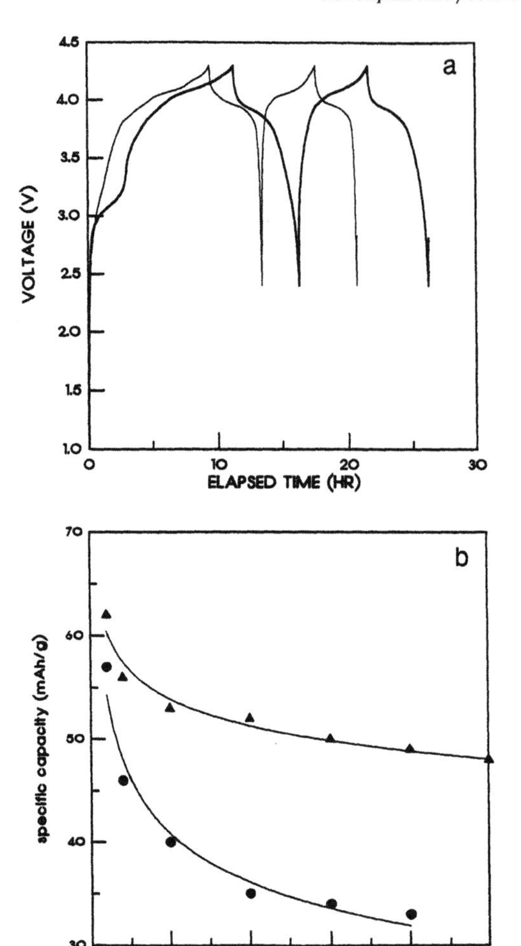

Fig. 5. (a) Voltage profiles of unoptimised $C/Li_xMn_2O_4$ cells. Thick solid line: carbon fibre anode, thin line: synthetic graphite anode. (b) Specific discharge capacity vs. number of cycles of the Li-ion cells shown in (a). Triangles: carbon fibre anode, circles: synthetic graphite anode.

balanced $LiMn_2O_4$/petroleum coke cells with a comparable electrolyte appearing in the literature [22].

4. Conclusions

The characterization of spinel $Li_xMn_2O_4$ cathode material, prepared at different synthesis temperatures, cooling rates and excess lithium as well as post-calcination grinding, resulted in the optimization of the cathode material for use in Li-ion battery applications. Thus, material prepared with 0.05 Li excess at 730°C with uniform, micron-sized particles shows high Li utilization, excellent cyclability and good rate capability. Different carbonaceous materials used show a good cycling efficiency as a result of compatibility with the electrolyte systems used. Further work is underway in order to optimise these carbon materials with respect to their reversible capacity. Preliminary results of unoptimised, with respect to electrode mass balance, $C/Li_xMn_2O_4$ Li-ion cells show that carbon fibres are promising anode materials.

Acknowledgements

This work is part of a project funded by the Greek General Secretariat of Research and Technology, Ministry of Development, under the contract 'ENVIBAT/664'.

References

[1] T. Nagaura, K. Tozawa, Prog. Batteries Solar Cells 9 (1990) 209.
[2] S. Megahed, W. Ebner, J. Power Sources 54 (1995) 155.
[3] K. Brandt, J. Power Sources 54 (1995) 151.
[4] M. Gauthier, A. Belanger, P. Bouchard, B. Kapfe et al., J. Power Sources 54 (1995) 163.
[5] D. Aurbach, B. Markovsky, A. Shechter, Y. Ein-Eli, J. Electrochem. Soc. 143 (1996) 3809.
[6] M.W. Juzkow, S.T. Mayer, 12th Annual Battery Conference on Applications and Advances, Long Beach, CA, January 14–17, 1997.
[7] T. Iijima, K. Suzuki, Y. Matsuda, Synth. Met. 73 (1995) 9.
[8] N. Takami, A. Satoh, M. Hara, T. Oshaki, J. Electrochem. Soc. 142 (1995) 2564.
[9] T. Ohzuku, M. Kitigawa, T. Hirai, J. Electrochem. Soc. 137 (1990) 769.
[10] Y. Gao, J.R. Dahn, J. Electrochem. Soc. 143 (1996) 1783.
[11] M.M. Thackeray, M.F. Mansuetto, D.W. Dees, D.R. Vissers, Mater. Res. Bull. 31 (1996) 133.
[12] J.M. Tarascon, W.R. McKinnon, F. Coowar, T.N. Bowmer, G. Amatucci, D. Guyomard, J. Electrochem. Soc. 141 (1994) 1422.
[13] D.I. Siapkas, I. Samaras, C.L. Mitsas, E. Hatzikraniotis, T. Zorba et al., accepted for presentation at the 1997 Joint International Meeting-192nd Meeting of the Electrochemical Society and the 48th Annual Meeting of the ISE, Paris, Aug. 31–Sept. 5, 1997.
[14] D.I. Siapkas, Ph.D. Dissertation, Univ. of Thessaloniki, 1969.
[15] J.B. Goodenough, A. Manthiram, A.C.W.P. James, P. Strobel, Mater. Res. Soc. Symp. Proc. 135 (1989) 391.
[16] G. Pistoia, D. Zane, Y. Zhang, J. Electrochem. Soc. 142 (1995) 2551.
[17] R.J. Gummow, A. de Kock, M.M. Thackeray, Solid State Ionics 69 (1994) 59.
[18] V. Manev, A. Momchilov, A. Nassalevska, A. Kozawa, J. Power Sources 43 (1993) 551.
[19] Y. Ein-Eli, S.R. Thomas, V.R. Koch, J. Electrochem. Soc. 144 (1997) 1159.
[20] J.R. Dahn, A.K. Sleigh, H. Shi, B.M. Way, W.J. Weydanz, J.M. Reimers, Q. Zhong, U. von Sacken, in: G. Pistoia (Ed.), Lithium Batteries: New Materials, Development and Perspectives, Elsevier, Amsterdam, 1994, pp. 1–47.
[21] D. Guyomard, J.M. Tarascon, Solid State Ionics 69 (1994) 222.
[22] D. Guyomard, J.M. Tarascon, J. Electrochem. Soc. 139 (1992) 937.

ELSEVIER

Journal of Power Sources 72 (1998) 27–31

ZEBRA battery meets USABC goals

Cord-H. Dustmann *

AEG Anglo Batteries, Söflinger Straße 100, 89077 Ulm, Germany

Abstract

In 1990, the California Air Resources Board has established a mandate to introduce electric vehicles in order to improve air quality in Los Angeles and other capitals. The United States Advanced Battery Consortium has been formed by the big car companies, Electric Power Research Institute (EPRI) and the Department of Energy in order to establish the requirements on EV-batteries and to support battery development. The ZEBRA battery system is a candidate to power future electric vehicles. Not only because its energy density is three-fold that of lead acid batteries (50% more than NiMH) but also because of all the other EV requirements such as power density, no maintenance, summer and winter operation, safety, failure tolerance and low cost potential are fulfilled. The electrode material is plain salt and nickel in combination with a ceramic electrolyte. The cell voltage is 2.58 V and the capacity of a standard cell is 32 Ah. Some hundred cells are connected in series and parallel to form a battery with about 300 V OCV. The battery system including battery controller, main circuit-breaker and cooling system is engineered for vehicle integration and ready to be mounted in a vehicle [J. Gaub, A. van Zyl, Mercedes-Benz Electric Vehicles with ZEBRA Batteries, EVS-14, Orlando, FL, Dec. 1997]. The background of these features are described.

Keywords: ZEBRA; Battery; Electric vehicle

1. Introduction

Driven by the unacceptable bad air quality in Los Angeles, the California Air Resources Board (CARB) has generated regulations to reduce emissions from traffic. This is justified because about 70% of the local emissions in the Los Angeles basin are generated by traffic. As a part of this Clean Air Act, zero emission vehicles (ZEV) are to be launched to the market. This is not an easy task, because new technologies for batteries and drive systems have to be developed and brought to production in competition to conventional internal combustion engines. In order to support and focus the necessary battery development, the United States Advanced Battery Consortium (USABC) has set goals that have to be met at least mid-term in order to develop electric vehicles acceptable for the market. This situation has triggered battery development world-wide in order to participate in this large emerging market for electric energy storage devices.

Fifteen years ago, the development of the ZEBRA battery was started [1]. The guiding idea from the very beginning was to achieve high energy density and performance as demonstrated in sodium sulfur batteries but avoiding the safety concerns which are caused by the sulfur content. This lead to the invention of the new electrochemical couple sodium and nickelchloride. The charming background of the system is that neither during production nor during recycling has the metallic sodium have to be handled. The raw materials used for the production of the ZEBRA cells is nickel and salt in combination with a ceramic electrolyte and a molten salt electrolyte. This battery system meets the USABC goals and is being produced in a pilot line (Table 1).

The specific energy density of 85 Wh/kg is without the additional 10 Wh/kg thermal energy available for cab heating. The liquid cooling system allows the withdrawal of up to 8 kW heat available instantaneously for wind screen defrosting. The ragone diagram (Fig. 1) also shows that the unavoidable losses inside the battery are available for heating as well. Thus the ZEBRA battery allows a range of 150 to 200 km independent of climate conditions. The double-walled vacuum insulated battery box in combination with the thermal management is designed for operation at −40°C to +70°C.

The specific power of 150 W/kg at 80% depth of discharge for a two third open circuit voltage at the end of a 30 s pulse is the requirement for a proper acceleration in

* Corresponding author.

0378-7753/98/$19.00
PII S0378-7753(98)00015-9

Table 1
ZEBRA meets USABC criteria

Parameter	Mid-term criteria	ZEBRA Z11
Price	< US$150/kWh	US$300/kWh production start
Cycle life	600 @ 80% DOD	1000 @ 100% DOD
Range @ life (urban miles)	100,000	70,000 demonstrated
Calendar life	5 years	5 years demonstrated, 10 years potential
Power density	250 W/l	256 W/l
Energy density	135 Wh/l	149 Wh/l
Specific power	150 W/kg (200 desired)	150 W/kg[a]
Specific energy	80 Wh/kg (100 desired)	86 Wh/kg[a]
Regenerative specific power	75 W/kg	200 W/kg[a]
End of life (EOL)	20% of rated power and capacity specification	20% of rated power and capacity specification
Operating performance	−30°C to +65°C	−40°C to +70°C
Normal charge	6 h, 20–100% SOC	6 h, 20–100% SOC
High rate charge	< 15 min, 40–80% SOC	36 min, 20%–70% SOC
Efficiency at EOL	75%	> 80%
Off-tether pack energy loss	Thermal loss < 3.2 W/kWh (< 15% in 48 h); Self discharge < 15%/48 h	Thermal loss 5.5 W/kWh compensated by operation; Self discharge zero

[a]System weight including BMS, IFB, cooling.

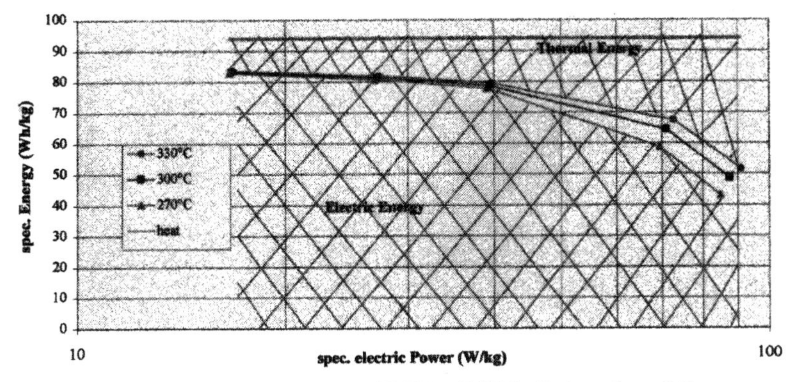

Fig. 1. Ragone diagram for ZEBRA batteries Z5, Z11, with ML3 cells thermally available energy.

normal traffic. By the redesign of the ceramic electrolyte shape and some other detailed work, this power is available nearly over all the battery capacity (Fig. 2) [2,3].

The cycle life of batteries is most important for the vehicle economy because the replacement of single modules or complete battery packs is always an expensive exercise. The ZEBRA battery has the important advantage in it's being maintenance-free for life because in the case of single cell failures no maintenance or anything else is necessary. The reason for this is the low resistant cell failure mode which means that in case of a cell failure the internal resistance of the failed cells is very low. In case of a crack in the ceramic, the liquid electrolyte reacts with the sodium to form aluminum and salt. This aluminum shortens the cell so that the ZEBRA cell has a looping element by its chemistry (Fig. 3). Therefore, up to 5% of failed cells can remain in the battery untouched before the battery would have to be taken out of service.

The performance of the ZEBRA battery is independent of the ambient temperature. All materials and the electronic parts of the battery system are designed for a temperature range of −40°C to +70°C. Besides the thermal energy storage capability, this is one of the main advantages of the hot battery because of its thermal management.

The ZEBRA battery has a very high level of safety incorporated in its features.

1.1. Chemical system

The chemical system has a built-in feature to reduce the heat burden in case of damage. The reaction of the liquid

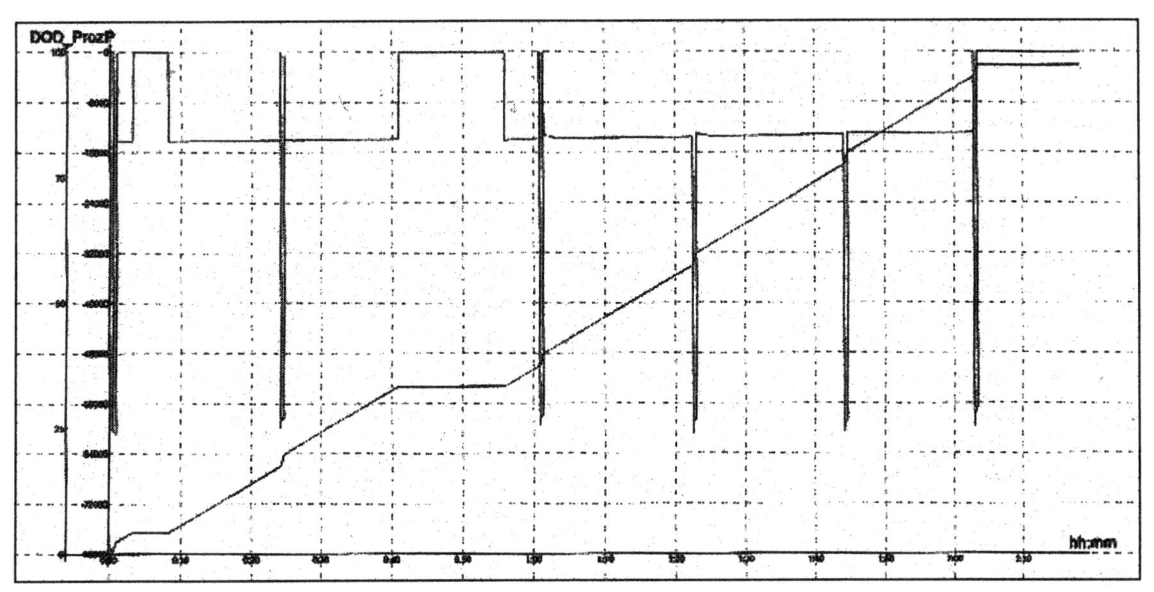

Fig. 2. Peak power independent of DOD.

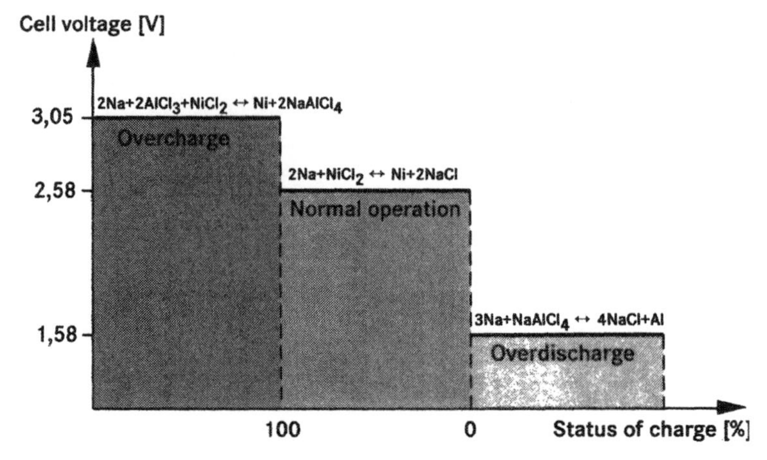

Fig. 3. ZEBRA cell reactions at 300°C.

electrolyte with the liquid sodium produces only two thirds of the energy of the main storage reaction, and the reaction products (salt, aluminum) passivate the cathode to suppress further normal discharge reactions (Fig. 4).

1.2. Three-fold steel enclosure

All reactants are encapsulated in the steel-made cell cans with welded tetrachlorobiphenyl (TCB) seals. The

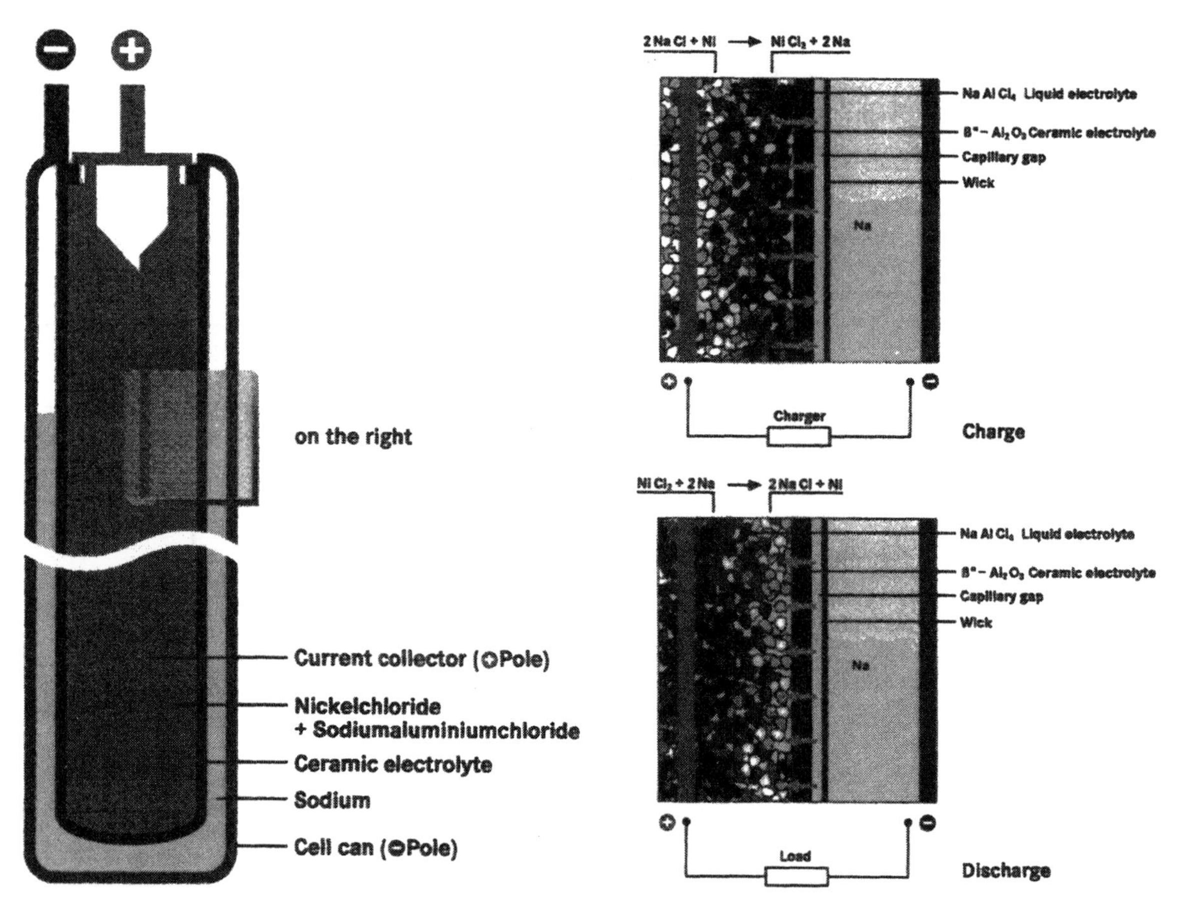

Fig. 4. ZEBRA cell design.

cells are ceramic coated for melt resistant electric insulation and are encapsulated in the double walled stainless steel battery box with inside high temperature resistant thermal insulation. This box encloses the reactants even at extreme conditions of 1000°C inside.

1.3. Internal short protection

In case of an internal short, a thermal runaway is prevented by the coated ceramic cell insulation, the partitions constraining any effects to battery segments and the built-in sump which absorbs melt in case of cell opening. The insulation resistance is permanently monitored by the BMS.

1.4. External short protection

The main circuit breaker is an integral part of the battery system and protects it from damage from an external short. It is operated by the BMS. In case of a vehicle crash, the crash sensor signal is used to disconnect the battery.

1.5. Crash and abuse tested

The qualification of each ZEBRA battery type includes on pole drop testing with 50 km/h vibration test for 24 h according to USABC specification, freeze/thaw cycling, overheating, overcharging and short circuit testing. In all tests the pass criteria is that the battery box remain closed except when it is mechanically opened from outside and no toxic components are emitted.

The ZEBRA battery passed all tests.

1.6. Thermal shut off

In contrast to ambient temperature battery systems, the ZEBRA battery can be passivated totally by cool down.

2. System design

The ZEBRA battery was developed and designed from the very beginning as a 'ready to mount' battery system with well defined interfaces to the car. The system includes the battery box with internal heating and cooling devices, the two pole main circuit breaker which is integrated in the interface box, the battery controller and the cooling box which contains the oil/water heat exchanger, the circulation pump and an expansion vessel for the cooling liquid. A logistic system was developed in order to supply the 'ready to mount' battery system to the production line of the cars.

3. Recycling

The ZEBRA battery is also being prepared for the recycling of worn out batteries. First investigations indicated that the recycling cost can be covered by the sale of the nickel inventory. The cost situation seems to be neutral.

4. Conclusion

The ZEBRA battery system is an advanced battery being tested very extensively with more than 1.5 million km on public roads.

The next important step now is to start series production which is solely dependent on the market.

References

[1] A.R. Tilley, R.J. Wedlake, ZEBRA—the sodium/metal chloride ceramic battery, Electric Vehicle Dev. 6 (4) (1987).

[2] C.-H. Dustmann, Power improved ZEBRA battery for zero emission vehicles, SAE Technical papers series 960445.

[3] A.R. Tilley, R.N. Bull, The ZEBRA electric vehicle battery—recent Advances, AutoFeds 97, Birmingham.

Journal of Power Sources 72 (1998) 32–36

Alkaline batteries for hybrid and electric vehicles

F. Haschka *, W. Warthmann, G. Benczúr-Ürmössy

DAUG—Deutsche Automobilgesellschaft, Emil-Kessler-Str. 5, D-73733 Esslingen, Germany

Abstract

Forced by the USABC PNGV Program and the EZEV regulation in California, the development of hybrid vehicles become more strong. Hybrids offer flexible and unrestricted mobility, as well as pollution-free driving mode in the city. To achieve these requirements, high-power storage systems are demanded fulfilled by alkaline batteries (e.g., nickel/cadmium, nickel/metal hydride). DAUG has developed nickel/cadmium- and nickel/metal hydride cells in Fibre Technology of different performance types (up to 700 W/kg peak power) and proved in electric vehicles of different projects. A special bipolar cell design will meet even extreme high power requirements with more than 1000 W/kg peak power. The cells make use of the Recom design ensuring high power charge ability at low internal gas pressure. The paper presents laboratory test results of cells and batteries.

Keywords: Alkaline batteries; Hybrid vehicles; Electric vehicles

1. Introduction

The compulsion to reduce noise and air pollution in the cities produced by combustion engine-driven cars, and to preserve individual and business mobility and flexibility, will force the development of hybrid cars. The hybrid system offers the advantage that the combustion engine can work with high efficiency in an optimal environmentally beneficial driving mode, and that the surplus of energy can be used for charging an energy storage system.

As energy storage systems, super capacitors, fly wheels and electrochemical storage systems are in competition. The advantage of the electrochemical storage systems are that the batteries are available today, while fly wheels and super capacitors are under development at the moment, and will be available in the mid-term or long-term future. Especially, the alkaline battery systems fulfil, besides the economic and environmental criteria, the technical requirements of high power input and output during charge and discharge, which are necessary for hybrid applications.

2. Recom cells in fibre technology

DAUG is involved in the development of all major alkaline storage systems especially nickel/cadmium-, nickel/metal hydride- and nickel/zinc- cells of different size and performance types. The cell electrodes are made from plaques of nickel-covered plastic fibres (Fig. 1) filled by a special process with active material. The fibre plaque electrodes provide mechanical strength and a certain degree of flexibility to compensate volume change of the active material during cycling. The fibre plate is a highly porous material of good electric conductivity.

Fig. 1. Fibre plaque.

* Corresponding author. Tel.: +49-711-17-66019; fax.: +49-711-17-66034.

0378-7753/98/$19.00
PII S0378-7753(98)00014-7

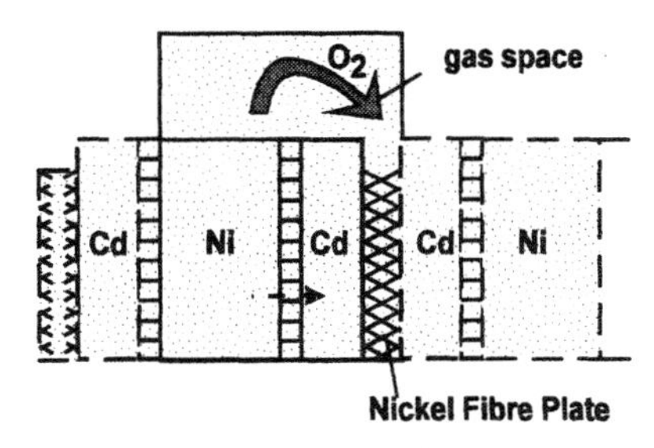

Fig. 2. Recom principle.

The manufactured cells are sealed, maintenance-free and of rectangular (prismatic) shape. The cells make use of the RECOM principle. The new and unique design is characterised by the use of a porous nickel fibre plaque 'sandwiched' between negative fibre electrodes (Fig. 2).

This composite negative electrode arrangement enhances the consumption of oxygen during recharge, so that there is a low pressure (vacuum) inside the cell even at the end of charge when heavy gassing would otherwise take place.

3. Nickel / cadmium system

The nickel/cadmium cells, manufactured in a pilot plant, are the most advanced system, and are now installed in about 100 EVs and hybrid EVs.

There are even buses powered by FNC-Recom batteries running in a normal time schedule for public transportation. These buses have been on the road for more than 18 months. Their batteries were recharged twice or three times in less than 50 min daily.

At Rügen island, the fast recharge ability of FNC-batteries has been demonstrated, which enables a daily cruising range of about 350 km. In fact, the fast charge ability is one of the most important properties of alkaline batteries to make them the battery of choice in hybrid cars.

For all the cars, DAUG is delivering the complete battery system including thermal and electrical battery management (Fig. 3). Battery data—voltage, current, temperature, accumulated charged and discharged capacity and energy, as well as the state of charge—are monitored on a display, and abuse conditions for the battery will be made visible.

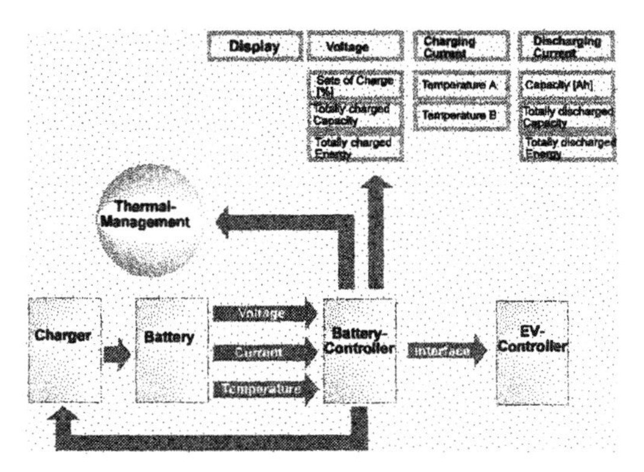

Fig. 3. Electrical and thermal battery management system.

The nickel/cadmium cells can be substituted easily by the nickel/metal hydride system, because the cell voltage is rather the same. In addition, the change from the cadmium electrode to the nickel metal hydride storage system will reduce the environmental problems.

4. Nickel / metal hydride system

Nickel/metal hydride cells have some advantages in comparison to nickel/cadmium cells. The energy density (W h/kg; W h/l) is considerably larger (30–40%). The electrochemical reaction during charge and discharge do not consume or produce water; therefore, the electrolyte concentration and composition do not change during the whole cycle. The cell reactions may be discussed as an 'ion shuttle' system with similarities to the advanced lithium ion cells (Fig. 4).

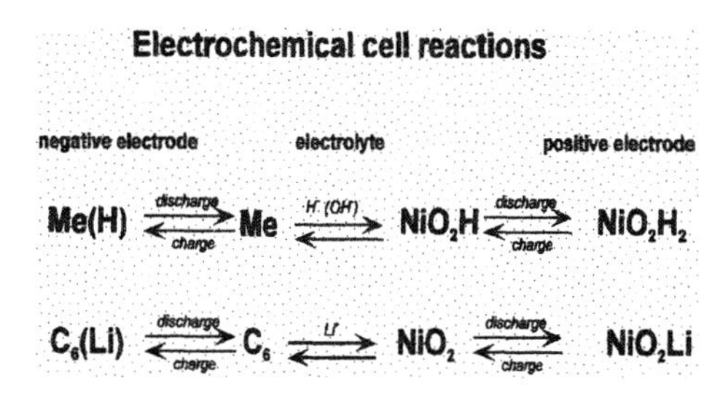

Fig. 4. Comparison of electrochemical cell reactions of lithium- and hydrogen-ion shuttle system.

The electrode processes are expected to proceed without hindrance, polarisation of the electrodes remain small, and a low internal resistance of the cells is assumed.

4.1. Characteristics of nickel / metal hydride cells

The cycle life of the nickel/metal hydride system is comparable to that of nickel/cadmium (Fig. 5).

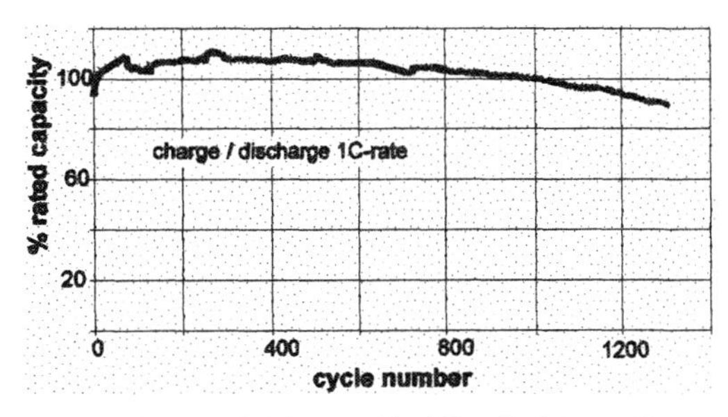

Fig. 5. Capacity of nickel/metal hydride cells during cycling.

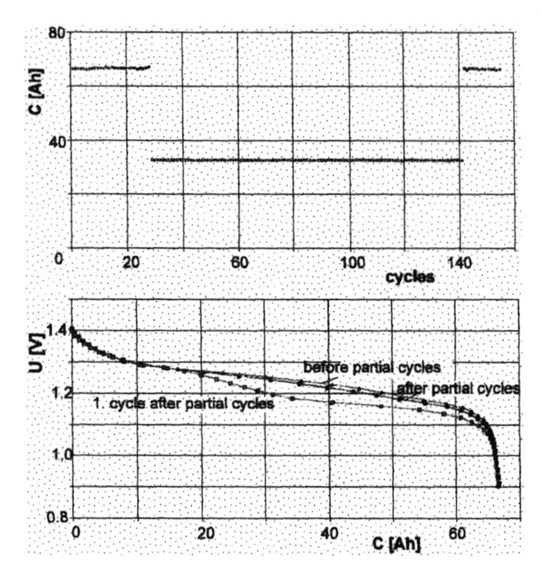

Fig. 6. Partially discharged nickel/metal hydride cells.

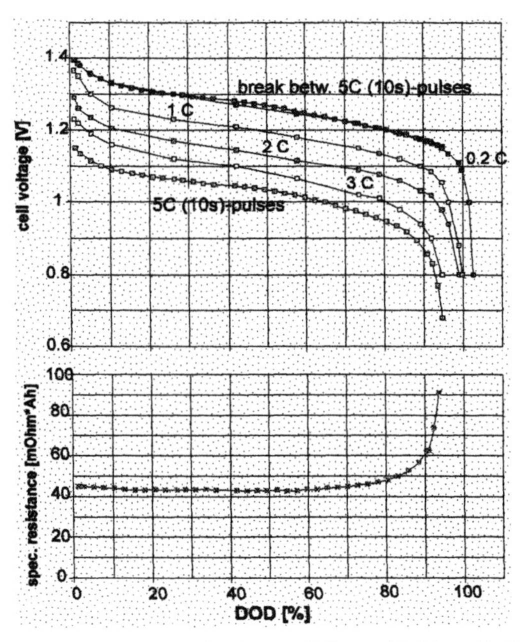

Fig. 7. Discharge curves and resistance of X-type nickel/metal hydride cells.

Experiments have shown that nickel/metal hydride cells do not have a capacity loss (memory effect) when they were partially discharged for several cycles (Fig. 6).

Typical discharge curves of a nickel/metal hydride X-type cell is shown in Fig. 7. The internal resistance determined by a pulse measurement remains approximately constant during the discharge, and is raised only at the end of discharge.

All alkaline systems will be produced at DAUG in the same cell case sizes, giving the advantage that the same thermal management can be used for all couples. There are three series of cell types, the H-type designed for pure EVs, the X-type for hybrid EVs (HEV) and additional the XX-type cells.

The XX-type cells will meet even extreme high power requirements for special HEV applications, e.g., demanded in the USABC/PNGV program.

In Table 1, the available cell sizes ($H \times W = 170 \times 115$ mm) and their characteristics have been compiled. Except the XX 11 A h ($H \times W \times T = 110 \times 59 \times 26.5$ mm) and the XX 22 A h ($H \times W \times T = 170 \times 59 \times 26.5$ mm) couples the cells differ only in their thickness (22 mm to 64 mm), which determines the cell capacity.

The cells can be used in a series of single cells or couples of two parallel cells allowing to build any required A h-capacity form 11 A h up to 280 A h.

The power available from a cell depends on the performance type (H-, X- or XX-type) and the state of discharge (% DOD) (Fig. 8).

Table 1
Cell dimensions and characteristics

	Dimensions									
Thickness (mm)	26.5	26.5	22	28	34	40	46	52	58	64
Width (mm)	59	59	115	115	115	115	115	115	115	115
Height (mm)	110	170	170	170	170	170	170	170	170	170
Volume (!)	0.17	0.27	0.43	0.55	0.68	0.78	0.9	1.02	1.13	1.25
	H-type cells for EVs									
Capacity (A h)	45	55	70	85	100	115	125	140		
Weight (kg)	1	1.15	1.4	1.7	2	2.3	2.6	3		
	X-type cells for HEVs									
Capacity (A h)	40	52	65	75	85	100	115	126		
Weight (kg)	1.06	1.25	1.55	1.8	2	2.4	2.75	3		
	XX-type cells for special HEVs									
Capacity (A h)	11	22	36							
Weight (kg)	0.38	0.72	1.15							

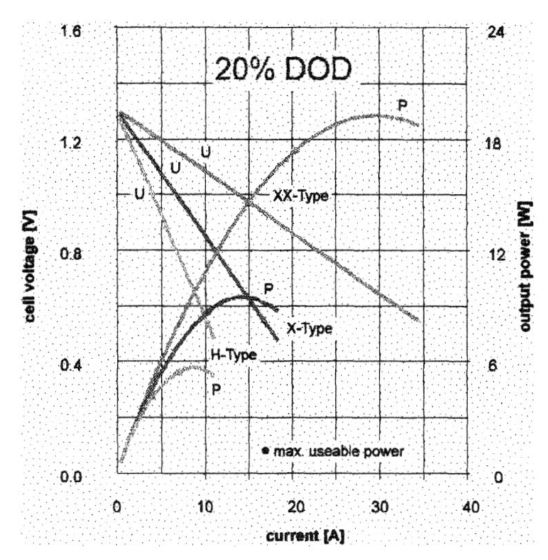

Fig. 8. Power output of nickel/metal hydride cells standardised to 1 Ah.

At a given current, the cell or battery voltage becomes reduced. The power output is a function of current, voltage and internal resistance. For practical applications, the discharge voltage should not drop down to lower values than 2/3 of the nominal voltage resulting in about 0.85 V for nickel/metal hydride cells, because the ratio of power output and heat generation becomes more and more unfavourable. The power output under this conditions for various cell-types as a function of depth of discharge is shown in Fig. 9.

4.2. Characteristics of nickel/metal hydride batteries

An example of results obtained from a complete X-type battery system is given in Fig. 10. It illustrates that large numbers of cells show the same behaviour as an individual cell, and prove that the battery and thermal management is well adapted to the needs of traction batteries.

Fig. 11 shows a sequence of current and voltage profile for a 4 kW h/11 A h XX-type battery (Fig. 12) during

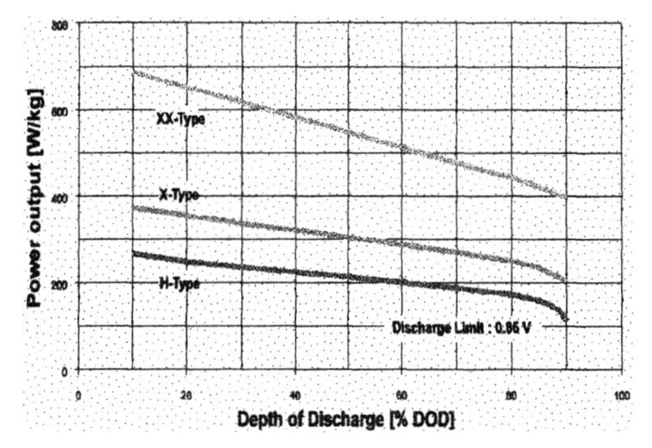

Fig. 9. Maximum power output of nickel/metal hydride cells at different state of charge.

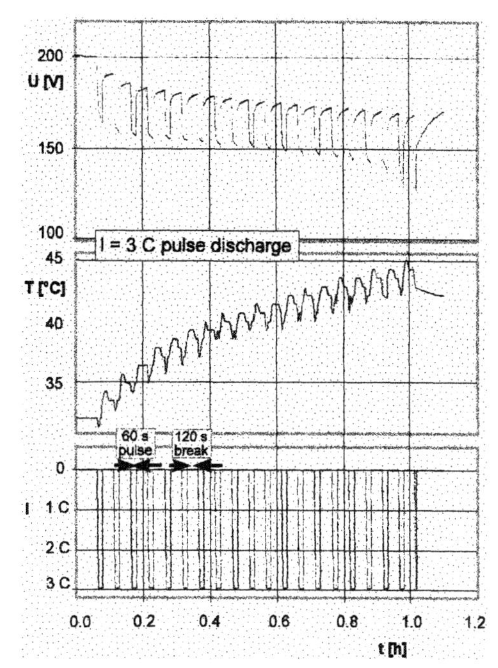

Fig. 10. Pulse discharge of a complete battery.

cycling. The battery consisting of 280 single cells was discharged with 40 kW from 90% state of charge (SOC) to 40% SOC and recharged with 20 kW.

During a 45-kW pulse-discharge the current rises up to 200 A at the end when the battery is fully discharged (Fig. 13).

5. Further developments

There are several ways to optimise a battery system. The choice of active electrode materials, cell design, thermal and electric management units influences the system performance. In addition, the improvement of the principal design of cells can help adapt the battery to the needs of car manufacturers.

DAUG is developing a new concept for bipolar nickel/metal hydride batteries (Fig. 14). The research

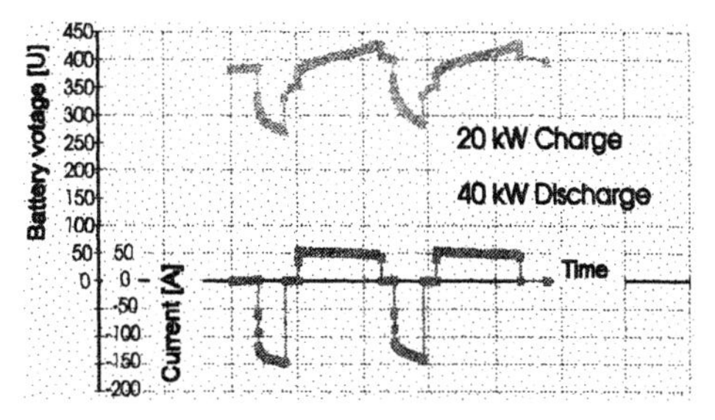

Fig. 11. Current and voltage profile during cycling.

Fig. 12. Four kW h XX-type high power battery.

activities are based on the current achievements in the monopolar cell design, and take full advantage of the features of the nickel/metal hydride system. In addition to the electro-chemical H-shuttle concept realised in such cells, there are additional measures for heat and hydrogen transfer that will enable an easier control of the battery devices.

The main advantages are expected in terms of energy density, lower costs, easier handling and better balance between individual sub-cells. It is expected that the resulting cells will have a considerably increased power density —even much higher than conventional XX-type cells—at larger energy density (Fig. 15).

Laboratory tests of bipolar nickel/metal hydride cells are very encouraging. The results confirm that the power output of more than 1000 W/kg and the energy density of app. Forty-five Wh/kg (Fig. 15) will be achieved.

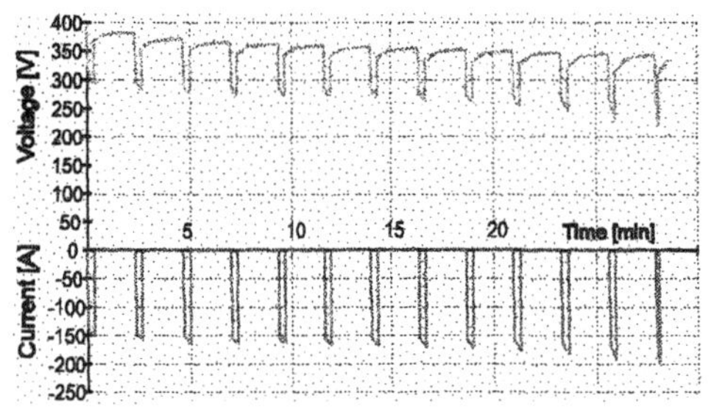

Fig. 13. Forty-five kW pulse discharge (20 s).

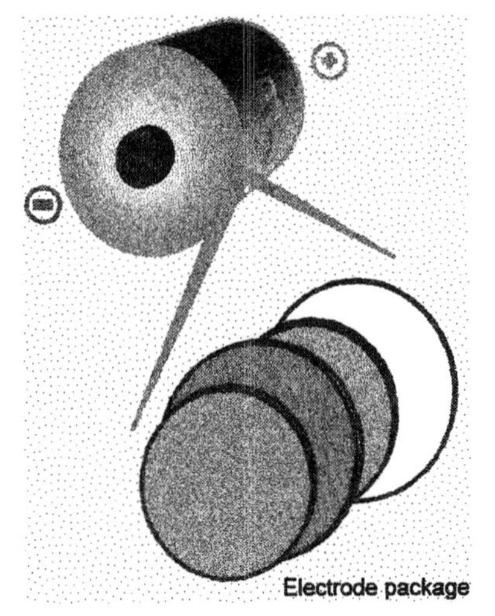

Fig. 14. Tubular bipolar battery design.

The power output of the bipolar battery system is similar, or higher than that of a super capacitor. The high energy density gives the battery system more advantage for hybrid applications than a super capacitor.

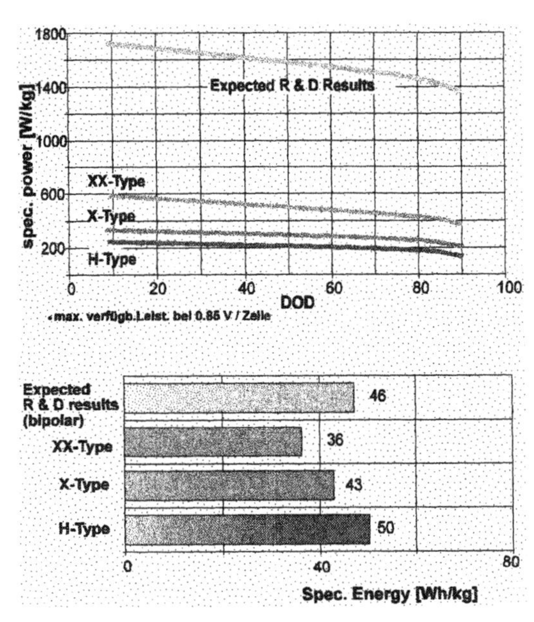

Fig. 15. Power output and energy density of new bipolar cell design.

ELSEVIER

Journal of Power Sources 72 (1998) 37–42

Real-life EV battery cycling on the test bench

Wolfgang Bögel *, Jean Pierre Büchel, Hélène Katz

RENAULT, Direction de la Recherche, 14 av. A. Einstein, Z.I. de Trappes Elancourt, 78190 Trappes, France

Received 26 August 1997; revised 23 September 1997

Abstract

When choosing a battery system for EV applications, there are many parameters we have to take into consideration: technical parameters, operational experience, economic factors, material availability, and the environment (A. Pellerin, Elec. Hybrid Technol. 96, 68–75). In this paper, we want to concentrate on one parameter only: battery life. The lifetime of traction batteries can be expressed either in terms of number of cycles or in terms of calendar life. Both are important and they strongly depend on the mission profile of the vehicle. A lot of work has been concentrating on bench testing to study the cycle life of traction batteries using standard cycle profiles such as ECE15 or DST. We have learned, however, that the results obtained show significant differences to the results obtained with vehicle trials. After a short introduction talking about the importance of the on-board energy, the EV mission profile and the time factor, we discuss the parameters influencing cycle life: rest periods, ambient temperature, depth of discharge (DOD) or peak power demand, etc. In the second part of the paper, we present a 'complex four-season cycle' integrating the previously mentioned parameters to approach real-life vehicle conditions. © 1998 Published by Elsevier Science S.A. All rights reserved.

Keywords: EV; Traction battery; Test procedure; Cycle life

1. Introduction

1.1. Importance of on-board energy

The objectives for different parameters of traction batteries are specified by organizations such as the United States Advanced Battery Consortium (USABC) [1], quoting target values for cycle life of 600 cycles for the midterm and 1000 cycles for the long-term.

The cycle life necessary for a vehicle depends on the on-board energy: a vehicle equipped with a battery of 15 kW h (useful energy under normal driving conditions) and an energy consumption of 150 W h km^{-1}, has a range of 100 km. For a vehicle life of 100,000 km, 1000 nominal cycles are needed (1000 times the nominal energy discharged), corresponding to a cumulated discharged energy of 15 MW h. The same vehicle equipped with a battery of 30 kW h and the same energy consumption of 150 W h km^{-1} (the extra weight of the battery is compensated by a lower average discharge rate and therefore, a higher available energy) has a range of 200 km, therefore, only 500 nominal cycles are needed for 100,000 km, the cumulated discharged energy remaining obviously the same at 15 MW h.

The same battery module might be used for the two-battery configurations, therefore our objectives for this module and the complete traction battery are not the same as far as cycle life is concerned. In order to be precise, cycle life has to be expressed in nominal cycles or in cumulated discharged energy for a given battery configuration (See Fig. 1).

1.2. Importance of the mission profile

The cycle life available for a given battery technology might be very much dependent on the mission profile of the vehicle. Let us consider the same vehicle as mentioned above, introducing the power characteristics of the drive systems. Let us assume a peak power of 30 kW from the battery. The power/energy ratio in the first case is 30 kW/15 kW h = 2.0; in the second case 30 kW/30 kW h = 1.0. For the same vehicle, the same battery technology might have a different discharge profile, e.g., maximum discharge power, as a function of the on-board energy.

On the other hand, for the same battery technology and the same on-board energy, the discharge profile varies

* Corresponding author.

0378-7753/98/$19.00 © 1998 Published by Elsevier Science S.A. All rights reserved.
PII S0378-7753(97)02775-4

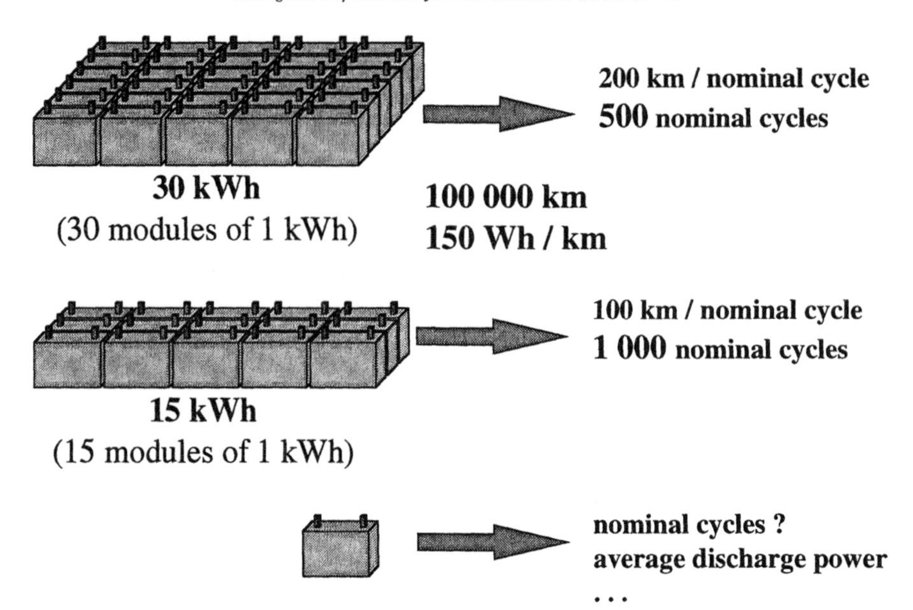

Fig. 1. Influence of the on-board energy.

from one vehicle to another. These differences might be due to different drive trains, aerodynamics, vehicle weight, vehicle performance characteristics such as acceleration and regenerative braking, etc.

In the first part of this paper, we are going to discuss the different parameters that have to be investigated to provide optimum battery integration and management, and therefore optimum battery performance on-board the electric vehicle.

1.3. Importance of the time factor

RENAULT has some first experience with its commercialized electric vehicles, with a typical driving distance

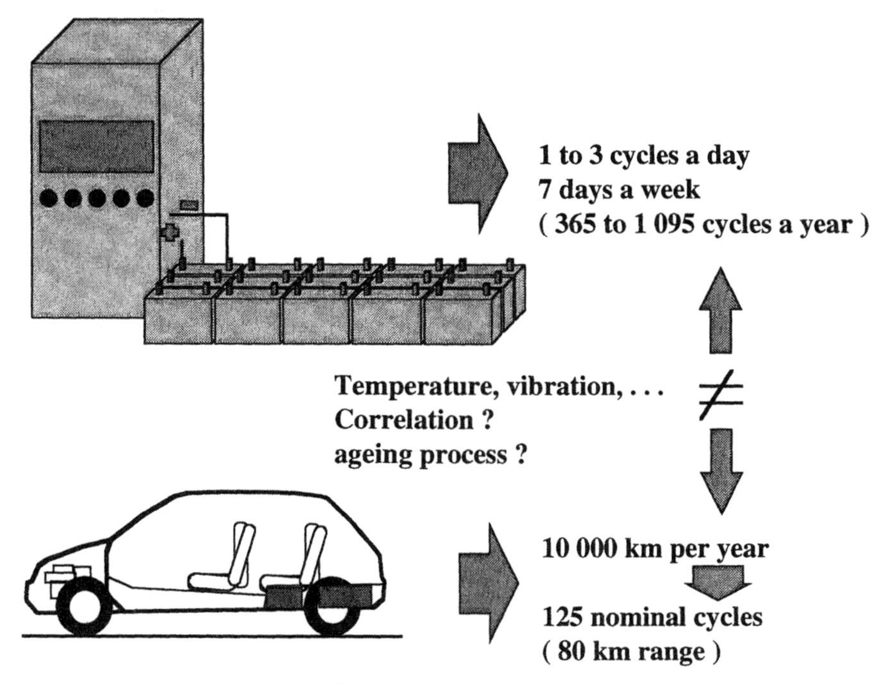

Fig. 2. Bench test–vehicle test.

rarely exceeding 10,000 km per year, corresponding to 125 nominal cycles only (typical vehicle range of 80 km). One might argue that the annual driving distance increases with an available driving range of more than 80 km. The USABC does also quote objectives for the calendar life of traction batteries: 5 yrs for the midterm, 10 yrs for the long-term, corresponding with our objective of 100,000 km and 10,000 km yr^{-1}.

For bench testing, most laboratories carry out 1 to 3 cycles per day, 7 days a week, resulting in some 365 to 1095 cycles per year. For a DOD of 80% normally used in this kind of test, we obtain 292 to 876 nominal cycles per year. As we can see, this is totally different from EV reality ($\leq$ 125 nominal cycles), where we observe long standstill periods, partial discharges, etc. As a consequence, the results obtained differ from vehicle testing (see Fig. 2).

However, it is clear that during research and development it is not possible to carry out cycling tests for prototypes at a rate of 125 nominal cycles per year only. We have the problem of not having the time available to verify the calendar life of a product that is expensive and that is undergoing continuous development. It is therefore, necessary to develop accelerated lifetime tests with proven correlation to forecast expected calendar life.

To conclude, we have to understand clearly the influence and correlation of accelerated bench test cycling and the ageing of a battery. A compromise has to be found between winning time on one hand, and obtaining all necessary information for the real vehicle application on the other.

In the second part of this paper, we discuss the approach of a 'complex cycling' profile to reach our objective.

2. Study of parameters influencing cycle life

2.1. Reference cycle

A lot of work has been concentrating on bench testing to study the cycle life of traction batteries using standard cycle profiles such as SAE 1227a [2], SFUDS [3], ECE15 or DST. Despite the continuous improvement of these cycles, we have learned, however, that the results obtained show significant differences to the results obtained with vehicle trials. Previous work with lead-acid batteries showed a cycle life on the bench of more than 700 cycles (TC69 WG3) [4], whereas only 300 cycles were obtained with the same battery on the electric vehicle.

In order to study different mission profiles, it is necessary to establish a reference cycle for further comparison. The cycle profile used is the TC69 WG3 (see Fig. 3). This dynamic discharge profile is very simple to use. It consists of a total discharge time of approximately 2 h and 30 min (for 100% DOD), and corresponds roughly to a suburban driving pattern. However, we have to accept the drawback of not having one step representing regenerative braking.

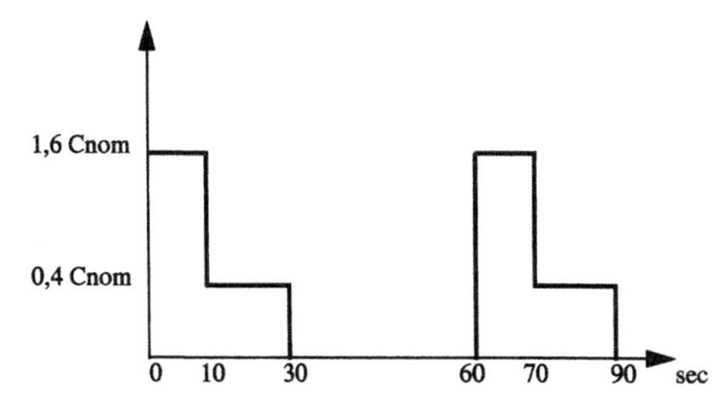

Fig. 3. Discharge with reference cycle TC69 WG3.

1. Normal charge (8 h)
2. Rest period of 1 h
3. Discharge TC69 WG3 (reference cycle), 80% DOD

$11 = 1.6\ C_{nom}$	10 s
$12 = 0.4\ C_{nom}$	20 s
$13 = 0$	30 s

4. Rest period of 1 h

This reference cycle can easily be modified to study the influence of one specific parameter, e.g., we examine the influence of the peak power demand by replacing the current step 11 by P_{max}. It is obvious that all test samples have to be as close in initial performance as possible to allow comparison and conclusion.

We establish a test plan, varying only one parameter at a time. The electrical data such as internal resistance variation, average discharge voltage, capacity evolution, energy efficiency, temperature, etc. have to be analysed and compared between different tests. The idea is not to obtain the life cycle capability of a system, we are more interested in studying the evolution of the identified parameters and the ageing process during testing. Once the cycling is completed, the active material has to be analysed in close collaboration with the battery manufacturer (post-mortem analysis).

2.2. Influence of discharge power

The available discharge power, continuous and instantaneous, is one of the most important parameters for the performance characteristics of the vehicle (acceleration, top speed). This specific discharge power for the battery might be a function of the on-board energy, the DOD, the temperature, battery age, cumulated discharged energy, etc.

In order to study the influence of the peak power demand, we have to take three parameters into consideration: the value of the peak power value, its duration and the frequency of the peak power demand. We modify our reference cycle and change the discharge step 11 to I_{max} (see Fig. 4).

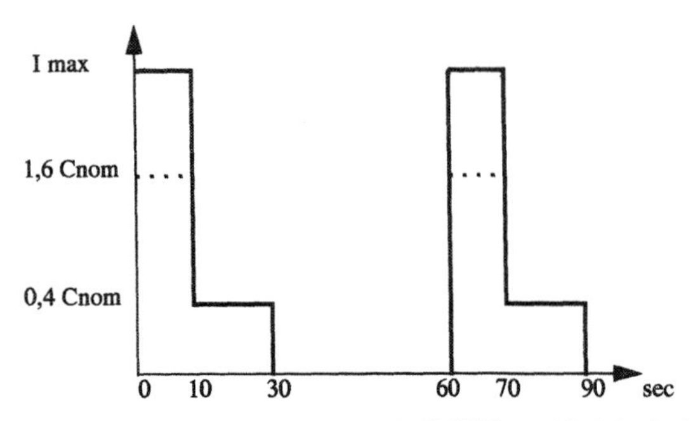

Fig. 4. Discharge with reference cycle TC69 WG3, modified for Peak Power value.

Example: peak power value
Fig. 3 Discharge with modified TC69 WG3, 80% DOD

$11 = I_{max}$, limited by U_{min}	10 s
$12 = 0.4\ C_{nom}$	20 s
$13 = 0$	30 s

Example: peak power frequency
Fig. 3 Discharge with modified TC69 WG3, 80% DOD

$11 = I_{max}$, limited by U_{min}	5 s
$12 = 0.4\ C_{nom}$	20 s
$13 = I_{max}$, limited by U_{min}	5 s
$14 = 0$	30 s

The value of I_{max} is provided by the battery manufacturer. For U_{min} we use 2/3 OCV, unless stated otherwise.

We know that this parameter is most important for advanced battery systems with high specific energy, e.g., Li-ion, NaNiCl2 (ZEBRA). To satisfy the proposed energy demand of 15 kW h, we require an approximate battery weight of 200 kg (ZEBRA, ≥ 75 W h kg^{-1}) and 125 kg (Li-ion, ≥ 120 W h kg^{-1}). For a given peak power of 30 kW, we require a specific peak power of 150 W kg^{-1} for ZEBRA and 240 W kg^{-1} for Li-ion (power/energy ration = 2). We notice the importance of the installed energy. The lower the installed energy, the higher the specific peak power. If we increase the peak power and/or reduce the battery energy, we approach the battery specifications of hybrid vehicles with a high power/energy ratio of 5–20. [5]

First initial tests of the systems show their peak power capability, compatible with our example. However, we have to study the influence of a repetitive power demand on cycle life to ensure the values provided by the battery manufacturer are truly useful peak power.

2.3. Regenerative power

It is well known that regenerative braking increases driving range, up to 25% for an urban driving cycle. Furthermore, regenerative braking reduces maintenance for the brake pads. However, little is known on the influence of strong regenerative braking on cycle life. We therefore introduce a high percentage of regenerative braking (up to 30% to emphasize the phenomena) into the reference cycle: 12 and 14, with $12 = 14 = -0.5\ C_{nom}$. In reality, we might have a regenerative braking current well above this value.

Example: regenerative power
Fig. 3 Discharge with modified TC69 WG3, 80% DOD

$11 = 1.6\ C_{nom}$	10 s
$12 = -0.5\ C_{nom}$, limited by U_{max}	5 s
$13 = 0.4\ C_{nom}$	20 s
$14 = -0.5\ C_{nom}$, limited by U_{max}	5 s
$15 = 0$	30 s

The power capability of a battery system has to be investigated in two directions: for discharge and for charge. The battery of a hybrid vehicle, for example, with a relatively low battery energy, has to be capable to accept high regenerative power and fast recharge.

The parameter 'regenerative braking' shows the importance of our overall approach: the peak power on discharge has to be specified within the product definition by marketing experts; in contrast, the management of regenerative braking is a more technical decision. The car manufacturer has to be aware of what is visible to the final customer (acceleration, available discharge power), and what is of less importance for the customer, but possibly as important for the optimal functioning of the battery system (regenerative power).

2.4. Depth of discharge

Concerning the depth of discharge (DOD), there are in general two possibilities: a fleet operator, with a daily use that is repetitive, therefore the battery is always discharged down to a similar DOD. The private user, however, has a mission profile which might vary at random from low to high DODs.

We have to use our reference cycle and go to a low DOD (e.g., 20%), thus accumulating a high number of cycles (but not nominal cycles!). The average battery voltage during discharge is higher than for a discharge at 80% DOD, and the battery is in the charge mode for a higher percentage of time, as the normal end of charge for all battery technologies consists of a low charging power.

Many papers have dealt with the so-called memory effect for alkaline and lead-acid batteries, which has been observed mainly on vehicle trials. This specific test allows us to verify this phenomenon more scientifically on the bench and to investigate its importance for the latest battery technologies: NiMH, NaNiCl2, and Li-ion.

2.5. Partial charge

As for the DOD, the vehicle battery might be fully charged regularly, or might be frequently charged without reaching a fully charged state. This might cause problems for the electrochemistry or the state of charge indication, and in consequence there might be an influence on battery life. We therefore have to modify our cycling profile, using some partial charges at various DODs and a complete charge only from time to time.

We believe that this kind of cycle profile has been used very little up to now, except for hybrid vehicle battery testing. In reality it might well happen that we use partial charging during the day to increase the daily driving range. This seems true especially for vehicles with a relatively low range, that is vehicles equipped with a battery energy of less than 15 kW h.

2.6. Rest periods after charge and discharge

As indicated in our reference cycle, we most often use a fixed rest period after charge and discharge: typically 1 h. The electric vehicle, however, will be most often 'used' in the rest mode, whereas on the test bench the charge mode is predominant (the charging time is normally longer (approximately 8 h) than the discharge time (approximately 2 h).

We have to modify the reference cycle and investigate long rest periods after charge. For alkaline and lithium batteries, this is important for the self-discharge or capacity loss (reversible or irreversible) and depends on the ambient temperature. For hot batteries we are more interested to measure the thermal discharge, i.e., the energy required to keep the battery at its operational temperature.

Concerning the rest periods after discharge, we have to investigate, for example, the effects of sulfatation for lead-acid batteries, or the cooling down of hot batteries (once the battery is discharged and not connected to the mains).

2.7. Deep discharge

Normal vehicle operation excludes the complete discharge down to 100% DOD, because this means vehicle breakdown. However, experience from vehicles equipped with lead-acid batteries showed that the driver might drive the vehicle until he comes to a standstill, waits for some moments and continues to drive until he comes again to a standstill (taking advantage of diffusion effects of the lead-acid battery, but severely damaging the battery). This shows the importance of well studied end of discharge strategy for the vehicle electronics and driver interface.

Deep discharge might also happen due to problems with the state-of-charge indication. Within our reference cycle, with a discharge down to 80% DOD, we integrate a deep discharge down to 100% and beyond (up to 0 V) every x cycles ($10 < x < 100$). In fact, this test combines the study of the influence of a deep discharge on cycle life, but also takes into consideration safety aspects under abnormal conditions.

2.8. Conclusion

We have to be very careful in the specification of our test procedures and the interpretation of the obtained results. When emphasizing the influence of one parameter at a time, we have to be cautious in choosing the right value in order to understand the effects it might have on the electrical test results during cycling. Furthermore, we have to be careful in assuring the overall test conditions to ensure the reproducibility of our test results.

Cycling is not necessarily carried out until the end of life criteria, but might be stopped once a certain number of cycles is obtained, for example 500 nominal cycles. At the end of the test a postmortem analyses is carried out in collaboration with the battery supplier, to analyse the ageing mode of the battery caused by the specific test.

The above list of parameters might not be exhaustive (e.g., fast charge or temperature might well be very interesting and are used in the 'complex four-season cycle', described below), but we believe that the parameters used in our test definition help to identify the most critical points of a battery technology for traction applications. It is the car manufacturer's role, in collaboration with the battery suppliers, to provide optimum control and management parameters for the battery system on-board the electric vehicle. These control parameters have to be integrated within the battery management system (BMS) or the vehicle management unit (VMU).

3. 'Complex four-season cycle'

This complex cycle consists of a number of charge/discharge cycles using no longer a fixed pattern as proposed in the reference cycle. We propose a combination of all the parameters, previously studied individually:

Rest periods	5 min → 72 h
Charging time	1 h (fast charge) → 8 h (normal charge)
DOD	20% → 100% DOD
Interrupted discharge	
Partial charge	
Ambient temperature	−5°C → +35°C

3.1. Ratio cycles / nominal cycles–calendar year / test year

Our complex cycle consists of 85 charge/discharge cycles for one given temperature, corresponding to 57 nominal cycles and a duration of approximately 6 weeks. The four seasons, thus one calendar year, is reduced to a duration of approximately 24 weeks, i.e., 6 months.

We obtain for one calendar year:
2 simulated years
680 charge/discharge cycles
456 nominal cycles

We notice that we do not achieve our initial objective of reducing significantly the number of nominal cycles per year. However, our complex cycle allows us to introduce some of the real life conditions of the electric vehicle and to save time (compromise between representative testing and time available). For a 15 kW h battery, we would have to cycle for approximately 24 months to obtain 1000 nominal cycles and thus 15 MW h (100,000 km). For the same cumulated energy (driving distance), and for a 30 kW h battery, the required time would be just over 12 months.

As mentioned before, we had difficulties in comparing the cycle life of lead-acid batteries obtained on the bench and on the vehicle. With the 'complex four-season cycle', for example, we are able to reproduce the reduced cycle life of only 300 cycles observed on the EV, mainly due to the longer rest periods of the complex cycle compared to conventional life cycle tests.

3.2. *Influence of the ambient temperature*

In order to simulate the calendar life, we introduce the four seasons of the year by modifying the ambient temperature of the test.

spring	(+20°C)
summer	(+35°C)
autumn	(+10°C)
winter	(−5°C)

We do not want to represent one specific geographical area, we rather try to integrate the variation of ambient temperature into a cycle life test. Traditional test procedures might propose to carry out life cycle tests at different temperatures. However, little is known about the influence of a regular temperature variation during the cycling test.

3.3. *Analysis of the complex cycle*

The complex cycle allows us to study in detail: (1) the influence of rest periods lasting up to 72 h (Monday effect, self-discharge, temperature, energy consumption for hot batteries, etc.), (2) the influence of fast charging (energy rechargeable, temperature evolution, etc.), (3) the influence of interrupted discharge (energy availability, temperature evolution, energy consumption for hot batteries, etc.), (4) the memory effect (available energy after frequent partial discharge), (5) the precision of a state-of-charge algorithm (if a gauge is available, the state of charge calculation is monitored by the bench), (6) the energy efficiency of the battery system,

$$\eta = \frac{\text{energy discharged}}{\text{charged energy} + \text{energy consumption of auxiliaries}}.$$

The importance of the above parameters might vary from one temperature to another, or their importance change with an increase of the cumulated discharged energy. The energy efficiency of the electrochemistry is surely a function of the applied cycle profile. However, for the overall energy efficiency of a battery technology, we have also to take into account the battery auxiliaries.

4. Conclusions

We propose in this paper a new approach for test procedures concerning battery bench testing for EV application. We have to clearly identify all possible parameters influencing the functioning of the cell electrochemistry or complete battery system. Once the parameters have been identified for a given vehicle specification, we have to study them individually to optimize the battery utilization on the EV.

In a second step, lifetime data have to be obtained by a complex cycle using real life vehicle data if possible together with a simulated driving pattern taking into account the parameters discussed before. Only this approach allows to obtain test bench data that are representative of vehicle data and to draw conclusions concerning the aptitude of a given battery technology for a specified vehicle utilization.

References

[1] K. Rajashekara, R. Martin, Electric vehicle propulsion systems—present and future trends, J. Circuits Syst. Comput. 5 (1) (1995) 109–129.
[2] Electric vehicle test procedure, SAE J227a, British Standard 1988, 28, Passenger cars, trucks, buses and motorcycles.
[3] A Simplified version of the federal urban driving schedule for electric vehicle battery testing, Idaho National Engineering Laboratory, DOE-Report No. DOE/ID-10146, August 1988.
[4] R. Wagner, W. Bögel, J.P. Büchel, Valve-regulated lead/acid batteries of the AGM design for electric vehicle service, 29th ISATA, Florence, Italy, 3–6 June 1996.
[5] J.P. Büchel, C. Hiron, H. Katz, W. Bögel, Utilization of traction batteries for EV application, EVT '95, Paris, France, 13–15 November 1995.

ELSEVIER

Journal of Power Sources 72 (1998) 43–50

Industrial awareness of lithium batteries in the world, during the past two years

Hélène Katz *, Wolfgang Bögel, Jean-Pierre Büchel

RENAULT, Direction de la Recherche, 14, Avenue Albert Einstein, ZA Trappes-Élancourt, 78190 Trappes, France

Received 26 August 1997; revised 22 September 1997

Abstract

The different requirements of lithium battery systems, as well as the structures of different lithium secondary systems, have been reviewed. The main research and development programs in the world are listed, in order to try to understand better the future directions for lithium systems development.

Keywords: Lithium battery; Lithium systems development; Electric vehicle

1. Introduction

This work has been carried out in the context of electric vehicle applications. We have attempted to draw a rough panoramic view of the actual situation of lithium batteries technologies in the world, with a view to gaining a better understanding of the present and future actors in the field. As lithium technology is still under development for most of its possible applications, the actors are generally involved in the following research and development areas: battery structure and composition optimization, manufacturing, use and arrangement, and recycling.

In this paper, we shall first review the main different kinds of lithium battery technologies (that could be suitable as electric vehicle traction batteries), and then the main national and trans-national R&D programs, in Europe, in the USA and in Japan. We shall then indicate the directions which seem to be actually taken and the most striking outstanding features. For this, we have focused on literature, congress proceedings, and US patents, all of them published within the last 2 years. We shall try to evaluate each step of the battery life (battery design, manufacturing and recycling).

2. Different uses for lithium technologies

Several kinds of lithium battery technologies have appeared, as well as several configurations. We can distinguish between primary and secondary batteries, and among secondary batteries, we can consider portable batteries, and traction or storage batteries, these differences in applications resulting in differences in structure and composition.

The most widespread technologies for primary batteries using lithium metal as anodes are commonly based on the use of manganese, copper or sulfur oxides as cathodes, soluble or solid, with aprotic electrolytes and lithium-based solute additives for electrolyte conductivity enhancement. The shape of cells may be button type, cylindrical or prismatic.

A thorough knowledge and a confirmed experience of the manufacturing and use of lithium primary cells may be a good basis for the development of secondary cells, providing an idea of the constraints existing on the choice, compatibilities and behaviours of the different components and additives.

Lithium secondary batteries are generally used in applications requiring high energy densities and high voltages. They are outstanding candidates for all portable applications like cellular phones, notebooks, or small computers, and the demand for them is presently in tremendous growth, particularly in Japan, where several battery manufacturers make them at an industrial level. Common points exist between portable applications and bigger devices for

* Corresponding author.

0378-7753/98/$19.00
PII S0378-7753(97)02783-3

home energy levelling, particularly the high cyclability (even if it is not practically the same order of magnitude, around 1000 for portable, and closer to 3000 for stationary), even though sizes and capacities are different in both cases, and if weight is less crucial in stationary devices.

The energetic characteristics of lithium secondary systems might make them quite attractive as electric vehicle traction batteries, the biggest differences between portable and traction applications lying mainly in size and consequently the thermal or safety behaviour of the batteries.

The above differences and common points give rough directions for systems optimization. It is clear that if one system could lead to high energy density, high cyclability, and still remain of small size, it could be of great interest in many fields of life and industry.

Some minor differences may play a part in the resulting structure and composition of lithium secondary batteries, such as, for example, the need for low costs, high safety, the types of mission profiles or the particular power demand levels.

Other applications of lithium batteries, that concern military systems that require high energy density but not necessarily high cyclability, low costs or small sizes will not be dealt with here.

3. Different compositions of lithium secondary batteries

Many particular factors may influence the configurations of lithium secondary batteries [1], depending on the requirements set by users, as mentioned in Fig. 1. However, setting aside requirements, lithium secondary systems have been developed in several directions, that appeared to diverge for a certain time, and that now seem to be converging once more.

We can distinguish between different kinds of lithium systems, related to several criteria, as listed in Table 1: on the composition of the negative electrode, of the positive electrode, on the physical state and composition of the electrolyte, or on some other criteria.

A major difference arose, in the 1980s, between what has been called lithium-ion (or lithium–carbon, or rocking-chair system), and lithium-polymer (-electrolyte) batteries, the difference lying mainly in the physical state of the electrolyte (liquid for lithium-carbon, and solid for lithium-polymer), and in the negative electrode composition (carbon with intercalation of lithium ions, as opppsed to pure lithium in the initial lithium-polymer systems).

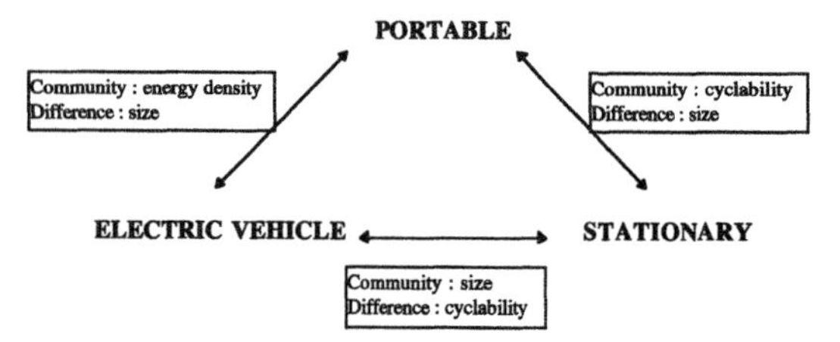

Fig. 1. Lithium battery systems applications and main requirements.

Table 1
Criteria on differences between lithium secondary systems

●Criterion on negative electrode composition:
Pure lithium
Carbon–lithium, graphite–lithium
Lithium alloy
Other alloy
●Criterion on electrolyte state:
Liquid electrolyte
Polymer-solid electrolyte
Gelified electrolyte
●Criterion on positive electrode:
Intercalation cathode ($LiCoO_2$, $LiNiO_2$, $LiMnO_2$, V_2O_5,...)
Polymer cathode
●Criterion on lithium based additive in electrolyte
●...

Both systems, lithium–carbon and lithium-polymer, were developed separately in different directions, until about 1992. Research was mainly directed to the improvement of the drawbacks of each system:

- For lithium-ion systems:
  - new materials of intercalation for the negative and the positive electrodes
  - new electrolytes and conductive salts
  - safety improvement
  - cost reduction
  - cyclability
- For lithium-polymer systems:
  - improvement of the conductivity of the electrolyte and additive salt
  - test of new materials for the positive and the negative electrodes
  - cyclability

The distinction between the two systems lost part of its meaning in 1992, when Bellcore (Bell Communications Research) laid patents showing the possibility for making cells using both solid polymer electrolyte and carbon as negative electrode, involving at the same time the intercalation phenomenon of lithium ions in the carbon electrode, and their transport through a salt in the solid electrolyte, thus combining the safety aspect of the lithium-polymer system and with the high potential energy density of the lithium-ion. Since this time, this 'hybrid' concept has been widely developed, particularly in Japan, and now it would be necessary to specify if 'lithium-polymer' is developed with carbon as negative or with pure lithium, or if 'lithium-ion' has a solid or liquid electrolyte. Unfortu-

Table 2
USABC funded programs in the US

Partners	Program duration and funds	Objectives
W.R. Grace (USA)	3 years (Beginning: 1992); $24.5 million	Develop lithium-polymer batteries
Saft America–Argonne Nat. Lab. (USA)	3 years (Beginning: 1992); $17.3 million	Develop LiAl/FeS2 batteries
3M (USA)–Hydro-Quebec (Canada)–Argonne Nat. Lab. (USA)	2 years (Beginning: 1993); $32.9 million	Develop lithium-polymer batteries
Varta (Germany) – Duracell (USA)	2 years (Beginning: 1995); $18.0 million	Develop lithium-ion batteries for electric vehicles
SRI International (USA)	1/2 years (Beginning: 1996); $0.8 million	Develop lithium-ion technology
Saft America	1/2 year (Beginning: 1996); $1.4 million	Develop lithium-ion batteries
3M (USA) – Hydro-Québec (Canada)	2 years (Beginning: 1996); $27.4 million	Manufacture lithium-polymer batteries for electric vehicles
Varta (Germany) – Duracell (USA)	2 years (Beginning: 1997); $14.5 million	Develop lithium-ion batteries for electric vehicles

nately, only few manufacturers give this kind of information.

Another system, the hot lithium–aluminium/iron sulfide (or disulfide), has been developed by the Argonne National Laboratory in the United States since the 1960's. Its running temperature is around 450 to 500°C. Even if potentially suitable for electric vehicle applications, because of its performances, this system does not offer the required safety guarantees (problems of corrosion at high temperatures).

Other systems using pure lithium as negative electrode have been mainly developed for military applications (high energy density) but offer low interest for EV applications, because of poor cyclability.

4. R&D programs

Many programs concerning secondary lithium batteries have been or are funded worldwide, by national or continental authorities, in the United States, in Japan or in Europe. We shall list the most important of them hereafter and deal with the different policies on lithium battery research and development in the different continents.

4.1. In the United States

The United States Advanced Battery Consortium (USABC) has already funded several big programs on lithium batteries, in order to power electric vehicles (Table 2). For years, the USABC has been funding only hot lithium batteries and lithium-polymer technologies, and it is only very recently, in 1995, that the USABC began to get interested in the lithium-ion system, allocating some funds to SRI, to Saft America, and to Varta and Duracell. The last contract Varta and Duracell have signed with the USABC is aimed at production ability, safety, life time and cost improvement of the lithium-ion battery.

Hot lithium systems have been particularly developed in the USA by the Argonne National Laboratory, since the 1960s, and by Saft America. This battery could have satisfied the USABC requirements in some respects, but needs further development.

The USABC has been investing a lot in lithium-polymer, as it has been claimed for years, in the United States, that it could reach much higher performances than the lithium-ion system. As the differences between the two systems have been diminishing, and as both systems have been improved, the USABC now begins to evaluate lithium-ion.

The total amount of money invested in lithium battery research by the USABC reaches $136.8 million, of which only part is really provided by USABC, the other part being supported by the battery companies themselves.

Table 3
LIBES objectives for lithium systems development

	High cyclability batteries	High energy batteries
Energy density		
Wh/kg	120	180
Wh/lit.	240	360
Lifetime	3500 cycles	500 cycles
Energy efficiency	90%	85%
Prototypes capacity	20 kWh	30 kWh
Companies involved	Hitachi	Asahi–Toshiba
	Mitsubishi Electric	Japan Storage
	Sanyo	Matsushita
	Yuasa	Nippon Denso

4.2. In Japan

In Japan, the most famous program concerning lithium batteries is the LIBES, that began in 1992, and should last for 10 years. The total Government funding reaches ¥14 billion (about US$118 million). This program gathers several big Japanese battery manufacturers, all specialised in batteries and lithium systems. Only SONY, one of the most advanced lithium-ion Japanese battery companies, has not been involved. The LIBES has structured research in two directions: those batteries that could offer good cyclability, and those that could offer good energy density, as these two features seemed conflicting. We can see in Table 3 the target values for these two systems, and the companies involved. Some other companies, like OSAKA GAS, CRIEPI, MITSUBISHI Petrochemical, are involved in the LIBES materials research too. Each company had a precise task in the development of batteries [2].

4.3. In Europa

Several programs, Joule or Brite-Euram, gathering European companies, for lithium batteries development, have been or are funded by the European Community [3] (Table 4).

The European Community has funded eight Joule programs concerning the development of lithium batteries, for a total amount of 12 338 kECU (about $14 million). We can see that, as in the USA, many programs concern the lithium-polymer technology, only one being dedicated to lithium-ion with liquid electrolyte.

Even if an important difference seems to appear in the level of the funds dedicated to lithium systems between Japan, the US and Europe, it must be remembered that some more national programs, not listed here, are helped by some of the European Governments.

5. Manufacturing and commercialization of lithium batteries

If we first have a look at the Japanese industry of primary lithium batteries, we can notice that output has

Table 4
European funded lithium batteries programs (Joule)

Partners	Duration and alloc. funds	Objective
VARTA (D)–HARWELL (UK)–Univ. Roma–Univ. Warsaw–Bulgarian Acad. Sc.	2 years (Beginning: 1993); 852 kECU	Lithium-ion polymer electrolyte of 2 kWh
DANIONICS (Dan)–Univ. St Andrews– Univ. Southampton–Univ. Uppsala	2 years (Beginning: 1993); 570 kECU	Develop lithium-polymer modules
DANIONICS (Dan)–Univ. St Andrews– Univ. Southampton–TU Delft		Develop lithium-polymer batteries
CEA (F)–CNRS (F)–EDF (F)–BOLLORÉ (F)–SADACEM (B)	3 years (Beginning: 1993); 560 kECU	Develop positive electrodes for lithium-polymer systems
Sonnenschein (D)–DANIONICS (Dan)–ISITEM Nantes (F)	2 years (Beginning: 1994); 500 kECU	Develop lithium-ion technology with polymer electrolyte
SAFT (F)–VARTA (D)–SOLVAY (B)–AAR (F)–EUCAR	3 years (Beginning: 1996); 6000 kECU	Develop lithium-ion batteries with liquid electrolyte for electric vehicles
Sonnenschein(D)–DANIONICS (Dan)–ISITEM (F)–AAR (F)–TU Delft–UK Univ.	x years (Beginning: 1996); 1900 kECU	Develop lithium-polymer batteries
DANIONICS (Dan)–TOBIAS JENSEN (Dan)–STENOVIST (Sw)–DB (D)	x years (Beginning: 1996); 1306 kECU	Develop lithium-polymer batteries for electric vehicles

Table 5
Monthly production of lithium-ion secondary portable batteries

	Million units manufactured per month	Date of forecasts
Asahi–Toshiba	1.6	mid-1996
Fujifilm	1.5	1998
Hitachi–Maxell	6	July 1997
Japan Storage	3	April 1997
Matsushita	20	1999
Matsushita US	1.5	1997
Moli Energy Canada	10	2000
Nippon Moli	2	January 1998
Sanyo	5	September 1996
Sony	10	March 1997
	20	1999

drastically increased from 1988, when it was 234 million units, to 1992, when it reached 451 million units, representing more than 38% of the total Japanese production of primary batteries [4].

The production of secondary portable batteries (of capacities ranging between 0.5 and 2 Ah) is increasing at a much higher rate, companies announcing every month that the output will be doubled soon, and that new production lines are to be built. We give hereunder the production of several big Japanese companies, as well as the dates forecast for these, as published in the newspapers during the first six months of 1997:

It appears that in 1996, the Japanese battery industry produced 120 million lithium based portable units. It should be 3 times more in 2000.

All the companies listed in Table 5, even if they do not all belong to the LIBES, have already been working a lot in the field of research and development concerning lithium secondary batteries, laying patents and sometimes funding their own studies.

We must emphasize that Japan is the only place in the world where the growth in lithium portable battery manufacturing is so significant. Some other companies in the world, like Saft or Varta, already make and commercialize lithium primary batteries since the 1990s, and since very recently, portable secondary batteries, but today in much smaller quantities than in Japan. The same phenomenon is happening in the United States, with a lower increase than in Japan.

Concerning lithium secondary batteries that could be EV suitable, only very few manufacturers really produce prototypes. In Japan, Sony makes lithium batteries for Nissan electric and hybrid vehicle applications, according to a special agreement. In Canada, Hydro-Quebec will manufacture big EV batteries next year, that will be reserved for USABC members' tests. In Europe, Saft and Varta have to manufacture EV size lithium-ion batteries in the frame of their Joule program with EUCAR, for bench tests.

So that today, lithium-ion batteries for EVs are not commercially available, unless by special agreement with battery manufacturers.

6. Recent and possible future developments in the lithium battery industry

Several areas have been developed in research and development on lithium systems. We have listed the main ones, as well as the companies known for their progresses in these areas, and the most striking events (Table 6). This progress may concern secondary portable as well as EV or stationary applications. If a company is not listed in a special area, it does not mean that this company has done nothing, but that we have found nothing about its activities. We did not list all the universities participating in these researches, although many of them do, in Japan, in the United States and in Europe.

Much work has already been done on the improvement of the different components of lithium ambient temperature systems.

Concerning the positive electrodes, many elements have been studied, as possible intercalation compounds: cobalt, nickel, manganese oxides, as well as many combinations of these elements, and use of additives in the positives. Vanadium oxides and titanium sulfide combinations have been studied alone or mixed as carbon composites. Recently, some other elements have been introduced as positives, like gallium by Sumitomo [5], or copper by Hitachi [6].

The negative electrodes have been tested in different forms: pure lithium, generally with polymer electrolytes, or carbon-based. When carbon-based, many kinds of carbons; graphites, pitches, fibers of different sizes and combinations, have been tried, as well as many pretreatments for these fibers or mixes [7–11]. Fuji patented new compositions for negatives, replacing the carbon by composite materials of mixes between metals of groups IIIA, IVA or VA of the Mendeleyev Table (B, Al, Ga, In, Tl, Si, Ge, Sn, Pb, P, As, Sb, Bi), with oxide, sulfide, selenide or telluride (of group VIA), underlining that the best results were observed with tin oxide composites [12]. Hitachi suggested the use of composites of lithium–lead–carbon [13]. Osaka University has studied negatives made of carbon composites mixed with conducting polymers [14].

The electrolytes have been the object of much work too; on their composition and on the nature of the salts to be dissolved in them for conductivity enhancement. Moreover the electrolytes have to be aprotic, in order to avoid explosions with lithium. They might be liquid organic, polymer organic or gelified. Recently, Nagoya University found a new kind of electrolyte, that is inorganic and aprotic. Conductive salts have been widely tested too, and several lithium-ion liquid electrolyte manufacturers seem to prefer LiPF6.

Table 6
Companies' involvement in lithium battery R&D

	Europe	US–Canada	Japan
Positive electrodes	AAR (F), Bolloré Technol, (F), CNRS (F), EDF (F), Saft (F), TU Graz (A). Univ. Sofia (Bu), Univ. Uppsala (S), Varta (D)	Argonne NL, Bellcore, Hydro-Quebec, Dalhousie Univ.	Aichi Steels, Asahi–Toshiba, Hitachi–Maxell, Japan Storage, Matsushita, Mitsubishi, Mitsui, Moli Energy, Nippondenso, Sony
Negative electrodes	AAR (F), Bolloré Technol, (F), Carbone Lorraine (F), CNRS (F), EDF (F), Saft (F), TU Graz (Au), Varta (D)	Bellcore, Livermore NL–Berkeley, Hydro-Quebec, Dalhousie Univ.	Asahi–Toshiba, Fuji, Hitachi–Maxell, Japan Storage, Matsushita, Mitsubishi, Moly Energy, Nippondenso Nippon Steel, Sanyo, Sony
Electrolytes	AAR (F), Bolloré Technol. (F), CNRS (F), EDF (F), Elf Atochem (F), Saft (F), Varta (D)	Arizona Univ., Hydro-Quebec, 3M, Dalhousie Univ.	Asahi–Toshiba, Fuji, Hitachi–Maxell, Japan Storage, Matsushita, Mitsubishi, Mitsui, Moli Energy, Nippondenso, Sony, Yuasa
Manufacturing	Bolloré Technol. (F), Saft (F), Tadiran (Is), Varta (D)	Bellcore, Hydro-Quebec, Dalhousie Univ., Valence	Asahi–Toshiba, Hitachi–Maxell, Japan Storage, Matsushita, Mitsubishi, Moli Energy, Nippondenso, Sony
Recycling		Greatbatch, Bellcore, National Technical Systems	Canon, Hitachi–Maxell, Miyiazaki Univ.

Concerning possible structures, Yardney has recently patented a lithium based bipolar battery [15].

7. Conclusions

Lithium battery technologies are still at a research and development stage, and the first EV suitable prototypes are actually, or will soon be, under test.

Many points have still to be improved for EV powering applications, like costs, cyclability, and safety. One of the biggest difficulties comes from the fact that these three points have to be optimized at the same time.

Anyhow, much has already been done since the beginning of lithium batteries, even if it is a very young technology, and it seems that further scientific advances, like on conducting polymers, or new micro-visualization processes, will be helpful. We can see either that some interesting trends have recently been observed and should allow some new advances. Moreover, the interest in these technologies is great worldwide, and the level of money invested is high. So, we could hope to see some applications to EVs... at the beginning of the next century...?

References

[1] D. Linden (Ed.), Handbook of batteries, 2nd edn., McGraw-Hill, 1995.

[2] T. Koyamada, H. Ishihara, Research and development program of lithium battery in Japan, Electrochim. Acta 40 (13–14) (1995) 2173–2175.

[3] Research and Technological Development Programme in the field of non-nuclear energy, Project Synopses, Joule II, DG XII, 1994, Project Synopses, Joule III, DGXII, 1997.

[4] A. Ikegami, A look at Japan's primary battery industry, Japan, January 21, 1994, p. 61.

[5] Lithium secondary battery having a cathode containing gallium, US Patent No. 5 595 842, 1997.

[6] Lithium cell with a cathode comprising a copper compound oxide, US Patent No. 5 470 678, 1995.

[7] F. Salver-Disma, J.-M. Tarascon, Effet unique du broyage mécanique sur l'intercalation électrochimique du lithium dans des carbones de morphologies différentes, Journée d'étude sur les accumulateurs au lithium organisée par la SFC, Paris, 1996.

[8] T. Kasuh, A. Mabuchi, K. Tokumitsu, H. Fujimoto, Recent trends in carbon negative electrode materials, 8th International Meeting on Lithium Batteries, Nagoya, Japan, 1996.

[9] T. Tamaki et al., Characteristics of boron doped mesophase pitch-based carbon fibers as anode materials for lithium secondary cells, 37th Japanese Battery Symposium, Kyoto, 1996.

[10] T. Nohma et al., Electrochemical behaviour of natural graphite electrodes in various electrolyte solutions, 37th Japanese Battery Symposium, Kyoto, 1996.

[11] T. Maeda et al., A long-life secondary battery using graphite–coke hybrid carbon negative electrode, 37th Japanese Battery Symposium, Kyoto, 1996.

[12] Non-aqueous electrolyte secondary battery, Japanese Kokai Patent, Toku-Kai-Hei 8-273 668, 1996.

[13] H. Miyadera, Hitachi's research on lithium secondary battery, Japan, August 21, 1994, p. 75.

[14] S. Kuwabata et al., Preparation of composite of poly(3-*n*-hexylthiofene) and carbon material and its charge–discharge properties as an anode active material for rechargeable lithium-ion batteries, 37th Japanese Battery Symposium, Kyoto, Japan, 1996.

[15] Bipolar lithium-ion rechargeable battery, US Patent No. 5 595 839, 1997.

Journal of Power Sources 72 (1998) 51

Author Index of the ICAM/E-MRS Conference

Elsevier Science S.A.
PII S0378-7753(98)00069-X

Journal of Power Sources 72 (1998) 52

Subject Index of the ICAM/E-MRS Conference

Elsevier Science S.A.

Printed and bound by CPI Group (UK) Ltd, Croydon, CR0 4YY

08/05/2025

01864803-0001